Landscape Architecture Framed from an Environmental and Ecological Perspective

Edited by Mustafa Ergen
and Yaşar Bahri Ergen

Published in London, United Kingdom

IntechOpen

Supporting open minds since 2005

Landscape Architecture Framed from an Environmental and Ecological Perspective
http://dx.doi.org/10.5772/intechopen.94789
Edited by Mustafa Ergen and Yaşar Bahri Ergen

Contributors
Frank Chibuzor Okenmuo, Bruno Marques, Jacqueline McIntosh, Hayley Webber, Lazaro Eliyah Mngumi, Yonggui Chen, Jon Bryan Burley, Chunqing Liu, Xiaowen Jin, Zhi Yue, Zhen Wu, Selma B Pena, Maria Luísa Franco, Manuela Raposo Magalhães, Mustafa Ergen

First published in London, United Kingdom, 2022 by IntechOpen
IntechOpen is the global imprint of INTECHOPEN LIMITED, registered in England and Wales, registration number: 11086078, 5 Princes Gate Court, London, SW7 2QJ, United Kingdom
Printed in Croatia

British Library Cataloguing-in-Publication Data
A catalogue record for this book is available from the British Library

Additional hard and PDF copies can be obtained from orders@intechopen.com

Landscape Architecture Framed from an Environmental and Ecological Perspective
Edited by Mustafa Ergen and Yaşar Bahri Ergen
p. cm.
Print ISBN 978-1-83969-698-5
Online ISBN 978-1-83969-699-2
eBook (PDF) ISBN 978-1-83969-700-5

Meet the editors

Mustafa Ergen graduated from the Department of Landscape Architecture, Abant İzzet Baysal University, Turkey, in 2000. In 2005, he completed his first master's degree in Urban and Regional Planning at the Gebze Institute of Technology, Turkey. He completed his second master's degree in Landscape Architecture at Anhalt University of Applied Sciences, Germany, in 2006. During 2007–2008, he studied Geographic Information Systems and Remote Sensing at the Mediterranean Agronomic Institute of Chania, Greece, and was granted a specialization diploma in Environmental Management. He also received a Dr.Eng from the Technical University of Dortmund, Germany. Currently, he works in the Department of Landscape Architecture, Faculty of Agriculture, Sakarya University of Applied Sciences, Turkey.

Yasar Bahri Ergen graduated from Istanbul State Academy of Engineering and Architecture, Department of Architecture, in 1973. In 1976 he completed his master's degree in Urban Planning at the Istanbul State Academy of Engineering and Architecture. From 1983 to 1984 he lived in the USA to conduct research in his field of expertise. He received a doctorate degree from Gazi University in 1987. His academic life started in 1982 and took a turn from the USA to Gazi University. In 1995 he established the Department of Urban and Regional Planning at Bozok University and in 2009 he established the Department of Urban and Regional Planning at Amasya University. In 2017 he established the Faculty of Fine Arts and Design at Siirt University In 2018 he received the title of associate professor. After that, he started to work at Aksaray University in 2021 and still works at Aksaray University.

Contents

Preface

Landscape architecture involves designing, planning, and managing natural and built environments. Its goal is to balance natural spaces with the human element. It is an important field for understanding how to live in harmony with nature. The development of human settlements is generally not friendly to the natural environment, thus new tools are needed to create healthy and user-friendly spaces in both urban developments and natural environments. This book is a guide to designing and planning within existing ecological environments.

Mustafa Ergen
Agriculture Faculty,
Landscape Architecture Department,
Sakarya University of Applied Sciences,
Sakarya, Turkey

Yaşar Bahri Ergen
Architecture and Design Faculty,
Architecture Department,
Aksaray University,
Sakarya, Turkey

Chapter 1

Introductory Chapter: Understanding the Ataturk Forest Farm from an Ecological Perspective

Mustafa Ergen, Bayram Cemil Bilgili and Yaşar Bahri Ergen

1. Introduction

The farm that has gained a significant place in the minds of the Turkish people and has been most well-known to them since the proclamation of the republic is undoubtedly Ataturk Forest Farm. Mustafa Kemal Ataturk laid the foundations of the modern Ataturk Forest Farm, initially called Gazi Forest Farm, in 1925, 5 years after the opening of the Grand National Assembly of Turkey and 2 years after the proclamation of the republic.

Following the years when Turkish people achieved epic triumphs, Ataturk Forest Farm was built during a challenging period of poverty. The Anatolian people were the main population of the Ottoman Empire and the basis of the Republic of Turkey, where 13,648,000 people [1] (76% of the population) lived in rural areas. Without a doubt, Mustafa Kemal Ataturk made the statement "Agriculture is the Basis of a National Economy" to impose the idea that the development of the Republic of Turkey should start in rural areas.

The mission of combining forestry and agriculture, two different applied sciences, was undertaken through the Ataturk Forest Farm. This mission was called the Ataturk Forest Farm because it was not solely based on farming activities and because forestry activities were more dominant than agricultural activities. Forestry and agriculture are similar disciplines but their purposes and methods are different, resembling General Surgery and Psychiatry, which both basically focus on the human body but use totally different methods and concepts. Furthermore, agricultural plans and activities are considered and conducted annually or seasonally, but forestry activities may cover a period of 100 years. Considering these differing perspectives and conditions of time, Ankara and Ataturk Forest Farm should be assessed collectively.

2. The evolution of Ankara and Ataturk Forest Farm, based on the development plan

Ankara was founded in an area that could fit into a single photo frame when it was declared the capital (**Figure 1**) [2].

Ankara is an arid city in the middle of Anatolia, the symbol of the modern Turkish Republic and its window to the world. Ankara's development plan and practices shaped the new form of the Turkish Republic, as well as the city itself.

Figure 1.
An old view from the Atatürk forest farm.

The development plan considered the impact on the city of biotic and abiotic factors, such as climate, fauna, flora, geology, geomorphology, and soil. Its approach also observed the impact of the city on people and the environment.

One of the most important factors here is undoubtedly the population projection, indicating how many people are estimated to be present in a certain place as well as the dimensions of this place. The number of people living in the Turkish Republic was 13,648,000 during those years, and the first census was performed for the first time after the proclamation of the republic was performed on October 28, 1927. Ankara's urban population was 74,784, of whom 49,439 were male and 25,345 were female [3]. Since the number of people living in rural areas of the Turkish Republic was 76% of the total population, a total population of 250,000–300,000 people was estimated for the whole Ankara area.

Carl Lörcher, a German architect and planner, had prepared two separate plans for the old and new city in 1924 and 1925 for Ankara's development between 1925 and 1930. However, the extent of his spatial planning studies and the housing conditions suggested in the plans were not adequate for forming a living area for 250–300,000 people, so a new development plan was needed. Exceeding the afore-noted population projection considered for Ankara's development plan between 1925 and 1930 forced the city to undergo spatial expansion, and new development plans were needed. Accordingly, an international contest was held and won in 1928 by Hermann Jansen in 1928 [4].

During these years, the farms called Orman, Balgat, Yağmur Baba, Macun, Güvercinlik, Ahimesud (Etimesud) and Çakırlar, constitute 80,000 decares of Ataturk Forest Farm land (102,000 decares in total), were purchased from the relevant owners using Ataturk's account number 2 in İş Bank [5]. By 1932, Hermann Jansen delivered the final development plans, and the surface area of Ataturk Forest Farm covered a larger area than the macro form of Ankara [6]. Ataturk Forest Farm reached its maximum size of 102,000 decares in the 1930s, but Ankara continued its urban development. This process can be described with the following comparison: Ataturk Forest Farm is greater than Ankara, and Ankara is greater than Ataturk Forest Farm from the period between 1930 and current times. Two important objectives of the Republic of Turkey: the agricultural development Mustafa Kemal conducted with Ataturk Forest Farm, and ensuring the urban development of Ankara that started with the plans of Hermann Jansen and that was based on modern and scientific grounds.

2.1 Agriculture and Ataturk Forest Farm

Agricultural development was one of the most important objectives of the Turkish Republic during the 1920s because most of the country's total imports consisted of main food elements, such as wheat. During those years, more than half of the total exports consisted of agricultural products, such as cotton, figs, grapes, hazelnuts, and tobacco [5].

Ataturk explained his purpose in founding Ataturk Forest Farm as follows: "The basis of the national economy is agriculture. Therefore, we attribute great importance to agricultural development. The planned and practical activities that will spread into the villages will facilitate adopting this objective. However, to gain success in this purpose, a research-based agricultural policy should be formed and an agricultural regime that every villager and citizen can easily understand and practice should be established" [7]. Ataturk reveals the notion behind the establishment of Ataturk Forest Farm with this statement.

Ataturk formed a group of Turkish agricultural experts and told them that *he wanted to establish a large farm in Ankara*, asking them to find a suitable location. One of the experts in the group stated: "We did not need to search through the entire city of Ankara and look for different natural characteristics. The reason was simple. A medieval city in the middle of an arid steppe... No trees, water, nothing... While looking for a place to establish a farm around Ankara, we focused on the current location of the farm the least. This place was not treated kindly by nature; it looked neglected and pale and made you feel pessimistic when you looked at it. The reeds in the swamp of the land through which trains pass poisoned the urban life made the people living around appear as pale and sick and acting as a source of malaria. Eagles and vultures made nests around the land, which had nothing but an adobe house back then. There was only a thin railroad as the sign of civilization and humanity on this land. When our assessments were complete, we submitted the result to the Great Chief. Ataturk showed the current location of the farm and asked: Did you visit there? We extended our common thought that the location he pointed out had none of the qualities needed for a farm and that it was only a swamp with arid sections. He replied as follows: This is what we want. A swamp, arid, poor location around Ankara. It is up to us, rather than anybody else, to physically improve that location" [5].

Ataturk's own words show that he targeted agricultural amendment before agricultural activities. Agricultural undertakings are accomplished by finding a suitable location for the pre-determined agricultural product and conducting agricultural activities in that location. However, as noted in the expert's report on Ataturk Forest Farm, Ataturk wanted to correct the land for agriculture first and only later to conduct agricultural activities.

The assessments of agriculture experts were based on the probability of performing agricultural activities on the farm. After determining the location of the modern farm land, some experts claimed it was unsuitable for any agricultural activity, but some admitted that, with stringent efforts, the land could be made suitable. For instance, Schmid, an expert from the Ministry of Agriculture, stated: "This is such a plan that either patience or money will be consumed on these arid lands and in these climate conditions" [5]. His statement is supported by the detailed soil analyses provided below, which were performed in 1926. These soil analyses indicate that the activities performed at Ataturk Forest Farm were conducted in the light of the most up-to-date data at the time.

ARID SOIL (Source [8]:)
Chemistry Laboratory
Number 32/42

1926 ANALYSIS REPORT
Sent by: Directorate of Forest Farm.
Analyzed: Soil sample.
Amount: Approximately 10 kilograms.
Result of mechanical analysis:

Main	% in the soil from which water vapor is dissolved
<0.002 millimeters	38.0%
0.002–0.02 millimeters	33.5%
0.02–0.25 millimeters	27.5%
0.25–2.0 millimeters	1.0%

Source [8].

Chemical analysis:	
Amount of carbon calcium	4.1%
Amount of fluid solution:	
Cations: Alkali ions (Na and K)	2.5138%
Calcium ion (Ca)	0.0337%
Magnesium ion (Mg)	0.0709%
Anions: Hydrocarbon ion (HCO4)	0.0419%
Chlorine ion (Cl)	0.2190%
Sulfate (So4)	2.2698%
Nitrate (No3)	0.0845%
The main carbon amount was calculated through Na_2Co_3	0.0056%
Concentrated hydrogen ion (Ph)	8.7%

Source [8].

Different experts analyzed the lands of Ataturk Forest Farm at different times, and the analyses revealed that the soil structure of the farmland was not homogeneous. Therefore, different agricultural activities were performed on the different soil types.

Agriculture is affected by all climate parameters but the amount and timing of rainfall is the most important parameter. Water should be supplied, or water resources should be accessed to supplement rainfall and sustain agricultural activities.

Accordingly, Atatürk Forest Farm interventions to access water included the following:

1. Building a barrage to irrigate the bottomlands with water from Ince Su and Bend Lake, and opening a 10-kilometer canal;

2. Building a large barrage to irrigate the other large section of the plain from Çubuk River, and opening a second canal, nine kilometers in length;

3. Building a barrage to irrigate the Tahar plain from Macun and the Çubuk River, and opening another canal, 1 km in length;

4. Building a concrete barrage to collect the surface water flowing through the Tahar strait in winter, and to irrigate the lands by collecting the underground water;

5. Collecting the underground water in Çorak River and building a pond;

6. Building an artificial pond on Kelek meadow to collect rainwater and underground water;

7. Building a dam in Istanbul that is 146 meters long, and opening irrigation canals;

8. Collecting the underground water in Çakırlar Farm through galleries, and irrigating crops in time; and

9. Purchasing five large transportable centrifuge pumps and irrigating crops throughout the farm [5].

Using irrigation increased productivity on Ataturk Forest Farm, and agricultural activities that suited the climate and land conditions were performed in certain parts of the Farm, although the infrastructure activities mentioned before were too expensive [5]. Ankara has been a model for agricultural activities. Due to the nature of agriculture, fruit productivity at Ataturk Forest Farm falls to minimum levels and significantly varies, year by year, due to climate issues, even though conservation actions are taken" [5].

3. Forestry and Ataturk Forest Farm

Ataturk Forest Farm did not focus solely on agricultural activities; it also played a key role in growing trees that were conserved under the conditions of that era. By building a forest, *a unity of life collectively formed* by trees, shrubs, bushes, and herbaceous plants of a certain height, characteristics and density, kelp, Boston ferns and fungi, microorganisms living under and over the ground, and various bugs and animals, all of which can create a specific climate in the borders of the forest, is created [9].

The afforestation of Ataturk Forest Farm was greatly assisted by the *Forestry School* [10] opened in İstanbul in 1857. The first afforestation activity conducted in Turkey dates back to 1892. Aleppo pine, cedar, black pine, gallnut, and ash trees were planted in a 20–25 decare area. The fatigue arising from World War I and after the Turkish war of independence, the rapid occurrence of reforms, and limited financial resources, all meant little importance was attributed to forestry and afforestation till 1937. Afforestation of Ankara's Ataturk Forest Farm was renewed after World War II when the Yalova-Termal afforestation and Tarsus-Karabucak eucalyptus forest were among the important afforestation activities [11]. Various laws enacted between 1923 and 1937 enabled performing afforestation at Ataturk Forest Farm, but no legal actions regarding the control of flooding were taken during those years [12]. The micro-climatic impact expected from the afforestation activities performed at Ataturk Forest Farm has finally been realized today, although conditions have changed since those early days. The necessity of a challenging study can still be mentioned if the soil analyses provided above are to be assessed by modern forestry and agriculture experts.

3.1 Ataturk Forest Farm being gifted to the treasury

Mustafa Kemal Ataturk and his farmworkers made great efforts for 13 challenging years. At the end of this period, Ataturk thought about granting his properties and goods to the Republican People's Party, which he founded, and he mentioned this idea during the congress of 1927 [5]. In 1937, Ataturk submitted a letter to the Speaker's Office at the Grand National Assembly of Turkey indicating his wish to donate the Ataturk Forest Farm, and he stated [5]: "As you know, I founded numerous farms in different parts of the country at different times to gain experiences in terms of agriculture. I give all these farms which hosted various agricultural arts as well as all sorts of agricultural products produced under relevant climates *during the challenging efforts lasting 13 years along with all equipment, animals, and fixtures to the treasury*. A brief list indicating the lands of farms, equipment, and fixtures is attached" [7].

Ataturk mentioned the 13 years of challenging effort spent building up the farm in the letter in which he donated Ataturk Forest Farm to the treasury—and the challenge and the effort remain even under modern conditions. Humanity's interest in (and opposition to, in certain cases) soil and climate continue to be manifested by forestry and agricultural activities. Nevertheless, it was the dedicated efforts of Ataturk and his farmworkers that turned this challenging process into a success story.

After 1937, the farmlands were managed by the government. In 1950, when Şemsettin Günaltay was the Prime Minister of Turkey, "The Law on the Foundation of Ataturk Forest Farm" was enacted. The 10th Article of the law states:

> *The transfer and handover of Ataturk Forest Farm as well as the real estates within the border of the farm during the time this law was enacted to natural and legal persons and expropriation of the aforenoted are subject to receiving permission through a special law. Prior to the enactment of this law, the aforenoted shall not be applied to the real estates that were sold to the official offices and organizations through the approval of the Administrative Board of State Agricultural Business Department and Ministry of Agriculture [13].*

Accordingly, some sections of Ataturk Forest Farm were sold, and thus the surface area of the farmland shrank. The farm has continued shrinking physically because land from Ataturk Forest Farm was generally transferred in four ways: through a special law, through a protocol, through renting, and without law or protocol. The percentage distribution of the transferred land is as follows [14]:

Distribution of transferred lands	**(%)**
National Ministry of Defense	65%
Various public institutions and organizations and universities	22%
Metropolitan Municipality of Ankara	8%
Worker houses, farmers, different building societies	5%

Parts of Ataturk Forest Farm continued to be transferred for approximately 100 years. This was because a population of 250–300,000 people was estimated when planning the development of Ankara in the 1930s, but the actual figures were much higher. The most unexpected result of this process is that Ankara now has a population of more than 5.5 million people. This unexpected population increase put pressure not only on Ataturk Forest Farm but also on other natural resources. What would Hermann Jansen, the person who created the development plan of Ankara in the 1930s, say when he was told that the population of Berlin where he resided would remain at 3.5 million even after a century but that Ankara, the arid city in the middle of Anatolia, would have 5.5 million people 100 years later?

Considering the conditions of the era, neither Jansen nor anybody else would believe such a claim. While this unexpected population growth expanded Ankara, it shrank Ataturk Forest Farm. The process started during the era of Ataturk's contemporaries after Ataturk passed on. This transformation was not directed against Ataturk Forest Farm; instead, it was conducted in conjunction with efforts to maintain and improve the Republic of Turkey.

3.2 Ecological impacts of Ataturk Forest Farm

The transformation of urban spaces undoubtedly causes changes in environmental processes. In a 2009 study by Bilgili, the temperatures at different areal and green locations of Ankara were measured by fixed and mobile climate stations. Ataturk Forest Farm was statistically shown to be cooler than the built-up areas around Altınpark, Gençlik Parkı and Kurtuluş Parkı and was also cooler than the other green areas. The scientific findings varied according to the green areas, the micro-scale climate change caused by tall trees with large petal leaves in and around Ataturk Forest Farm, and the spatial (size), structural (pattern and plant cover), and temporal (phenological periods) traits of green areas. Therefore, the contribution of green areas to an urban ecosystem is a dynamic process caused by the collective impact of these areas' spatial, structural and temporal traits, and it changes over time. According to the Normalized Difference Vegetation Index, the plant cover of Ataturk Forest Farm, and of the forests surrounding the agricultural areas there, have a linear form, distinctively separate from the agricultural areas [15].

The trees planted approximately a century ago still have an impact on the climate parameters of Ankara in current times. Because it is in the center of the city, Ataturk Forest Farm's spatial traits that will create Ankara's green area system for the next century should be reassessed at this point.

Conflict of interest

The authors declare no conflict of interest.

Author details

Mustafa Ergen[1]*, Bayram Cemil Bilgili[2] and Yaşar Bahri Ergen[3]

1 Sakarya University of Applied Sciences, Sakarya, Turkey

2 Çankırı University, Çankırı, Turkey

3 Aksaray University, Aksaray, Turkey

*Address all correspondence to: mustafaergen2002@yahoo.com

References

[1] TURKSTAT. General Censuses. Turkish Statistical Institute; 2020. Retrieved from: http://www.tuik.gov.tr/PreTablo.do?alt_id=1047 [Accessed: 26 May 2020]

[2] Memlük Y, Keskinok Ç, Dinçer G, Kılıç C, Şenoğlu K, Erkan K. Bir Çağdaşlaşma Öyküsü Cumhuriyet Devriminin Büyük Eseri Atatürk Orman Çiftliği. Koleksiyoncular Derneği: Ankara; 2007

[3] Toprak Z. Cumhuriyet Ankara'sında İlk Nüfus Sayımı (Tecrübe Tahriri - 1927). Ankara Derg. 1991;**1**(2):57-66

[4] Burat S. Yeşilyollarda Hareketle İstirahat': Jansen Planlarında Başkentin Kentsel Yeşil Alan Tasarımları ve Bunların Uygulanma ve Değiştirilme Süreci (1932-1960). İdealkent Cilt. 2011;**4**:101-102

[5] Öztoprak İ. Atatürk Orman Çiftliği'nin Tarihi. Ankara: Atatürk Araştırma Merkezi. 2006;**14**:25-35

[6] Kimyon D, Serter G. Atatürk Orman Çiftliği'nin ve Ankara'nın Değişimi Dönüşümü. Cilt. 2015;**25**(1):51

[7] Anonim. Atatürk ve Tarım. Antalya: Akdeniz Üniversitesi Ziraat Fakültesi; 2020

[8] Anonim. ODTÜ Mimarlık Fakültesi AOÇ Araştırmaları. Ankara: ODTÜ; 2020. Retrieved from: http://aocarastirmalari.arch.metu.edu.tr/belge-yapitlar/.s.12

[9] Anonim. Türkiye Orman Varlığı. Ankara: Orman Ve Su İşleri Bakanliği Orman Genel Müdürlüğü; 2014

[10] Anonim. İstanbul Üniversitesi Orman Fakültesi Tarihçe. İstanbul: İstanbul Üniversitesi Orman Fakültesi; 2020

[11] Arslantaş M. Atatürk Orman Çiftliğinde Kızılçam (Pinus Brutia), Karaçam (Pinus Nigra), Sedir (Cedrus Libani) Türleriyle Yapılan Ağaçlandırma Çalışmalarının 6 Yıllık Sonuçlarının İrdelenmesi. T.C Isparta Uygulamalı Bilimler Üniversitesi. 2019;**1**

[12] Mustafa Dal, "Türkiye'de Yandere Islahı Çalışmalarında Örgütsel Yapı", İstanbul Üniversitesi Orman Mühendisliği Anabilim Dalı Orman İnşaatı ve Transportu Programı Yüksek Lisan Tezi. 2011. s. Viii

[13] Türkiye Cumhuriyeti Resmi Gazetesi. 5659 sayılı Atatürk Orman Çiftliği Kurulması Kanunu. Ankara; 1950

[14] ATAK E, ŞAHİN SZ. Atatürk Orman Çiftliği'nin 79 Yılı ve Çiftliğin Korunmasına Yönelik Politika Arayışları. Planlama. 2004;**3**:80-88

[15] Bilgili BC. Ankara Kenti Yeşil Alanlarının Kent Ekosistemine Olan Etkilerinin Bazı Ekolojik Göstergeler Çerçevesinde Değerlendirilmesi Üzerine Bir Araştırma. Ankara Üniversitesi Fen Bilimleri Enstitüsü Doktora Tezi; 2009. pp. S.1-165

Chapter 2

Erosion Quantification and Management: Southeastern Nigeria Case Study

Frank Chibuzor Okenmuo

Abstract

Soil erosion in Southeastern Nigeria is assuming an unusual dimension despite efforts by successive governments to control the phenomenon. Agronomic activities on eroding surfaces can give rise to landscapes much different from the original. Research activities in erosion quantification, the findings and how their applications have contributed to soil erosion management are highlighted. A key factor is the community efforts which have been relegated to a top-down approach occasioned by land use, land tenure and technological changes. The system is often a preventive management approach which achieves ecological and economic benefits. This chapter also discusses the indigenous methods of soil conservation and proposes their inclusions for sustainable management. To manage soil erosion in the region, emphasis must be placed on preventive management rather than crisis-management. Such approach will ensure that fewer resources are expended and land is appropriately conserved. To this end, soil can play its many environmental roles adequately.

Keywords: soil erosion, indigenous knowledge, soil conservation, erosion quantification, land use

1. Introduction

Methodologies for sustainable management of land degradation, economic growth and poverty reduction have become topical issues in present African research activities because of the danger posed by their neglect [1]. Land degradation especially the soil erosion aspect has been recognized as a serious threat to environmental sustainability. It impacts life on earth through degradation of land resources, loss of farmlands, decline in soil fertility due to top soil losses, contributes to climate change due to a compromise in soils C-sink potentials. In lowlands, eroded soils are often deposited as sediments on both land and river bodies. Thus further impoverishing rural communities who are often ill- equipped to manage the threat on land and water resources. As a result of the many implications of soil erosion on the environment, many efforts have been made to adequately understand the phenomenon so as to better manage it. Many of such efforts have failed due to little consideration of the several factors and their environmental peculiarities. The factors include: rainfall, soil properties, topography, and land-use and management. Since soil erosion begins in the farmer's field, scientific results could

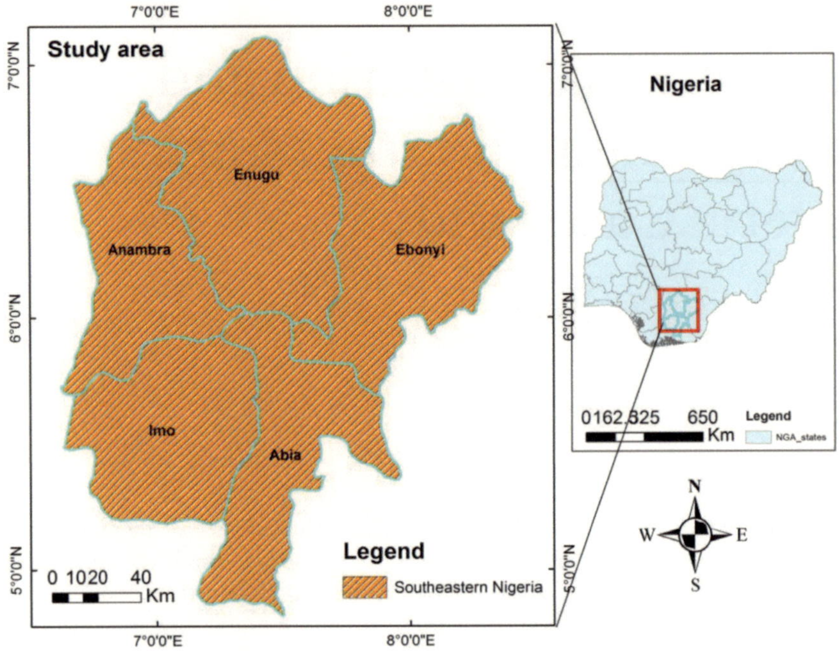

Figure 1.
Location of the reviewed area showing the states

be complemented by indigenous knowledge interventions in order to achieve better results. Local people have not only found nomenclatures for their soils, the indigenous knowledge has also extended to practices that have been through longterm observation of their interaction with the environment and transferred from generation to generation. Ezeaku and Salau [2] defined indigenous soil knowledge system as "the knowledge of soil properties and management possessed by people living in a particular environment for a long period of time. Soil conservation practices applicable to the North- Western zone of Nigeria include: contour farming, ridge tying, strip cropping, crop rotation, planted fallows, conservation pits, crop livestock farming and adequate fertilizer use [3, 4]. Howevever not much studies have been carried out in the Southeastern Nigeria (**Figure 1**) which is particularly an erosion prone zone.

2. Soil erosion and landscape evolution

Landscapes evolve under the influence of a complex suite of natural processes, many of which may be either directly or indirectly influenced by land use [5]. For example, a long history of cultivation can leave significant footprints on the original landscape. Under an unfavorable land-use condition, soil particles are moved from one position to another through agents of erosion such as wind, water and gravity. The series of particle movement consisting of detachment, transportation and deposition contribute to the evolution of landscapes. This is more so when the phenomenon occurs at accelerated dimensions as a result of continued anthropogenic activities. The effect of the denudation is a drop in soil surface level [6]. Future landscape evolution driven by soil erosion is expected to be exacerbated by land-use change, agricultural intensification and climate change [7, 8], coal mining and exploration [9], terracing [10], other mining activities, road networks and agricultural practices [11]. With current efforts to make resources available to the ever increasing population, the impact of these events in terms of magnitude and impact is expected to increase and destabilize geomorphic systems. In Southeastern Nigeria, Nwajide [12] observed that a major exogenic geological hazard is soil loss due to sheet and gully erosion. Of which the sheet type occurs as more or less removal of topsoil by flood but does not appear to threaten agricultural production or human habitation. Nevertheless, during major floods, sheet erosion may threaten small holders' food production for a period of time (**Figure 2**). However, its impact is often considered to be obliterated by the rapid rate of soil regeneration. The gully types have been observed to be the most obvious because of the remarkable impression they leave on the surface of the earth [13]. A gully is a distinct channel carved by running water into an unconsolidated substratum, and through which water flows only during and immediately after heavy rains. They are also a visible manifestation of the physical loss of the land due to erosion (**Figure 3**). Idike [13] observed that most studies of soil erosion in Southeastern Nigeria had strong focus on gully incision and gully prone regions but less on the little noticed sheet erosion. Such erosion contributes to land degradation but often in slower dimensions that land users fail to notice and are they often occur alongside with gullies in erosion prone areas. It is quite clear that soil erosion alters hydrology and landscape and connectivity patterns therefore necessitating efforts towards its better quantification.

Figure 2.
Inspection of lowland sites affected by sheet erosion following the 2012 flood at Odekpe, Southeastern Nigeria.

Figure 3.
A gully incised landform at St. Francis Enugwu-Ukwu, Southeastern Nigeria.

3. Soil erosion quantification techniques

In Nigeria, the detailed history of erosion quantification evolution may have been lost. However, there are remarkable timeline of events that are observable from literature. Earlier (1930–1955), soil erosion studies were descriptive, involving much field surveys and subsequent mapping. It was dominated by geographers and geologists who made efforts to understand the soil and its environment at regional scales (for example, see [14–16]). The second phase of soil erosion researchers (1955–1985) were mainly agriculturally-inclined (agronomists and soil scientist) with fewer geomorphologists, who established runoff plots, simulated soil loss using desurfacing approaches and a host of other techniques which were experimented in order to understand soil-water, erosion-productivity interactions in fields [17, 18]. The transitional period (1985–2000) focused on attempts to integrate technology towards broadening the scale of soil erosion research [19–24]. Presently (2000-date), soil erosion research has been multidisciplinary, multidimensional (local, regional and global) and with strong links to global issues such as social inclusion, sustainability and climate change. It is worthy to note that researchers continue to apply different methods depending on their objective and no method is obsolete Per se but a compromise of the other. The methods that have been applied by researchers in Southeastern erosion quantification are summarized in **Table 1**. The common erosion quantification methods are discussed below.

3.1 Mapping and direct field observations

Photographs can be used for detecting morphological change at varying scales and for recording the spatial relationship of landforms in order to provide three-dimensional information that can be used to construct Digital Terrain Models (DTMs) (for example, see [31]). It is also carries supplementary details useful for interpreting erosion rates or patterns. Maps produced as a result of detailed reconnaissance land resources surveys generally also contain information on the erosion hazard and on evidence of past erosion. Areas affected by sheet-rill-and gully- erosion can be recognized on aerial photographs and the growth of the erosion affected areas or the effects of conservation measures can be be traced from available maps or photographs and additional information collected in the field. Information obtained in the field may include using a simple scoring system to rate the severity of the erosion from e.g. the exposure of tree roots, the surface crusting, the thickness of the A horizon, erosion forms and shapes etc. Onweremadu [32] used field sampling aided by morphological landscape changes to identify erosion units for conservation treatments. However, field surveys can be time consuming but with the development of remote sensing techniques, more efficient methods of obtaining spatiotemporal erosion information are emerging. The disadvantages of using mapping as a tool for assessing soil erosion are the needs for cartographic skills, challenge of ascertaining whet difficulties in interpreting whether the current situation of the erosion phenomenon time constraints and variations in map quality.

3.2 Runoff plot studies

Runoff-plot methods are designed by using artificial boundaries to define a plot area and sediments are collected from a receptacle downslope. They could also be closed plots systems which uses rainfall simulators to study erosive events or open systems. Runoff plots are valuable research tools in soil erosion and surface runoff (soil loss) studies, evaluating conservation measures, effect of different crops and

DOI: http://dx.doi.org/10.5772/intechopen.99551

Location	Erosion Type	Quantification method	Result	Management recommendation	Author(s)
Abia State	gully	GPS and GIS	False bedded sandstone formed the major area for gully dispersion especially where slope was greater than 15 degrees	Proper land use	[25]
Onitsha, Port Harcourt, Owerri, Enugu, Uyo, Calabar, Ikom and Ogoja	Sheet	USLE, Rainfall erosivity	Severe erosion causing rains are associated with the rainy seasons; Calabar, Owerri and Port Harcourt had very high erosivity index	Monitoring hydrologic and climate related factors as well as land-use management	[26]
Anambra	Sheet	GIS and RUSLE-based	Mean value of estimated soil loss of 214.82t ha-1.;High rainfall erosivity combined withhigh slope factor and decreasing vegetal cover	Comparison between estimated and measured soil loss	[27]
Anambra and Enugu States	Sheet, interill and rill erosion	SLESMA and USLE erosion models	Coincidental and significant relationship between the USLE estimated maps and extent of actual gullying on ground	assesment of individual soil erodibility and not rating based on soil taxonomy	[24]
Orlu-Okigwe Asix of Imo State	Gully	Field and Landsat data	Surface phenomenon of washing away of loose top soils is not the only factors responsible for gullies but also deep-seated near-surface structural weakness along gully axis.	Emergency environmental monitoring	[28]
Anambra Basin	Gully	Field, Remote sensing and Geotechnical analysis	Development of gullies on steep slopes and non-vegetated areas are facilitated by cohensionless and very permeable nature of Ajali and Nanka sandy formations	Agronomic and engineering practices could help mitigate formation and expansion of gullies	[29]
Orashi Catchment, Anambra State	Not known	Field, expert judgment and remote sensed data	Vegetation and slope are the main factors governing erosion susceptibility	Change in landuse to a more integrated basin development system and public awareness on soil conservation and strategic planning	[30]

Table 1.
Summary of some erosion quantification approaches and management recommendations in Southeastern Nigeria.

management practices. They are commonly used to monitor hillside erosion but the design of runoff plots (in terms of plot dimension, runoff and erosion collection system, methods to monitor sediment concentration etc) are not standardized, making their results technique dependent [33]. Li *et al.* [34] anticipated that runoff research will tend to be more precisely location and model-inclined, technologically advanced and quantitatively precise in future. Iwara and Ewa [35] constructed erosion plots on natural fallow vegetation varying ages in southeastern Nigeria. They observed that July to September experienced highest amount of runoff and sediment losses. A better performance of the 10 and 3-year old fallow over the 5-year old fallow lead them to the conclusion that surface cover type and extent had greater influence over erosion processes than the age of fallow. The use of runoff-plot often alters the natural hydrology of fields due to their artificial boundaries and therefore may not accurately represent the actual erosion conditions. Extrapolation of the plot scale experiment beyond the area of observation may also be erroneous.

3.3 Erosion pin technique

The erosion pin method is a simple and feasible approach for soil erosion monitoring by inserting rods or nails into surface of slopes and using the basis of length of pin exposed or movement of washer placed on the pin. The technique has been successfully been modified and its photo-electronic erosion pin (PEEP) modification was efficiently used to monitor stream bank erosion by Lawler in 1989 [34, 36]. Erosion pin can aid in dynamic monitoring of the initial stage of gullying by identifying surface roughness, detachment and deposition. It can also conveniently monitor bank collapse and other short-term field monitoring. Some of its limitation include: susceptibility to environmental and human interference, need for close contact with assessed land and small range of observation. It can also be used to monitor gullies and landslides.

3.4 Erosion marker technique

Erosion markers allow carrying out analyses at larger temporal and spatial scales than those that are achieved through experimental plots. Bio-markers such as tree ring characteristics have been used to estimate the rates of soil erosion from decennial to millennium time scales by applying dendrogeomorphology [37]. The original landscape in relation to exposed roots can be a marker of soil erosion processes. However, as it is not always easy to identify the original land surface level, the vertical distance from an exposed root to the present ground surface may represent an underestimation of the total depth of the material [38]. The use of biomarkers is useful for long-term erosion quantification but it is subject to errors due to the natural variability of plants.

3.5 Radionuclide tracer method

Over the last few decades, geochemical methods have also been used to quantify erosion rates at different temporal scales. Examples of radionuclides which have been used as erosion tracers include ^{137}Cs, ^{210}Pb and ^{7}Be. The application of environmental radionuclides in soil erosion surveys is based on the premise of adsorption and redistribution of fallout by soil and sediment particles following erosion and sedimentation [39]. Radionuclide observations showing losses compared to the reference value indicate erosion. Observations greater than the reference value shows deposition. Unfortunately, this approach is yet to be applied to Southeastern Nigeria. Its first application in Ibadan, Southwestern Nigeria was reported to be a valuable alternative to conventional methods for soil erosion for obtaining quantitative data on soil erosion and deposition [40].

3.5.1 ^{137}Cs tracer method

^{137}Cs tracer technology rapidly developed and became the major means for monitoring soil erosion, determining soil erosion and sedimentation rates, quantitatively analyzing soil net loss, and other applications in the field [34]. The principal limitations of the ^{137}Cs approach include the costs of analytical equipment and the difficulties experienced in interpreting medium term estimates of average soil redistribution rates in the absence of complementary information on land use patterns and intensity. The method also requires a long measurement time and a high cost of laboratory analysis.

3.5.2 $^{210}Pb_{ex}$ tracer method

According to Li *et al.* [34], the $^{210}Pb_{ex}$ tracer method can distinguish the changes in atmospheric particles and human causes of trace elements, the reconstruction of pollution sources, and the history of river deposition and erosion in the past 100 years. However, its limitations include complex sample processing, high accuracy requirements, and difficulty in obtaining the flux of deposition for a particular year. In future, development of this technique is to further improve the quantitative relationship between the amounts of $^{210}Pb_{ex}$ loss and soil erosion. Its combination with other tracers and models is also likely.

3.5.3 ^{7}Beryllium tracer method

Li *et al.* [34] observed that this method can be applied to evaluate soil erosion under a particular intensity of land use, thereby providing an important basis for the monitoring and control of soil erosion. However, the application of ^{7}Be tracing still has some problems. For example, the shallowness of ^{7}Be distribution complicates sampling. It is, however, important to recognize that the use of ^{7}Be measurements is best suited to situations where significant erosion events are separated by 5 months in order to minimize the effect of previous erosion [39]. Since ^{7}Be could reflect the effects of soil erosion factors. Therefore, the scope of applications of the ^{7}Be tracer method can be broadened to explore the comprehensive effect on specific small watersheds based on the hydrological and meteorological conditions of soil erosion [34].

3.5.4 Magnetic tracer method

The application of magnetic tracers has two aspects: (i) to trace sediment sources using magnetic minerals in the environment, (ii) to indicate environmental change in basins. Hence Li *et al.* [34] synthesized that the magnetic tracer method can reflect the history of land use pattern, vegetation succession, and soil erosion in a watershed. It can also identify the soil distribution and the erosion rate for certain period. Therefore, this method can be used to provide a theoretical basis for soil erosion prediction and monitoring, and a history of the development of small watersheds. The advantages of this method are the transportability of the equipment, the methods simplicity; meeting the need of large samples and non-destructive nature. The method is however constrained by inability to trace magnetic properties and depth of soil erosion or deposition. Presently, magnetic tracers have been used to study soil formation, classification of soils, and the quantitative description of evolution, occurrence, and development of erosion.

3.6 Soil erosion models

Soil erosion models are quantitative approaches in study of soil erosion. Based on literature searches, the application of models in Southeastern Nigeria is still minimal. Models can be classified into three groups viz. Empirical, Physically-based and Conceptual (partly empirical/mixed) [41].

3.6.1 Empirical statistical model

Empirical models are based primarily on observation and inductive logic from the environment. Empirical models such as the Universal Soil Loss Equation (USLE), Revised Universal Soil Loss Equation (RUSLE), the Unit Stream Power Based Erosion/Deposition model (USPED), Erosion Productivity Impact Calculator (EPIC). Empirical models are mainly based on the USLE and remain widely used till date even in regions with limited data. Li *et al.* [34] documented the advantage and disadvantage of these models to include: (1) the formula is concise and the meaning of each factor is clear. (2) The calculation method of the factor has been basically mature and the parameters are easy to obtain for the continuous improvement and perfection of the model. (3) After several years of verification and testing, the accuracy of the model meets the needs of the application. In the tropics, the incompetence of the USLE in its rainfall erosivity component has been overcome by the incorporation of rainfall erosivity values in the RUSLE [27]. Applying the equation in the Anambra area, they observed that about 1804.39 km^2 (39.49%) of the area had slight erosion rate of 0–10 t/ha/yr., while rates of erosion in 746.60 km^2 (16.34%), 1025.38 km^2 (6.28%) and 45.59 km^2 (1.02%) of the area are 10.6–85.3, 85.4–235.2, 235.3–608, 608.1–2200 and > 2200.1 t/ha/yr. respectively. They noted that high rainfall erosivity, moderate to high slope and decreasing vegetal cover were the major factors driving soil loss in the area. In an earlier study, Igwe *et al.* [24] compared the USLE and the Soil Loss Estimation Model for South Africa (SLESMA) in producing soil erosion working maps in Anambra and Enugu States, South-East Nigeria. They found out that the USLE model reflected better the actual field situations except for its high values, absolute values compared to the global scale. The values were categorized into very slight (<50 Mg/ha/yr); slight (50–150 Mg/ha/yr); moderate (151–500 Mg/ha/yr); severe (501–1500 Mg/ha/yr) and very severe (>1500 Mg/ha/yr). A similar high value of above 200 t/ha/yr. was reported in Uyo metropolis, Nigeria by Fashae *et al.* [42] who observed that the values corresponded with areas with active gullies and altered vegetation cover. Obinna *et al.* [43] applied RUSLE model on the entire Southeastern Nigeria and observed that the results corresponded with known areas of gully menace in the region. Most of the erosion hotspots were located around the north-eastern part of the region covering most parts of Ebonyi State, some parts of Enugu State (Northwest axis), Anambra State (South East and Central axis), and most parts of Abia State. It could, therefore, be concluded that the high number of active gully occurrence may translate into the likelihood of other forms of erosion in the tropics.

3.6.2 Physical process model

The physical process model is based on the study of the processes and mechanisms of soil erosion e.g. stream flow or sediment transport. Examples of physical models include Water Erosion Prediction Project (WEPP), European Soil Erosion Model (EUROSEM). The WEPP model can simulate soil erosion, non-regular steep slope, and soil, tillage, and management measures by calculating the temporal and spatial distribution of soil erosion and predicting the movement of sediment in the slope and basin [34]. The WEPP model reflects the applicability and ductility of the temporal and spatial distribution of erosion and sediment; thus, numerous scholars still use this method. Although the physical model greatly compensates for the defects of the empirical model, this approach also has some shortcomings. (1) The physical mechanism of soil erosion is relatively complex and unclear. Some parameters in the physical process model are still dependent on the empirical model. (2) The large range of the study area is the major obstacle that hinders the use of the model because of the exacting demand of the model parameters. (3) The structure of the physical process model is complex and may change because the form has not been unified.

3.6.3 Conceptual model

Conceptual models lie somewhere between physically-based models and empirical models, and are based on spatially lumped forms of water and sediment continuity equations. Parameter values for conceptual models have typically been obtained through calibration against observed data, such as stream discharge and concentration measurements [44]. ANSWERS (Areal Nonpoint Source Watershed Environment Response Simulation), CREAMS (Chemicals, Runoff and Erosion from Agricultural Management Systems), and MODANSW (MODifiedANSWers) are basically conceptual and event based models [41].

3.7 Remote sensing

Remote sensing allows detection of erosion in large areas using aerial photographs and satellite remote-sensing data without disturbing the studied land area. The assumptions in the use of this method include, first, that the geomorphic processes of interest produce detectable changes in the spatial or temporal pattern of electromagnetic radiation and, secondly, that any geometric distortions arising from the sensor can be discriminated from real changes in landscape features [39]. Quickbird, SPOT 5 and IKONOS are very promising for identifying erosion features, such as individual gullies. NigeriaSat-1 image data and Landsat ETM data have been applied for a comparative classification of landuse patterns and gully development in southeastern Nigeria [45]. There are also opportunities in obtaining data using cheaper sources such as unmanned aerial vehicles (UAVs), ordinary cameras, and smartphones. Due to the capability of this method to provide spatial and often real-time data over large areas, it is adjudged the best method suitable for Southeastern Nigeria. It is hoped that its potential will be more explored in future.

4. Factors of soil erosion

Ofomata [19] viewed the factors of soil erosion as two major components: physical (geological or "natural") and anthropogenic (human or "accelerated"). Highlighting that the human component is often exaggerated and the physical component underestimated, he divided the physical factors of soil erosion into four: climate (mainly rainfall), surface configuration (relief/slope), surface materials and vegetation. Igwe *et al.* [24] recognized rainfall, topography/relief, soil factors (geology and soil characteristics), vegetation, land use and management as the main agents that determine the extent of soil erosion hazard. The factors affecting soil erosion by water is commonly expressed in the Universal Soil Loss Equation (USLE) (Eq. (1)) as a multiplicative equation counting six environmental factors:

$$A = R \times K \times L \times S \times C \times P \quad (1)$$

Where A is the mean annual soil loss (metric tons per hectare per year), R is the rainfall and factor or rainfall erosivity factor (mega joule millimeters per hectare per hour per year), K is the soil erodibility factor (metric tons hours per mega joules per millimeter), L is the slope length factor (unitless), S is the slope steepness factor (unitless), C is the cover and management factor (unitless), and P is the support practice factor (unitless).

4.1 Rainfall erosivity factor R

The R-factor is the sum of individual storm *EI*-values for a year averaged over long time periods (>20 years) to accommodate apparent cyclical rainfall patterns [46]. The *EI* term is an abbreviation for energy multiplied by the maximum intensity in 30 minutes. In the humid tropical environments, rainfall amounts and intensities often exceed the infiltration rate of excessive runoff. Ojo-Atere *et al.* [47] opined that the phenomenon was common in cultivated fields where at the peak of the rainy season, intensities of rainfall often exceed the infiltration rate of 25 mm/hr. and the soils are also nearly saturated throughout the rainy season. In fact, an earlier study by Roose [48] attributed the severe erosion damage of bare soils in the tropics to the special erosivity of the tropical rainfall rather than the ferrallitic or ferruginous soils. Salako [49] evaluated the temporal variation of rainfall erosivity between sub-humid zone (Ibadan) and the humid zone (Port Harcourt) of Nigeria. He observed a strong positive relationship between rainfall erosivity and rainfall amount.

According to Salako [50], data required are such that rates of rainfall at short-intervals (preferably ≤15 minutes) must be known, and these are very rare in many developing nations. Note that although EI_{30} is recommended by RUSLE, E_{15} was recommended for the tropics to avoid underestimation of the R-factor. The trends in rainfall erosivity have been generally evaluated using commonly available annual rainfall amount data. Lal [17] postulated a combination of daily rainfall (A) amount and maximum intensity (I_m), expressed as AI_m as a reliable index for evaluating index of tropical rainfall. Obi and Ngwu [51] observed that Lal's index of AI_m had an advantage over other indices of KE > 1 and EI_{30} in Southeastern Nigeria. Extensive studies by Igwe *et al.* [24] applied a method proposed by Arnoldus [52] to calculate the R-factor of USLE because autographic rainguage was not present in the study location and this gave the equation an advantage over the other equations. This method used monthly rainfall data to construct sub-annual R factors and then aggregated the R factors to an annual scale (Eq. (2)). It was modified from Fournier [53]'s map of the theoretical risk of erosion in Africa based on the damaging effect of precipitation.

$$FI = \frac{pi^2}{P} \quad (2)$$

pi is the average precipitation in the wettest month of the year, *P* is the mean annual total of rainfall.

Due to unsatisfactory results in West Africa, the Fournier index was modified. The modified index is given as Eq. (3):

$$MFI = \sum_{i=1}^{12} \frac{pi^2}{P} \quad (3)$$

Where MFI is the modified Fournier index, *pi* is total monthly rainfall and *P* is the total annual rainfall.

In West Africa, Eq. (4) is best to determine rainfall erosivity.

$$R = 5.44MFI - 416 \tag{4}$$

where R is the rainfall erosivity factor (mega joule millimeters per hectare per hour per year) and *MFI* is the modified Fournier index.

4.2 Soil erodibility factor K

This factor relates to the rate at which different soils erode, due to inherent properties Generally, soil properties which affect detachability include; particle size distribution, organic matter content, soil moisture, presence of cementing material such as Fe and Al oxides, stability of aggregates, clay mineralogy, rock fragments and balance of cations on the exchange complex, permeability, soil structure and strength [54]. In southeastern Nigeria, clay content, level of soil organic matter (SOM) and sesquioxides such as Al and Fe oxides, clay dispersion ratio (CDR), mean-weight diameter (MWD) and geometric-mean weight diameter (GMD) of soil aggregates were observed to influence soil erosion hazards [55]. Different parent materials were studied by Obi *et al.* [56] using four (4) methods: wet-sieving method, the Wischmeier nomograph, portable rainfall stimulator and runoff plot measurements. They recommended that the nomograph approach were unsuitable for soil erodibility studies in Southeastern Nigeria. The influence of geology on soil erodibility has been noted. For example [55] reported that sites with the worst catastrophic gullies in the classical gully sites the whole of sub-Saharan Africa exists in Southeastern Nigeria on sandy geological formations of False-bedded sandstone, Coastal Plain sands, Nanka Sands and the Bende Ameki compare to their Shale formation counterparts. Nwajide [12] observed that most soils in Nigeria bear the property of the underlying parent material from which they were formed. This follows the behavior of the soil under erosive conditions. For example soils formed on limestone, dolomite and igneous rock were more resistant than soils of sandstone and clay sedimentary formations [24]. However, information on erosion categories of various sedimentary formations of South East Nigeria is rather scanty.

According to [47], soils in the tropics with high sand contents (>60%) and low silt and clay values (<12%) and (<40%) respectively are highly erodible. Also, the weak, fine crumb surface horizon and weak subangular subsurface horizons of the former increases its vulnerability to erosion. In contrast, [24] noted that both large and fine particles were more resistant to transport because greater forces were required to entrain the former and the resistance due to cohesiveness of the latter.

4.3 Topograpy factor LS

LS reflects the influence of length and steepness of slope on soil erosion, it determines the behavior of the surface runoff. It is defined as the distance from the point where overland flows starts to the point where either the slope steepness decreases to such an extent that deposition occurs, or where surface runoff enters a well-defined channel. According to Ojo-Atere *et al.* [47], topography modifies soil profile development in three ways: (1) by influencing the quantity of precipitation absorbed and retained in the soil, thus affecting soil moisture relations, (2) by influencing the rate of removal by soil erosion and (3) by directing the movement of materials in suspension or solution from one area to another. The thinness of the solum, less organic matter and less distinct horizons than soils on level or undulating topography has been attributed to erosive exposure of the lower horizons due to slope steepness. In Southeastern Nigeria, soil erosion can occur even at slope of 5% as highly friable sandstones from the upland yields to detachment due to concentrated runoff [19]. Even in highlands or cuestas with somewhat stable lithology and erosion resistance, aggressive runoff from them devastates the lowland areas especially at the toe slopes and river head-waters [57].

LS is expressed as a unitless ratio with soil loss from the area in question in the numerator, and that from a standard plot (9% slope gradient, 22.13 m slope length) in denominator. Although L and S factors can be determined separately, the problem has been simplified by causing the L and S factor and considering the two as a single topographic factor [58]. Eq. (5) below considers the effect of L and S factors:

$$L = (\lambda / 22.13)^m [65.4 \sin^2 Q + 4.56 \sin Q + 0.065] \quad (5)$$

where λ is the slope length (in meters) and m is an exponent factor equivalent to 0.5 for slopes steeper than 5%, 0.4 for slopes between 3–4%, 0.3 for slopes between 1–3% and 0.2 for slopes less than 1% (based on a Wischmeier' nomograph) and Q is the slope angle.

4.4 Crop management factor C

Erosion and runoff are markedly affected by different types of vegetative cover and cropping system (vegetation type). The factor is defined as the ratio of soil loss from a field with a particular cropping and management to that of a field with a bare, tilled soil. The factor ranges from 0 to 1.0, a value of 0 indicating a 100% protection of the soil against erosion and 1.0 where there is little soil cover (e.g. freshly graded bare soil on construction site) [59]. Vegetation intercepts raindrops by facilitating infiltration of water, improving organic matter soil composition, thereby ensuring minimal erosion. The stage of growth of the crop will influence the management need (e.g. fertilizer), ability to hold soil together and canopy protection. Landuse activities that deprive soil surface of its vegetation, contributing directly to sliding, slumping, sheet and gullying include; road construction, sand mining, urbanization, industrialization and general infrastructural development [19].

4.5 Erosion control practice factor P

The erosion control practice factor P is the ratio of soil loss under a particular practice compared with the soil loss occurring under normal tillage. It therefore accounts for the positive impacts the support practice. Control practices reduces erosion potential by influencing drainage patterns, runoff concentration, runoff concentration, runoff velocity and hydraulic forces exerted by runoff on soil [60]. This factor ranges from 0 to 1 and is 1 where there are no support practices an 0 under good conservation practice. The conservation measures usually included in this factor are contouring, contour strip cropping, grassed waterways, terracing and surface mulching. Conservation measures like conservation tillage, crop rotations, residue management etc. are incorporated in the C-factor [59]. The effectiveness of conservation practices and thus the value of the P-factor generally depends on the slope steepness.

5. Landscape management in focus

In colonial times, the British Government worked on natural resource management as interest was high in expanding commercial farming enterprises. The practices were often implemented without consideration of the natural schemes used by local people to protect their soils from erosion and fertility declines. Trees were exploited without ecological considerations and conservation approach was a top-down type. Most farmers quickly abandoned such conservation model. However with increasing resource demands, most farmers are presently using unsustainable farm practices. As such, soil erosion occurrence is exacerbated. However, little study has attempted to understand occurrence of erosion in farmer's field yet sheet erosion, rills and gullies occur there. Local people with experience have recognized the peculiarity of their soils and adopted practices that are suited to their soils. The practices identified include; ditches, water harvesting 'umi', ridging, agroforestry, manure application, mulching, soil stabilizing stones, multiple cropping and embankments. Others are fallowing, conservation tillage etc.

5.1 Some soil and water conservation practices observed in Southeastern Nigeria

5.1.1 Traditional drainage ditches

These are structures constructed by digging deep within the farm or outside so as to divert runoff water and debris before reaching the farmland (**Figure 4**). It also serves as a watercourse, channeling runoff into ground water with minimal sheet erosion. This practice reduces the distance over which runoff travels over farmland and reduces water logging conditions. Morgan [61] reported that such drainage practices are constructed along the slope, often covered with grass to prevent destruction, and primarily installed in areas with high rainfall rates. The depth of this structures are variable depending on the slope of the land and need of the farmer. Locally, farmers in the study area call the traditional ditches "Umi".

5.1.2 Ridges

Ridge preparation is carried out by hoe with minimal disturbance which is a practice of conservation tilllage. Ridges are constructed across the slope so as to reduce the velocity of water movement (**Figure 5**). However, in some areas farmers

Figure 4.
Traditional drainage ditches.

Figure 5.
Broken ridges across the slope allows water movement with minimal soil loss.

tend to ignorant of implication of slope direction. The ridges are low and with large spaces between. This is to facilitate ease of water flow along the ridges. Also broken ridges are observed to be conservation measures against erosion.

5.1.3 Embankment structures

Barriers made of vegetation (dried or live), short walls (earthen or concrete) (**Figure 6**) are used to reduce the velocity of runoff water which could lead to water erosion. Embankments can function as sediment filters, aid in runoff velocity reduction, infiltration facilitator and could serve as boundaries. Farmers' in the study area view this as a very important measure of checking sheet and gully erosion.

5.1.4 Manuring/residue mulch

Farmers shred parts of trees, leave the vegetative remains after harvest and sometimes add animal wastes to the soil (**Figure 7**). Eventually, the faster decomposing part serves as nutrient to restore soil physical, chemical and biological

Figure 6.
(a) Dried vegetative barrier (b) Wall barriers

Figure 7.
Animal wastes applied to enhance soil properties.

properties while the woody part is utilized as firewood. FAO [62] reported that the practice of manuring serves conservation needs by; protecting soil surface from adverse weather conditions, increasing infiltration and reducing runoff velocity, increase organic matter supply, reduces evaporation, nutrient recycling.

5.1.5 Multiple cropping/agroforestry

Most farmland in the study area consist of woody perennials and annuals in the same land unit. Examples of annuals observed include; Mango, Citrus, Oil palm, Banana, Plantain etc. (**Figure 8**). Young [63] attributed the potential of agroforestry as an erosion control measure to its capacity to supply and maintain a good soil surface cover by the tree canopy and the pruning material. The deep roots will helpto stabilize the soil and increase moisture absorption through transpiration by the trees and crops [62].

5.1.6 Rainwater harvesting

Reservoirs-like constructions, constructed for the purpose of storing the surface run-off, generated from the catchments area. In the study area, this practice began a long time ago and is still being practiced with modifications such as larger collection reservoirs. The harvested water does not only serve irrigation purposes and

Figure 8.
Agroforestry practices common in the study area.

Figure 9.
Stones incorporated into the soil so as to stabilize it against erosion.

erosion prevention, it also saves cost for the farmer's family. Other uses include; storing nutrients and rich soil materials as well composting etc. The reservoirs used for water harvesting vary in size depending on the need of the farmer.

5.1.7 Surface roughening

Farmers incorporate stones, charcoals, kernel shells, pieces of wood etc. so as to roughen and stabilize the soil against the effect of water erosion (**Figure 9**).

6. Conclusion

Soil erosion and other anthropogenic activities can leave remarkable footprints on earths' surface. In erosion prone landscapes, the earth's surface is continually evolving due to events and processes such as (i) road and foot tracks, (ii) land use and land cover changes, (iii) hydrogeological dynamics (iv) soil erodibility etc. To understand soil erosion, efforts have been made to assess and quantify the phenomenon. However, literature on soil erosion research in Southeastern Nigeria is still minimal and some promising techniques are yet to be tested. This may be partly the reason why soil erosion related land degradation remains critically high amidst scarce and often inaccurate data. Another reason is the non-inclusion of the sustainable home-breed sustainable solutions into erosion management plans. Such knowledge could assist in both research and practice in developing robust environmental conservation approaches; that is cognizant of cultural and socioeconomic situation of the area. Local people knowledge of the soil has helped to protect soil from erosion while ensuring that its fertility is maintained. When improved, better conservation outcomes in Southeastern Nigeria can be achieved.

Acknowledgements

The research leading to this chapter has received funding from the African Union, under the Pan African University Doctoral Scholarship.

Author details

Frank Chibuzor Okenmuo
Chukwuemeka Odumegwu Ojukwu University, Igbariam, Nigeria

*Address all correspondence to: fcvokenmuo1@gmail.com

References

[1] Ezeaku PI, Davidson A. Analytical situations of land degradation and sustainable management strategies in Africa. J. Agri. Soc. Sci. 2008:**4**:42-52

[2] Ezeaku PI, Salau, ES. Indigenous and scientific soil classification systems: A case of differences in criteria in some soils of Northcentral Nigeria. Production Agriculture and Technology Journal:2005:**1**(1):54-66.

[3] Junge B, Deji O, Abaidoo R, Chikoye D and Stahr K. Assessment of past and present soil conservation initiatives in Nigeria, West Africa. Conference on International Agricultural Research for Development; October 9-11, 2007. Tropentag. University of Kassel-Witzenhausen and University of Göttingen; 2007.

[4] Odunze AC. Soil Conservation and Management Practices. Department of Soil Science, IAR/A.B.U Zaria; 2008.

[5] Doe WW, Harmon RS. Introduction to soil erosion and landscape evolution modeling. In: Harmon RS, Doe WW, editors Landscape Erosion and Evolution Modeling. Springer, Boston, MA; 2001. https://doi.org/10.1007/978-1-4615-0575-4_1

[6] Kaffas K, Hrissanthou V. Introductory Chapter: Soil Erosion at a Glance, Soil Erosion- Rainfall Erosivity and Risk Assessment, IntechOpen; 2019. DOI: 10.5772/intechopen.89773. Available from: https://www.intechopen.com/books/soil-erosion-rainfall-erosivity-and-risk-assessment/introductory-chapter-soil-erosion-at-a-glance

[7] Walling DE and Fang D. Recent trends in the suspended sediment loads of the world's rivers. Global Planet Change: 2003:**39**:111-126.

[8] Panagos P *et al.* Global rainfall erosivity assessment based on high-temporal resolution rainfall records. Scientific Reports: 2017:7(4175): 1-11 DOI:10.1038/s41598-017-04282-8

[9] Agwu J. Gender and climate change in Nigeria. A study of four communities in north-central and South-eastern Nigeria. Heinrich Böll Stiftung (HBS); 2009. Available from: www.boellnigeria.org.

[10] Tarolli P, Preti F, Romano N. Terraced landscapes: From an old best practice to a potential hazard for soil degradation due to land abandonment. Anthropocene: 2014. Available from: http://dx.doi.org/10.1016/j.ancene.2014.03.002

[11] Tarolli P, Sofia G. Human topographic signatures and derived geomorphic processes across landscapes, Geomorphology: 2015. Available from: doi:10.1016/j.geomorph.2015.12.007

[12] Nwajide CS. Geology of Nigeria's Sedimentary Basins. CSS Press, Lagos, Nigeria; 2013; 507.

[13] Idike FI. On Appraising Soil Erosion Menace and Control Measures in Southeastern Nigeria Soil Technology: 1992:5:57-65.

[14] Stamp LD. Land utilization and soil erosion in Nigeria. Geographical Review: 1938:**28**:32-45

[15] Skyes RA. History of anti-erosion work in Udi. Farm and Forestry: 1940:**1**:3-6.

[16] Grove A T. Land Use and Soil Conservation in Parts of Onitsha and Owerri Provinces. Geol Surv. Nigeria; 1951; Bull No. 21;74.

[17] Lal R. No-tillage effects in soil properties under different crops in Western Nigeria. Soil Science Society of America Journal:1976:**40**(5):762-768.

[18] Mbagwu JSC, Lal R, Scott TW. Effects of artificial desurfacing on Alfisols and Ultisols in Southern Nigeria: Changes in soil physical properties 1. Soil Science Society of America Journal:1984:**48**(4):1-6.

[19] Ofomata GEK. Soil erosion. Southeastern Nigeria: the view of a geomorphologist, Inaugural lecture series University of Nigeria Nsukka; 1985

[20] Akamigbo FOR. Erosion menace in Anambra state: Government efforts in combating it. A paper presented at the International symposium on erosion in South East Nigeria, Erosion research unit, April 12-14, 1988; Federal university of technology, Owerri.

[21] Nwajide CS. Gullying in the Idemili River catchment, Anambra State, Nigeria: theory and cure. In Freeth SJ, Ofoegbu CO, Onuoha KM, editors. Natural Disasters in West and Central Africa. Vieweg and Sohon Veralagsgesellschaft, Wiesbaden; 1992; 149-162.

[22] Igwe CA. The applicability of SLESMA and USLE erosion models on soils of Southeastern Nigeria [thesis], Department of Soil Science, University of Nigeria, Nsukka; 1994.

[23] Igwe CA. Land use and soil conservation strategies for potentially high erodible soils of central eastern Nigeria. Land Degrad. Develop. 1998:**10**:425-434.

[24] Igwe CA, Akamigbo FOR, Mbagwu JSC. Application of SLESMA and USLE erosion models for potential erosion hazard mapping in South-eastern Nigeria. Int. Agrophysics. 1999:**13**:41-48.

[25] Nwilo P, Olayinka D, Uwadiegwu I,Adzandeh A. An assessment and mapping of gully erosion hazards in Abia State: A GIS approach. Journal of Sustainable Development. 2011:**4**(5):112-122.

[26] Ezemonye MN, Emeribe CN. Rainfall erosivity in Southeastern Nigeria. Ethiopian Journal of Environmental Studies and Management, 2012:**5**(2):112-122

[27] Fagbohun BJ, Anifowose AYB, Odeyemi C, Aladejana OO and Aladeboyeje AI. GIS-based estimation of soil erosion rates and identification of critical areas in Anambra sub-basin, Nigeria. Model Earth Syst. Environ. 2016:**2**(159):1-10

[28] Ibeneme SI, Okoro C, Ezetoha NO, Ahiarakwem CA, Nwankwor GI Integrating Field And Landsat Data In Structural Mapping Of Gully Erosion Sites Within Orlu-Okigwe Axis Southeastern Nigeria. Journal of Earth Sciences and Environmental Studies. 2019:**4**(4):696-707

[29] Ocheli A, Ogbe OB, Aigbadon, OG. Geology and geotechnical investigations of part of the Anambra Basin, Southeastern Nigeria:implication for gully erosion hazards. Environmental Systems Research. 2021:**10**(23)1-16

[30] Odunuga S, Ajijola A, Igwetu N, Adegun O. Land susceptibility to soil erosion in Orashi Catchment, Nnewi South, Anambra State. Proc. IAHS. 2018:**376**:87-95

[31] Chiemelu N, Okeke F, Nwosu K, Ibe C, Ndukwu R, Ugwuoti A. The role of surveying and mapping in erosion management and control: Case study of Omagba erosion site, Onitsha Anambra State, Nigeria. Journal of Environ. and Earth Science. 2013:**3**(11):129-137.

[32] Onweremadu EU. Assessment of mined soils in erosion-degraded farmlands in southeastern Nigeria. Estud. Biol. 2006:**28**(65):59-67.

[33] Lal R. Soil Degradation by Erosion. Land Degradation and Development. 2001:**12**:519-539.

[34] Li Y, Bai X, Tian Y, Lou G. Review and future research directions about major monitoring method of soil erosion. IOP Conf. Ser.: Earth Environ. Sci. 2017:**63**:1-15.

[35] Iwara AI, Ewa EE. Seasonal variability in runoff, sediment and nutrient losses in vegetation fallows in the Rainforest zone of Southern Nigeria. Ghana Journal of Geography 2018:**10**(1):42-57

[36] Lawler DM. Some New Development in Erosion Monitoring: The Potential of Optoelectronic Techniques [R]. School of Geography, University of Birmingham Working Paper 47, 44; 1989.

[37] Prosdocimi M, Jordan A, Tarolli P, Keesstra S, Novara A, Cerda A. The effectiveness of barley straw mulch in reducing soil erodibility and surface runoff generation in Mediterranean vineyards. Science of the Total Environment: 2016:**547**:323-330

[38] Casalí J, Giménez R, De Santisteban L, Alvarez-Mozos J, Mena J, Del Valle de Lersundi J. Determination of long-term erosion rates in vineyards of Navarre (Spain) using botanical benchmarks. Catena. 2009:**78**:12-19. http://dx.doi.org/10.1016/j.catena.2009.02.015

[39] Collins AL and Walling DE. Documenting catchment suspended sediment sources: problems, approaches and prospects. Progress in Physical Geography. 2004:**28**(2):159-196.

[40] Junge B, Mabit L, Dercon G, Walling DE, Abaidoo R, Chikoye D, Stahr K. First use of the ^{137}Cs technique in Nigeria for estimating medium-term soil redistribution rates on cultivated farmland. Soil and Tillage Research. 2010:**110**:211-220

[41] Ganasri BP, Ramesh H. Assessment of soil erosion by RUSLE model using remote sensing and GIS-A case study of Nethravathi Basin. Geoscience Frontiers. 2016:7(6):953-961.

[42] Fashae OA, Ubom O, Adeyemi O. Estimating soil loss from soil erosion-induced land degradation in Uyo metropolis, Nigeria using remote sensing techniques and GIS. Ife Research Publications in Geography. 2013:**12**(1&2):120-137.

[43] Obinna CD, Anejionu PC, Nwilo PC, Ebinne ES. Long term assessment and mapping of erosion hotspots in South East Nigeria. FIG Working Week, Environment for Sustainability, Abuja, Nigeria, 6-10 May, TS03B-Remote Sensing for Landuse Planning; 2013;6448;1-19.

[44] Sorooshian S. Parameter estimation, model identification, and model validation: conceptual-type models In Bowles DS, O'Connell PE, editors. Recent Advances in the Modeling of Hydrologic Systems. 1991: 443-467.

[45] Igbokwe JI, Akinyede JO, Dang B, Alaga T, Ono MN, Nnodu VC, Anike LO. Mapping and monitoring of the impact of gully erosion in southeastern Nigeria with satellite remote sensing and geographic information system; The International Archives of the Photogrammetry, Remote Sensing and Spatial Information Sciences. 2008:XXXVII:865-872.http://www.isprs.org/proceedings/XXXVII/congress/8_pdf/9_WG-VIII-9/03.pdf

[46] Renard KG, Freimund JR. Using monthly precipitation data to estimate the R-factor in the revised USLE. Journal of Hydrology. 1994:**157**:287-306.

[47] Ojo-Atere JO, Ogunwale JA, Oluwatosin GA. Fundamentals of Tropical Soil Science. Evans Brothers (Nigeria Publishers) Limited; 2009.

[48] Roose EJ. Use of the Universal Soil Loss Equation to Predict Erosion in West

Africa. Soil Conservation Society of America, Ankeny, Iowa, 1976; 60-74.

[49] Salako FK. Temporal variation of rainfall erosivity in Southern Nigeria, ASSET Series A, 2007:7(1):190-202.

[50] Salako FK. Rainfall variability and kinetic energy in southern Nigeria. Climate Change. 86:151-164.

[51] Obi ME, Ngwu OE. Characterization of the rainfall regime for the prediction of surface runoffand soil loss in southeastern Nigeria. Beitr. Trop.Landwirtsch. Veterin~irmed. 1988:**26**:39-46.

[52] Arnoldus HMJ. An approximation of the rainfall factor in the Universal Soil Loss Equation. In De Boodt M, Gabriels D, editors. Assessment of Erosion. John Wiley and Sons: Chichester; 1980:127-132.

[53] Fournier F. Carte du danger d'erosion en Afrique au sud du Sahara. 1962; CEC-CCTA, Avril.

[54] Das DK. Introductory Soil Science. Kalyani Publishers: New Dehli, India; 2017:749.

[55] Igwe CA. Gully erosion in Southeastern Nigeria: Role of soil properties and environmental factors. In Godone D, Stanchi S, editors. Research on Soil Erosion. IntechOpen. 2012;8:157-171. DOI: 10.5772/51020. Available from: https://www.intechopen.com/books/soil-processes-and-current-trends-in-quality-assessment/development-of-topographic-factor-modeling-for-application-in-soil-erosion-models

[56] Obi ME, Salako FK and Lal R. Relative susceptibility of some southeastern Nigeria soils to erosion. Catena. 1989:**16**:215-222.

[57] Ofomata GEK. Soil Erosion. Nigeria in maps, Eastern States. Ethiope Publishing House, Benin City, Nigeria. 1975.

[58] Singh G and Panda RK. Grid-cell based assessment of soil erosion potential for identification of critical erosion prone areas using USLE, GIS and remote sensing: A case study in the Kapgari watershed, India. International Soil and Water Conservation Research. 2017:**5**:202-211.

[59] Brady NC and Weil RR. The Nature and Properties of Soils. 13th ed, New dehli: India Pearson Education Inc.; 2002.

[60] Rellini I, Scopesi C, Olivari S, Firpo M, Maerker M. Assessment of soil erosion risk in a typical Mediterranean environment using a high resolution RUSLE approach (Portofino promontory, NW-Italy), Journal of Maps. 2019:**15**(2) 356-362. DOI:10.1080/17445647.2019.1599452

[61] Morgan RPC. Soil Erosion and Soil Conservation. Essex: Longman; 1995

[62] FAO. Manual on integrated Soil Management and Conservation Practices. Methods of Soil Conservation held at IITA, Ibadan, Nigeria 21 April- 1 May 1997. FAO, Rome, Italy; 2000

[63] Young A. Agroforestry for soil conservation. Exeter: Wheatons; 1989.

Chapter 3

Therapeutic Landscapes: A Natural Weaving of Culture, Health and Land

Bruno Marques, Jacqueline McIntosh and Hayley Webber

Abstract

Current concepts of therapeutic landscape combine landscape with principles of holistic health and the interaction of social, affective and material factors. As social tensions widen the gap between the places of emotional retreat and healing from those of everyday sociability, concepts of therapeutic landscape are evolving to reflect society's current values. This chapter examines how cultural place-based values affect and maintain physical, emotional, mental and spiritual health and well-being in the context of a therapeutic landscape. Five case studies from Australasia, Africa, Middle East and Latin America are analysed to understand better the interrelationships between land, culture and health that make an environment therapeutic. The case studies were selected based on their engagement with the cultural traditions of landscape architecture and how the boundaries of these cultural traditions are negotiated within a modern context. The chapter contributes to the knowledge base of landscape architects and academics interested in the role of culture in producing and maintaining therapeutic landscapes by presenting a cross-cultural analysis to illustrate a range of strategies for incorporating cultural traditions and customs into modern landscape architectural contexts to promote health and well-being.

Keywords: therapeutic landscapes, therapeutic environments, culture, health, well-being, Indigenous knowledge, landscape, landscape architecture

1. Introduction

It has always been known that humans find reassurance in nature. Therapeutic sanctuaries have been sought out through the ages by individuals who have a sense of disorientation or likely future ambiguity. For example, those diagnosed with acute and terminal illness often search for a space of solitude, needing time and space for reflection on significant life and health changes [1]. Explanations for this phenomenon find that the brain is capable of two types of attention: directed attention, belonging to the higher cognitive centres, and soft fascination, linked to the old parts of the brain [2]. In natural environments, the higher cognitive centres of the human brain can rest and reset, lending some of the first scientific evidence to the healing powers of nature.

Most recently, as social tensions associated with the Covid-19 pandemic, climate change and racial inequity deepen the need for places of emotional retreat and healing, therapeutic landscapes have become an increasing topic of interest [3, 4]. However, while more and more research [5–7] finds that contact with nature plays a critical role in psychological well-being, and for those in need, a faster recovery from psychological trauma or stress, Ulrich & Gilpin found when researching hospital gardens, an example that had a measurable negative effect on patient's health [8]. Similarly, following an extensive review of healing gardens, Stigsdotter & Grahn found that not all gardens are healing gardens [9], leading to the question, what is a therapeutic landscape or a healing garden?

The term therapeutic landscape was originally defined in 1992 by the geographer William Gesler as a place "where physical and built environments, social conditions and human perceptions combine to produce an atmosphere which is conducive to healing" [10]. This definition was contested, critiqued and elaborated on over the next two decades, extending from a literal relationship between health and place with the acknowledgment of extraordinary places to a much expanded and refined characterisation. The extended literature reflected the growing interest in the relational and situated approach to well-being, which acknowledged the therapeutic nature of places in the context of those social, cultural, material, affectional and sensual aspects of human and non-human factors [1]. Such therapeutic encounters are defined differently by others, such as 'networks' [11], 'experiences' and 'environments' [12], 'taskscapes' [13], 'mobilities' [14], 'assemblages' [15] and 'enabling places' [16], with the latter gaining particular purchase amongst health and cultural geographers alike.

This theoretical turn saw a much greater emphasis on qualitative and ethnographic methodologies in the study of health designed to reveal the histories, discourses, and lived experiences of a place. Building on the contention that different people experience therapeutic landscapes differently or at different times, the potential healing outcome of the landscape can be seen as a relational process, and therapeutic landscapes become socially and culturally responsive. People's behaviour is deeply embedded within a place, particularly around health [17]. It is not just the space that is healing but the intention of those using the space [18]. Places should not only be defined by the fact they are conducive to healing but also places that are conducive to the maintenance of health and well-being [19, 20].

Successful therapeutic spaces generally reflect a society's current values and aspirations, and as such, become part of its identity. Situated in the current context, 'place' becomes even more challenged and implies that such spaces may need to be flexible to changing interpretation to remain therapeutic over time. Foley & Kistemann [21] considered therapeutic spaces as emergent through a set of embodied experiential practices linking affects, emotions, and bodily sense that arise from being immersed in such therapeutic environments. This form of assemblage can therefore be better understood through its material, metaphorical, and inhabited dimensions [15, 22]. The material component would contain the tangible aspects of landscape that people experience due to their therapeutic qualities. The metaphorical component comprises the ethnographic and cultural values expressed through narratives, myths, and stories that are crucial in defining site-specific rituals and cures. Inhabitation brings together mind, body, and spirit because it draws from lived, experiential, and performative health dimensions.

To this concept, Andrews [17] proposed two streams of application. One relates to the impact of landscape on human experience. The other pertains to how therapeutic landscapes are shaped by the influence of different belief systems, leading to the cultural specificity of the therapeutic landscape concept. In this way, culture and requirements for health create an assemblage of layers that combine in bespoke combinations depending on the individual and the time. The implication is that a healing environment cannot be achieved by a set of design requirements or details you need to meet, it must facilitate a healthcare service that is patient-focused and centres on the diverse needs of all [23]. This extends the therapeutic landscape concept to encompass both tangible and intangible values, supporting the proposition that health, well-being, and place are intricately intertwined and emergent through the layering of architecture and material; with practices and responses in a narrative of individual and communal history where people are simply seeking well-being.

Indigenous methods of restoring health and well-being offer new opportunities for understanding the complexity of health and well-being. Most of the Western understanding around therapeutic landscapes has come from the healing properties of the physical space and adopting a bio-medical approach, translated into spas, mineral springs and mountain retreats [24]. For example, in areas of natural springs, much was made of the unique chemical make-up of the water until scientists uncovered the lack of chemical difference [11]. With this narrow view, what was overlooked was the non-physical dimension to these spaces [24], as scientists failed to engage with and quantify the practices and traditions associated with these landscapes that enabled the healing process [25]. Similarly, what has not been translated into Western culture is the deep-rooted connection with the land and its relation to self-identity on an everyday basis. The associated values and practices embedded in culture and practice are often the foundation of the healing nature of the therapeutic landscapes. Practices in Indigenous cultures are able to be embodied and translated into everyday places and landscapes.

In fact, only a small portion of the literature on therapeutic landscapes discusses the inclusion of Indigenous methods and the relationship between the environment and health and well-being that most of their spiritual connections are grounded by. However, 80% of people in developing countries still rely on traditional medicines or methods. They are inexpensive and easily accessible, they are also believed to be stronger and more effective than treatments offered at a healthcare facility [26]. Indigenous methods and practices often combined with everyday health and well-being, mainly when connected to a sense of place and the sense of identity and symbolic healing the environment can enable [24].

This chapter explores therapeutic landscapes and landscape architectural approaches that can be applied to creating or understanding therapeutic landscapes that are culturally and socially responsive. It addresses the gap in the literature regarding therapeutic landscapes to understand how different social and cultural contexts can affect physical, mental, emotional, and spiritual health. Our intent is to illustrate landscapes that include cultural practice and Indigenous healing methods to promote well-being at both individual and community levels. Examining a series of case studies from around the world, it explores landscape architectural responses to healing. The case studies have been selected based on their success in catering to the social, mental, physical and emotional needs of the user. The chapter unpacks the social and cultural contexts that have shaped these environments and discusses how this can be applied to the universal design of everyday therapeutic landscapes.

2. Method

To investigate the ideas underlying therapeutic environments, a literature review was undertaken using electronic databases such as Scopus, Google Scholar and ProQuest, as well as landmark book publications. Search terms included therapeutic landscapes, therapeutic environments, healthcare landscapes, and restorative outdoor spaces to identify design criteria. Following the first selection and exclusion of non-relevant materials, twenty-two peer-reviewed publications were evaluated.

The articles were critically assessed by conducting a strengths and weaknesses analysis of each study and considering their relevance in promoting health and well-being. As causal relationships between therapeutic environments and human health are difficult to establish, this critical literature review focussed on studies that dealt with association rather than causation. To better understand the theoretical constructs and obtain an international perspective of approaches to health and well-being, 32 case studies were initially selected during the scoping of the literature to demonstrate a breadth of scales. To get an understanding of how different places with differing social and cultural contexts responded through their therapeutic environments, samples were selected from five different countries, namely Kopupaka Reserve in Aotearoa-New Zealand, Lions Park in Queensland, Australia, Freedom Park in Pretoria, South Africa, Jiyan Healing Garden, in Chamchamal, Iraq, and Jardín Etnobotánico de Oaxaca in Mexico, for closer investigation.

3. Kopupaka reserve

Kopupaka Reserve is a large urban wetland in Auckland, New Zealand. It is unique as it combines an expansive natural environment, which has been identified as having unique healing qualities, with Indigenous Māori cultural values. Māori call themselves the *tāngata whenua* (people of the land), which places importance on the connection between land and sea and respect and preservation of the natural environment and its people [27]. Māori collective cultural orientations brings with them a much more holistic approach to health and well-being, combining sustainable and restorative methods with genealogy (*whakapapa*).

In an effort to enhance the presence, visibility, and participation with design, a set of design principles for the creation of designed environments were developed based on Māori cultural values [28]. The values that underpin these design framework are *rangatiratanga* (chieftainship), *kotahitanga* (unity), *kaitiakitanga* (guardianship), *wairuatanga* (spirituality), *manaakitanga* (hospitality), *whanaungatanga* (relationship, sense of family connection), and *mātauranga* (knowledge and wisdom) [29]. Seven principles can be applied to all aspects of design thinking and require an ongoing engagement with Māori tribes. These principles are: *mana* (status of *iwi*, tribe, and *hapū*, sub-tribe, is recognised and respected), *whakapapa* and *mahi toi* (Māori names are celebrated by the recognition of traditional place names on signage and wayfinding and cultural landmarks are acknowledged), *taiao* and *mauri-tu* (the natural environment is protected, restored and/or enhanced), and *mahi toi* (Māori narratives are captured and expressed creatively and appropriately) [28].

The Auckland City Council commissioned the Kopupaka Reserve as a new public space for the emerging Westgate shopping centre. The Kopupaka Reseve is an ancestral name for the area and relates to the meeting of the Tōtara Creek and Waiteputa Stream. The reserve is of a linear shape that follows the riparian corridor of Tōtara Creek and Waiteputa Stream, which had become extremely degraded from the area's long history of growing market produce and most recently a large-scale strawberry farm. Once rich with *māhinga kai* (traditional Māori foods), the waterways had become overloaded with nutrients and infested with weeds. The design restored the native species that once would have grown in the area, including harakeke flax, traditionally used for weaving [28].

Wetland waterways provide an abundance of life with food and purification, contributing to the well-being of the people as Māori maintain that a healthy landscape makes for a healthy individual. Views over the wetlands surrounding the confluence of the two freshwater streams, display the varied birdlife and biologically diverse horticulture. Hard landscape forms are inspired by the abundance of *kai* (food) found in wetlands and guide the boardwalk design (**Figure 1**). Open spaces for communal activities are contrasted with private places for emotional retreat (**Figure 2**). The body is engaged through sitting or walking, the senses are engaged through sights, sound and smells. The park is a celebration of the inseparable bond of water and earth. This is deemed vital to sustaining and balancing the natural environment and which is reflected through wairua and which refers to the spiritual plane. Healthy water and rich resources are deeply embedded in Māori cultural and spiritual landscape values as water is sacred.

A distinct and defining feature of the park is a series of interlocked curving timber structures that frame the wetland ponds and a woven together with vegetated paths. The timber frames are a symbolic representation of the *hīnaki waharua* (eel baskets) that were used to collect the *tuna* (eel) that gathered at one of the historical wetlands in the area [28].

What makes Kopupaka Reserve a successful therapeutic landscape its ability to encourage community engagement and establish a strong sense of place and place identity. These themes are threaded throughout most of the literature that has been written on therapeutic landscapes. As research shifted from its focus on the big event or extraordinary healing spaces to the everyday, the method of healing also shifted from an internal singular process to a communal and inclusive process. Urban park and community gardens emerged as having healing qualities and were celebrated for their unique ability to bring communities together and improve overall well-being. The 22-hectare reserve is located within the centre of Westgate, a new town centre that has emerged as a result of urban sprawl. The park connects together the surrounding suburbs and is a central point for many people. While people are not directly forced to engage with each other or participate in rehabilitative activities, the opportunity is given. Signs and cultural motifs placed around the park tell a narrative of the history of the reserve, people feel more attached to the park as they can understand its significance. The diversity of spaces allows locals to find what they need from the park and offers opportunities to improve their personal health and well-being. The park aimed to restore the mauri (lifeforce) of the area, which was mainly achieved through improved storm water management and the revegetation of key areas improving the water quality of the area and

Figure 1.
Kopupaka reserve and the wetlands, where timber frames are deeply inspired in traditional Māori eel baskets.

Figure 2.
Recreational activities at Kopupapa reserve.

enhancing habitats and ecological corridors [28]. Kopupaka park demonstrates how urban growth and community engagement can be successfully paired with ecology and engineering [30].

4. Lions Park

Lions Park is a small urban park located in Gladstone, Queensland, but in spite of its size, it is considered a revolutionary park for Australia, being one of the first intergenerational and therapeutic environments located within the public realm. It is an inclusive space that has been established for intergenerational socialisation and as a safe space for those with disabilities. Building from the evidence that a strong link exists between the natural environment and improved health and well-being, these spaces provide moments of relaxation, social interaction and physical activity [31]. The park is a destination for locals and it also acts as a thoroughfare between the commercial zone, suburbs and natural environment. The Leonardo Da Vinci inspired playspace incorporates sensory features that can accommodate children and adults with disabilities. The designer, Playscape Creations, worked closely with local communities and those with impairments to create a space that accommodates a diverse range of needs [32]. Due to the urban location, Crime Prevention Through Environmental Design (CPTED) design principles were used throughout the entire design to enhance active and passive surveillance as well as making sure the park was a multi-use space. The increased surveillance also minimises the complexity of maintenance and ensures the park features last longer [32].

The form of the park is inspired by Da Vinci's cogs of interaction. The cogs create islands of play and recreation linked together with wide wheelchair-friendly paths and tactile water features. Each cog is unique with special elements that engage, inspire, and incorporate user preferences for each impairment. For example, those with sensory processing disorders like to spin and swing, but also enjoy shared play and like spaces to hide away in and just observe from. Those with auditory processing disorders engage well with tactile and visual elements but prefer less busy spaces and bold signage. Those with visual impairments are especially sensitive to light so shade is needed along with tactile environments that offer a contrast for navigation. Those with Down syndrome require muscle stimulating equipment, individual and group play opportunities and vestibular play elements. Those with autism can be sensitive to change, so enjoy tactile mapping and colour coding, as well as repetitive play equipment and spaces they can hide away in. For muscular dystrophy, spaces are required to provide shelter from light and heat and have a handhold that can be used with a closed fist. Those with cerebral palsy need rest spaces, handholds that can be used with a closed fist and tactile interactive panels. To accommodate spina bifida, wide wheelchair friendly pathways and transfer points to get on equipment are necessary but also are places for social rest.

Cog one is the entry node at the beginning of the park, which also contains the car park. The purpose of this area is predominantly a gathering space with arranged seating and shelter. From there, the user can move through into the body of the park. Cog two is the linking cog and social hub, connecting either to the nature cog and action cog or the commercial zone cog. In cog three, there is play equipment, informal water play, soft fall mats, turf, and walls that are intended to double as informal seating and shelters. Cog four is the swing, spin and discovery zone and a five-way swing stands in the centre. The play zone is enclosed by a dense vegetation buffer due to its proximity to the road. Cog five is an informal cog dedicated to nature play and exploration play (**Figure 3**). Activities are less structured in this area and there is more opportunity for imagination to take control [33].

In Lions park, play elements and natural features are thereby designed for all of the seven senses; hearing, sight, smell, touch, vestibular and proprioception. Specific play equipment has been included in the design of the playgrounds to ensure that users of all abilities have been included. These play elements include tactile water features for formal and informal water play, small and large shades for those with heat and light sensitivity, safe havens and small hideaways for those who want to escape, hand hold-on equipment that can be used with a closed fist for those who have limited grip, and spinning and full-body vestibular play features. In addition, consideration has been given to stimulating cognitive, social, psychological and physical senses while in the act of playing on a playground [34]. This playground accommodates a variety of bodily movements that stimulate the mind and body in much the same way as the activity of gardening.

Figure 3.
Lions park and the different cog areas.

5. Freedom Park

Freedom Park is located in one of South Africa's capital cities Pretoria. Situated on the top of a hill overlooking the city of Salvokop, the park was mandated by President Nelson Mandela to communicate a narrative of South Africa's journey to freedom following the end of the apartheid. It is a part of the rebuild seeking to address social fragmentation between races and significant disparities in access to education, health care and employment [35]. Freedom Park is a therapeutic landscape in the form of an urban park that tells a narrative of South Africa's journey to freedom and celebrates the 3.6 billion years of physical change in environment as well as those who fought for its freedom [36].

Freedom Park addresses the physical, mental, spiritual, societal and environmental elements of the landscapes of South Africa and embodies them into one space for all. The park supports South Africa's newfound identity, attempts to create new social networks, and supports the once fractured society. It acknowledges the holistic approach of traditional African culture focusing on the collective and bringing community interests to the forefront of any decisions [37]. One African principle that was not lost with colonialism was 'ubuntu', meaning that 'I am what I am because of who we all are', which follows on from the Zulu saying 'umuntu ngumuntu ngabantu', which means 'a person is a person through other persons' [37]. This principle values people first and foremost with a view of the collective whole [38].

One of the most vital features of the park is its engagement to a sense of place. The development of a sense of place is a social and cultural process that depends upon the user's response and cannot be derived from location alone [39]. Creating a sense of place is affirmed in Freedom Park by the inclusion of the historical narrative and the use of traditional practices. When entering through the //Hapo external space the users make their way through traditional and native medicinal plants. The journey educates the user about traditional practices and sparks conversation (**Figure 4**). Sense of place is not only an embedded relationship with the natural

Figure 4.
Use of traditional materials.

environment but directly links to the people within the place [39]. The inclusion of the eleven languages around the park aids in establishing place identity. For example, the //Hapo is from the Khoisan language, S'Khumbuto (the memorial) is Swati, Uitspanplek (A picnic area on the northern side of the hill) is an Afrikaans word, and Isivivane is a Zulu word meaning 'to throw your stone upon the pile' [36].

//Hapo is the first point of arrival to the park. This space is an interactive exhibition telling the story of South Africa over the course of 3.6 billion years. The name //Hapo has been drawn from the phrase '//hapo ge//hapo tama/haohasib dis tamas ka i bo', which means 'a dream is not a dream until it is shared by the entire community'. Within //Hapo there are three external spaces. The first space is the Garden of Indigenous Knowledge/the Healing Garden. In this garden, water channels lead visitors into a quiet and contemplative space where they are immersed in medicinal plants and gain knowledge of Indigenous healing methods. The second space is Sentihaga, the children's area. The landscape has been terraced with a series of cascading walls. Children and others are able to play with water features and plant material. Totems, an amphitheatre and small passageways are features of the garden to spark imagination. The third space is called the Boulders. They are metamorphosed layers of rock composed in a circular formation and surrounded by savannah vegetation to tell the story of South Africa's creation. Indigenous vegetation surrounds //Hapo blending the building into the landscape and preserving the natural terrain of the hillside (**Figure 5**) [36].

Exiting the //Hapo, visitors continue on a contemplative journey up the Salvokop hill along a pathway named the Vhuwaelo. Along the Vhuwaelo there are a series of small gardens that you can weave in and out of, exploring the small private spaces comprised mainly of natural materials and plants. Dramatic views overlook the city of Pretoria and the Union Buildings. At the final moment before you descend down the hill is a space named the Mveledzo. It is made from heavy stones and concrete, providing a final resting spot. Then the visitor descends the hill to arrive at the S'khumbuto, a Swati word for remembering or memorial. The memorial is for those who fought for freedom and showed great leadership. Water has

Figure 5.
Layers of rock surrounded by the savannah vegetation.

been used throughout the memorial because of its importance to African culture and its relation to healing and purification [36].

6. Jiyan healing garden

The Jiyan Healing Garden is a treatment facility and therapeutic garden that was established by the Jiyan Foundation for Human Rights in Chamchamal, Iraq in 2016. The Jiyan Foundation began its work in 2005 by building rehabilitation centres throughout Kurdistan and Iraq for those who were survivors of torture. The healing garden is an animal-assisted trauma therapy centre for those who suffered at the hands of Saddam Hussein's regime and, more recently ISIS. Many children in Iraq suffer from problems associated with this trauma, including behavioural and speech disorders.

A crumbling public infrastructure, including the water and sanitation network, results from years of political unrest. The Jiyan Foundation for Human rights wanted to create a space that was self-sustaining and supported the surrounding environment. The team of architects worked in collaboration with BORDA (Bremen Overseas Research and Development Association) to create a decentralised wastewater system for the garden [40]. This was deemed necessary as the Kurdistan region continually suffers from a rainfall deficit and low groundwater levels. The decentralised wastewater treatment system taps into the sewage pipes that run under the facility and is able to clean 100 cubic meters of dirty water each day, enough water to provide the entire garden with water. The water system also processes animal and human waste, turning it into biogas used for heating the therapy rooms in the winter [41].

ZRS, the architects of the treatment facility and healing garden, wanted to create a space that represented trust, identity and well-being. The facility is an example of traditional and sustainable architecture. The buildings are constructed from local clay material that has been historically used in the area (**Figure 6**). The walls are built from the air-dried brick blended with the thin layer of earth-straw mix. The roof is also constructed from the thin earth straw mix. The earth materials have climate cooling properties that concrete and glass cannot provide. The earth keeps

Figure 6.
Illustration of the building's layout and materials used.

the interior space cool during the hot summers and then keeps the interior dry and warm during the wet and sharp winter months [42].

The treatment facility is situated in a recreated traditional village environment typical to that in Chamchamal. In total, there are 11 individual dwellings that provide therapy rooms and quiet spaces that users can retreat to. A courtyard and shade device connect the dwellings together and allow access to the central healing garden and the animal pens. The garden and buildings create a unique space to heal and the garden is full and flourishing compared to the barren terrain of Chamchamal. Those at the facility participate in group-based garden therapy, which gives them a support network and a new way to overcome their ordeal [42]. The treatment facility also allows them to participate in group activities in the garden and direct therapeutic work in the huts. The technique they tend to use with the children is 'mud therapy'. The kids are encouraged to play outside in the garden and in the mud. They are prompted to get dirty and interact with the many animals before coming back inside for a discussion when they feel safe and comfortable and are often more willing to open up (**Figure** 7) [40].

A distinguishing feature of this therapeutic landscape is the animal-assisted therapy. Those who have been subject to trauma caused by acts of war, torture, violence, or displacement have strong feelings of vulnerability. They are often automatically and subconsciously drawn to the animals. They will try to get close to the animals to feel safe and comfortable. It is also said that the practice of petting, brushing, walking and caring for the animals helps give people something else to focus on, increases self-esteem and reduces feelings of anxiety, grief and isolation. The animals introduced to the Jiyan Healing Garden are a mixture of local animals that many would find in the area as well as exotic animals. Those who use the facility have been subject to generations of pent up trauma and psychological abuse and most are having to adjust to a new life away from their home and loved ones.

The Jiyan Healing Garden adopts place-making strategies and familiarisation through structured individual and communal activities making this space an internationally award winning 'therapeutic landscape'. The gardens provide relief and help to those who are suffering trauma from torture and acts of war and it gives the community access to adequate green space [40]. Green space that is inclusive and contains structured activities is crucial for creating stronger and resilient communities. This is the grounding essence of a therapeutic landscape.

Figure 7.
Outdoor spaces and access to interact with animals.

7. Jardín Etnobotánico de Oaxaca

Jardín Etnobotánico de Oaxaca is deemed one of the world's most original public botanic gardens, it is an ethnobotanical garden on the grounds of the San Domingo monastery in Oaxaca Mexico. Mexico is one of the most biologically rich countries globally, and out of the 31 states in Mexico, Oaxaca is the most biodiverse. Oaxaca has a unique botanical history as it contains evidence of plant domestication of squash and corn in the Americas dating back to the beginning of the new world. The ethnobotanical garden has been designed to reflect and celebrate this unique diversity. Each section of the garden displays a different ecology. The northeast section of the garden is dedicated to the tropical forest. Species that have been included in this section are balsa, ceiba and huaie. The east section of the garden reflects the wet regions of Oaxaca, including species of cacao, vanilla and achiote. The west of the garden is dedicated to the dry lands and includes plant varieties; cacti and agaves [43]. A planter bed planted with squash has been raised above the rest of the garden and faces towards Guilá Naquitz [43].

The garden tells a strong narrative of the natural history of Mexico, as well as exploring the cultural and artistic traditions of Oaxaca. During the site excavation, structures more than 400 years old were discovered and incorporated into the final design. A number of other artists, scientists, anthropologists and horticulturalists collaborated on the design of the planting, hardscape and water systems [43]. The use of materials and hardscaping is unique to the garden and emphasises the narrative of Oaxaca's history (**Figure 8**). A distinct feature of the garden is the green-hued pathway that mimics the formation of a step-fret zigzag. This pattern is common in many pieces of pre-Columbian carving and art. Another distinct feature of the garden is a large water fountain that seeps a distinct red water. The fountain is made from the wood of the Montezuma cypress tree and the water has been dyed red from ground up cochineal coating the wood [43].

The designers and director of the garden choose to include minimal signage around the garden to allow for a more seamless aesthetic. Visitors who want to get a richer experience of the garden and learn about the specific varieties of plant

Figure 8.
Use of softscape and hardscape materials typical to Oaxaca.

Figure 9.
Horticultural practices at the site.

species are encouraged to take a 2-hour tour. During this tour, they are taken through the medicinal and ceremonial plants, including mesquite, copal used in incense, alebriies for carving, and agave plants traditionally used to make mezcal. Visitors are also taken through a section of the garden that is dedicated solely to traditional food crops. The crops include hierba de conejo, corn, beans, chepil (the herb used in tamal de chepil), jicama, amaranth, tomatoes, and chia. Many of these plants were domesticated in Oaxaca. Although, there are around 1000 different varieties of plants in the garden, they are all wild except for the traditional food crops (**Figure 9**) [44].

What makes this space a successful therapeutic environment is the strong sense of place that is created by place attachment and narratives. There are many repetitive and empowering cultural and historical motifs that feature in the garden and speak to Oaxaca's history as one of the more important sites of archaeological history for the Americas, showing the first signs of plant domestication [43]. In the 1970s, seeds of squash, bottle gourd, beans, agave, and chilli peppers were all excavated and included in the ethnobotanical garden design [44]. Visitors to the garden and locals can learn about the ancient history of Oaxaca and the rest of the Americas to understand the role the area played in the development of the new world, creating a strong sense of place identity. But what is also successful about this place is that it began as a passion project from the many locals in the community who wanted to see the preservation of art, culture and the environment. This developed into the current ethnobotanical garden that allows the community to explore the relationship between plants and people and work in harmony with the Cultural Centre. Local designers, artists, scientists, horticulturists, and anthropologists all worked together on the design, construction and infrastructure within the garden. As such, this 'community' garden provides unique opportunities to participate in communal activity and interaction.

8. Discussion

Successful therapeutic landscapes address the social and cultural contexts of their location as well as respond to the environmental, mental and physical needs of each user. In defining therapeutic environments as reliant on a series of encounters, networks, and associations, Duff [16] classified these enabling resources into three categories: social, affective, and material. Social resources are grounded in relationships but are also directly linked to place, since they are a product of it and simultaneously enable everyday experiences. Affective resources are the fusion of individual or collective feelings that define or restrict orientations or actions. Material resources characterise the way in which relationships are established through the materiality of place and how they impact our access to goods, services, and information. Therapeutic landscapes can strengthen and improve the overall well-being of an individual, a community or even a population. In the case studies, a variety of processes were used to foster therapeutic environments, bringing together the relational, affective, emotional and cognitive skills that create and maintain social networks and promote meaningful experiences, closely weaving together land, culture and health.

8.1 The land

Landscapes are experienced in different ways by different people. Different cultural orientations may also lead to different experiences of the same space; however, this does not negate its therapeutic qualities. Similarly, the experience of therapeutic landscapes will vary based on previous experiences or attachments. In the literature, sense of place is discussed as being related to the inclusivity of the design and the intention of the user, but it can also refer to the historical connection with the space, which can enrich the 'place identity'. Placemaking enables the restorative and healing abilities of therapeutic landscapes, aiding those seeking the social inclusion or home comfort and attachment that is often formed through activities and the communal associations made [39]. The spaces that were the most successful in creating a holistic healing landscape for the community, were those environments that were welcoming and established a strong sense of place for all users. Hereby, the sense of place in a therapeutic landscape can result from place identity and place attachment and is often felt by a user when they have an emotional or historical connection to that place.

Two case studies, in particular, illustrate how the land grounds the therapeutic nature of the place. The Kopupaka Reserve in Aoteroa-New Zealand guides visitors along constructed boardwalks. The surrounding waterways provide an abundance of life with views over the wetlands and freshwater streams to the varied birdlife and biologically-diverse horticulture. Open spaces for communal activities are contrasted with private places for emotional retreat, providing a human experience that engages with all the senses through sight, sound and smell. With the attention shift from intimate perspectives to distance views, the approach also changes from a very singular and internal healing process to a communal and inclusive one. The design is permeable and flexible, allowing the materials to flex with the movement of the land and using materiality to will enable the ecology to flow.

Freedom Park in South Africa also uses the land to bring people together. One of the strongest features of this park is its engagement with the sense of place. Bringing together formerly alienated groups in a neutral landscape that offers both distant views over a cityscape and intimate pocket spaces, it is concerned with the physical, mental, spiritual, societal and environmental landscapes of South Africa, embodying them into one space for all. Emotions typically run high when the historical injustices of apartheid are situated in a specific place. The park is designed to seamlessly disappear into its natural context as the materiality of the built environment emulates and draws inspiration from the surrounding nature.

8.2 Culture

Therapeutic landscapes can provide an opportunity to embed cultural and traditional practices that do not often get conveyed in everyday contexts. For example, the Jardín Etnobotánico de Oaxaca (Mexico) began as a passion project for many locals in the community who wanted to see the preservation of art, culture and the environment. The garden tells a story about the cultural and artistic traditions of Oaxaca as well as its place in the natural history of Mexico. There are many repetitive and empowering cultural and historical motifs that feature in the garden and speak to Oaxaca's history. From a social perspective, it brings together artists, horticulturalists, scientists and anthropologists as well as the community at large to reflect the purpose of the garden as an exploration of the relationship between plants and people. The use of materials and hardscaping is also unique to the garden and emphasises the specific narrative of Oaxaca's history. In this project, the health of individuals reflects the strength of the community and collective triumphs are celebrated over the personal gains of the individual.

Another important example of a successful therapeutic environment that celebrates cultural values and includes Indigenous and traditional materials and practices within the design of the landscape is the Kopupaka Reserve (New Zealand). Importance was given to creating a place identity and understanding of how the land was used pre-colonisation. The New Zealand case study adopted a series of principles, Te Aranga framework, to ensure a cultural approach was taken to design that encouraged designers to engage with local Māori tribes in all aspects of the design and construction processes and to ensure that plants and materials had been chosen correctly. Cultural values hereby affect and maintain physical, emotional, mental and spiritual health and well-being. Our findings align with those of Andrews [17], whereby a successful therapeutic landscape has been shaped by the influence of a belief system/culture and emotionally engages with human experience. If these are embedded within the design, the environment will inherently respond to the needs of each individual user in a holistic sense and provide the foundation for a more affluent, healthier lifestyle. For many, connection through place or land includes not just the individual but the entire community of beings, living and non-living, putting the need for connection at the forefront.

8.3 Health

The literature review highlighted some clear themes relating to health that are thread throughout the discussion of therapeutic landscapes and what a successful holistic healing space should provide. Therapeutic environments and healing environments need to be designed to the needs of the user; which includes the needs of the marginalised. Two landscapes that are specifically designed for unique human experience are the Jiyan Healing Garden (Iraq) and Lions Park (Australia).

In the Jiyan Healing Garden, the healing garden is designed for victims of trauma and to address the social tensions that had widened the gap between the places of emotional retreat and healing from those of everyday sociability. Here the healing garden serves as an emotional retreat that facilitates an 'extraordinary' healing experience. For this reason, it takes a unique approach and involves access to animal therapy and pays attention to the specific needs of the user. In addition, it offers the chance for patients to work on the upkeep of a garden actively, activities they may have done at home [23]. Participation at this level creates a sense of ownership and sense of place in the facility, improving the productivity of healing and rehabilitation. Wellness and health of a community do not rely solely on the physical gains from gardening but also the mental gains. Feelings of loneliness and isolation can be mitigated by involvement in communal gardening as well as the practice of plant propagation and cultivation [34]. Concerning the material, a diverse range of spaces and activities are essential to the inclusivity of a space. For this garden, the choice of plants was vital, which targeted the user and their preferred activities [23]. While the garden's main intention is to provide relief and help those suffering trauma from torture and acts of war, it also gives the community access to adequate green space. Green spaces that are inclusive and contain structured activities are crucial for creating more robust and resilient communities. This is the grounding essence of a therapeutic landscape.

Similarly, the Lions Park (Queensland, Australia) has been designed for participation and socialisation on a much smaller scale. With its design focus on intergenerational socialisation and inclusion for those with disabilities, interactive areas provide for various body movements that include vestibular and proprioception and sound, sight, smell, and touch. Social inclusion is achieved through the creation of spaces that meet the needs and desires of all potential users. There are also spaces of refuge that can be used to hide away, for relaxation or contemplation. Playgrounds provide various bodily movements that stimulate the mind and body, similar to the activity of gardening. In particular, play equipment was included within the design of the playground to ensure that users of all abilities were included. Wheelchair-friendly paths, shelters for those who are light-sensitive, tactile panels, visually stimulating signs and planting combinations all combine to create an inclusive and restoratively productive environment.

9. Conclusion

Our understanding of what makes up a therapeutic landscape is evolving. Social issues including; quality of life, access to health care, social inclusion and supportive environments were all addressed in most literature, but overall, we found a lack of engagement with the important cultural context of the therapeutic landscapes and in particular, the inclusion of Indigenous practices. Our research finds that Indigenous methods can provide a holistic approach to health and well-being in our communities and a finer grained understanding of a therapeutic environment.

Our case studies varied in scale from large and expansive natural landscapes to small manmade interventions, but all shared social inclusion as a recurrent theme and all acknowledged aspects of land, culture and health. The successful therapeutic environments were tied to the inclusivity of the design and the intention of the users, but also the historical connection to enrich the place identity. Most of these spaces have come about as projects as a response to passion and request from the local community or as a response to social or cultural demands. This has resulted in the inclusion of the community's needs in the designed spaces and their physical input in the landscape.

The concept of therapeutic landscapes has evolved with research and theory to reflect society's current values. Most recently, this has shifted with the desire to have healthy and strong communities. A stronger sense of place is considered to create stronger communities and improved community-wide health and well-being. The success of a community cannot be separated from the success of its place, including the natural settings, the local culture and wider surroundings. In this way weaving land, culture and health is essential to the concept of the therapeutic environment.

Conflict of interest

The authors declare no conflict of interest.

Author details

Bruno Marques*, Jacqueline McIntosh and Hayley Webber
Wellington School of Architecture, Victoria University of Wellington, Wellington, New Zealand

*Address all correspondence to: bruno.marques@vuw.ac.nz

References

[1] Bell SL, Foley R, Houghton F, Maddrell A, Williams AM. From therapeutic landscapes to healthy spaces, places and practices: A scoping review. Social science & medicine. 2018;196:123-130.

[2] Kaplan R, Kaplan S. The experience of nature: A psychological perspective. Cambridge University Press; 1989. 152 p.

[3] Ugolini F, Massetti L, Calaza-Martínez P, Cariñanos P, Dobbs C, Ostoić SK, Marin AM, Pearlmutter D, Saaroni H, Šaulienė I, Simoneti M. Effects of the COVID-19 pandemic on the use and perceptions of urban green space: An international exploratory study. Urban forestry & urban greening. 2020;56:126888.

[4] Honey-Rosés J, Anguelovski I, Chireh VK, Daher C, Konijnendijk van den Bosch C, Litt JS, Mawani V, McCall MK, Orellana A, Oscilowicz E, Sánchez U. The impact of COVID-19 on public space: an early review of the emerging questions–design, perceptions and inequities. Cities & Health. 2020:1-7.

[5] Thompson CW. Linking landscape and health: The recurring theme. Landscape and urban planning. 2011;99(3-4):187-195.

[6] Marques B, McIntosh J, Hatton W. Haumanu ipukarea, ki uta ki tai:(re) connecting to landscape and reviving the sense of belonging for health and wellbeing. Cities & health. 2018;2(1):82-90.

[7] Marques B, McIntosh J, Kershaw C. Therapeutic environments as a catalyst for health, well-being and social equity. Landscape Research. 2021;19:1-6.

[8] Ulrich R, Gilpin L. Healing arts. Nursing. 1999;4(3):128-133.

[9] Stigsdotter U, Grahn P. What makes a garden a healing garden. Journal of therapeutic Horticulture. 2002;13(2): 60-69.

[10] Gesler WM. Therapeutic landscapes: medical issues in light of the new cultural geography. Social science & medicine. 1992;34(7):735-746.

[11] Smyth F. Medical geography: therapeutic places, spaces and networks. Progress in Human Geography. 2005;29(4):488-495.

[12] Conradson D. Landscape, care and the relational self: therapeutic encounters in rural England. Health & place. 2005;11(4):337-348.

[13] Dunkley CM. A therapeutic taskscape: Theorizing place-making, discipline and care at a camp for troubled youth. Health & place. 2009;15(1):88-96.

[14] Gatrell AC. Therapeutic mobilities: walking and 'steps' to wellbeing and health. Health & place. 2013;22:98-106.

[15] Foley R. Performing health in place: The holy well as a therapeutic assemblage. Health & place. 2011;17(2):470-479.

[16] Duff C. Networks, resources and agencies: On the character and production of enabling places. Health & place. 2011;17(1):149-156.

[17] Andrews, G. Landscapes of Wellbeing. In: Brown T, Andrews G J, Cummins S, Greenhough B, Lewis D, Power A, editors. Health geographies: A critical introduction. London: John Wiley & Sons; 2017. p. 59-74.

[18] Bingley, A. Health, People and Forests. In: Convery I, Corsane G, Davis P, editors. Making Sense of Place:

Multidisciplinary Perspectives. Boydell and Brewer; 2012. p. 107-116.

[19] Khachatourians AK. Therapeutic landscapes: A critical analysis [Doctoral dissertation]. Department of Geography: Simon Fraser University; 2006.

[20] Williams A. Therapeutic landscapes as health promoting places. A companion to health and medical geography. 2010:207-223.

[21] Foley R, Kistemann T. Blue space geographies: Enabling health in place. Health & place. 2015;35:157-165.

[22] Foley R, Wheeler A, Kearns R. Selling the colonial spa town: The contested therapeutic landscapes of Lisdoonvarna and Te Aroha. Irish geography. 2011;44(2-3):151-172.

[23] Mei C. Planting design and its impact on efficacy in therapeutic garden design for dementia patients in long-term care facilities in North Texas [Doctoral dissertation]. The University of Texas at Arlington; 2012.

[24] Wilson K. Therapeutic landscapes and First Nations peoples: an exploration of culture, health and place. Health & place. 2003;9(2):83-93.

[25] Bell SL, Phoenix C, Lovell R, Wheeler BW. Seeking everyday wellbeing: The coast as a therapeutic landscape. Social Science & Medicine. 2015;142:56-67.

[26] Gold CL, Clapp RA. Negotiating health and identity: Lay healing, medicinal plants, and indigenous healthscapes in highland Peru. Latin American Research Review. 2011; 93-111.

[27] Marques B, McIntosh J, Hatton W, Shanahan D. Bicultural landscapes and ecological restoration in the compact city: The case of Zealandia as a sustainable ecosanctuary. Journal of Landscape Architecture. 2019;14(1): 44-53.

[28] Auckland Design Manual. Kopupaka Reserve Wetland Park [Internet]. 2017. Available from: http://content.aucklanddesignmanual.co.nz/resources/case-studies/kopupakareserve/Documents/Kopupaka%20Case%20Study.pdf [Accessed: 2020-11-17]

[29] Jacqueline Paul. Exploring Te Aranga Design Principles in Tāmaki. National Science Challenges Building Better Homes, Towns and Cities Ko ngā wā kāinga hei whakamāhorahora [Internet]. 2017. Available from: https://www.buildingbetter.nz/publications/urban_wellbeing/Paul_2017_exploring_te_aranga_design_principles.pdf [Accessed: 2021-04-12]

[30] Isthmus. Kopupaka Park, Landezine [Internet]. 2017. Available from: http://www.landezine.com/index.php/2017/04/kopupaka-park-by-isthmus-group/ [Accessed: 2021-03-21].

[31] Carter M. Wetlands and health: how do urban wetlands contribute to community wellbeing?. In: Finlayson C, Horwitz P, Weinstein P, editors. Wetlands and human health. Dordrecht: Springer; 2015. p. 149-167.

[32] Playscape Creations. Lions Park, Gladstone [Internet]. 2017. Available from: https://www.playscapecreations.com.au/lions-park-gladstone/ [Accessed: 2021-04-04]

[33] Gladstone Council. Lions Park, Gladstone [Internet]. 2017. Available from: https://www.gladstone.qld.gov.au/parks [Accessed: 2021-04-04].

[34] Ferrini F. Horticultural therapy and its effect on people's health. Advances in Horticultural Science. 2003;77-87.

[35] Curtis N. South Africa: The Politics of Fragmentation. Foreign Affairs. 1972;50(2):283-296.

[36] GREENinc, The Freedom Park by GREENinc, Landezine [Internet]. 2012. Available from: http://landezine.com/index.php/2012/09/the-freedom-park-by-greeninc/ [Accessed: 2021-04-11].

[37] Maluleke MJ. Culture, tradition, custom, law and gender equality. Potchefstroom Electronic Law Journal/Potchefstroomse Elektroniese Regsblad. 2012;15(1).

[38] Gade CB. The historical development of the written discourses on ubuntu. South African Journal of Philosophy. 2011;30(3):303-329.

[39] Lynch KA. The land tells our story: urban native place-making and implications for wellness [Doctoral dissertation]. Boston University; 2016.

[40] Alex MacDonald. Kurdistan 'healing garden' aims to help Iraqis deal with trauma, Middle East Eye [Internet]. 2017. Available from: https://www.middleeasteye.net/news/kurdistan-healing-garden-aims-help-iraqis-deal-trauma [Accessed: 2021-05-10].

[41] BORDA. Nothern Iraq: building a Healing Garden with a sustainable water supply, BORDA WesCA [Internet]. 2018. Available from: https://borda-wesca.org/healing-garden-providing-a-sustainable-water-supply/ [Accessed: 2021-04-28].

[42] ZRS Architekten Ingenieure. Jiyan Healing Garden / ZRS Architekten Ingenieure, Archdaily [Internet]. 2017. Available from: https://www.archdaily.com/883358/jiyan-healing-garden-zrs-architekten-ingenieure [Accessed: 2021-03-22].

[43] Jeff Spurier. Oaxaca's Ethnobotancial Garden, Garden Design [Internet]. 2011. Available from: https://www.gardendesign.com/mexico/oaxaca-ethnobotanical.html [Accessed: 2021-03-25].

[44] Rebecca Bailey. Jardín Etnobotánico de Oaxaca, Que Pasa Oaxaca [Internet]. 2018. Available from: http://www.quepasaoaxaca.com/jardin-etnobotanico-de-oaxaca/ [Accessed: 2021-03-26].

Chapter 4

Community Attachment and Environmental Stewardship: A Peri-Urban Perspective

Lazaro Eliyah Mngumi and Yonggui Chen

Abstract

This Chapter questions the negligence of attachment scholarship in the context of environmental stewardship with a specific focus in peri-urban areas. This Chapter has illuminated the imperatives of considering place attachement as an important factor in realizing environment stewardship in peri-urban areas. Three selected hamlets (*Nzasa, Kisarawe and Pugu-Kibaoni*) constitute the study area. A standard closed-ended questionnaire for assessing the extent of attachment of the community was deployed. Literature review on the other hand was used to map baseline information of the study area including the historical significance of the Pugu and Kazimzumbwi forest reserves. Three attachment attributes were explored;, community knowlegiability levels of the area; level of thoughts and feelings of the area; and the extent of community connection to natural resources in the area. It was revealed that the extent of community connection to the forest reserves are relatively strong. The study revealed considerable contrast on forest knowledgiability levels among men and women in the Pugu and Kazimzumbwi forest reserves. Males are generally revealed to be more knowledgiable of the Pugu and Kazimzumbwi forest reserves as compared to their female counterparts. The study revealed that there was substantial relationship between residence status and the level of thoughts and feelings on the forest reserves. The study has shown that natives have more thoughts and feelings of the present and the future of the forest reserves as compared to those who migrated from other parts of the country. The findings suggest that community attachment is of considerable importance in influencing environmental behavior either positively or negatively. Whilist the empirical evidence are drawn from the peri-urban areas of Pugu and Kazimzumbwi forest reserves of Dar es Salaam city, the message thereoff is representing a broad reality in the peri-urban areas of the Global South. The inclusion of community attachment perspective in negotiating environmental stewardship is advocated for as it might contribute in addressing the growing degradation of natural resources in peri-urban areas which has been increasingly declining.

Keywords: natural resource degradation, global south cities, extent of community connection, knowledgiability levels, thoughts and feelings

1. Introduction

There has been encreasing encroachment of the Pugu and Kazimzumbwi forest reserves in recent years. The encroachment has been alarming to the point of threatening the depletion of the forest reserves in the near future. At the same time, studies on land cover changes at the said forests have shown substantial decrease of land cover over the years [1]. This has led to the substantial decline of forest ecosystem goods and services including decrease of water volume on rivers such as nyeburu, decrease of water level on dams (Minaki), decrease of forest products such as honey, wild fruits; rainfall and temperature variability and decrease in crop yields (*Ibid*). Notably, increasing number of studies have recommended interventions geared at altering livelihood strategies of the forest adjacent community as part of the solution to the problem of forest enchroachment and its associated ill-effects. While the issue of livelihood is crucial in addressing the problem in question there are other underlying behavior related concerns to be taken on board for realizing a more holistic solution. Place attachment is one of the behavior related concept linked to community attitudes and behavior which affects positively or negatively community actions towards their surrounding environment. Notably, place attachment concept has been increasingly applicable both theoretically and empirically in rural settings as opposed to urban and peri-urban contexts [2]. While the amount of time one has stayed in a given locale partly explains the limited attachment research in peri-urban areas, this is an area worth schorlarly attention. The need for urgent attachment scholarship in peri-urban areas owes to the increasing degradation of natural resources especially in the Global South. Peri-urban being inhabited by a mixture of migrant populations and pressured by urban externalities poses unique attachment characteristic features in the context of environmental stewardship unlike the urban counterparts. This chapter therefore questions the negligence of attachment scholarship in the context of environmental stewardship in peri-urban areas. The chapter illuminates the imperatives of attachment scholarship in peri-urban areas by examining the commmunity attachment to the Pugu and Kazimzumbwi forest reserves and so establish weather the increasing encroachment is partly attributed to the extent of attachment.

2. Theoretical perspectives

2.1 Place, place attachment and environmental stewardship

Place concept is an increasingly contested concept across disciplines including positivistic scholars, social constructivism and sychometrics. In a review of place literature Low and Altman [3], established that as space is an integrating concept, there is no systematic theory of place and scholars have increasingly echoed luck of conceptual coherence in place research. The concept of place broadly connote to the subjective experience of embodied human experience in the material world [4]. Morgan argues further that place concept is paradoxical, with meaning that is not readily understood, but difficult to define. While understanding place is a step towards comprehending place attachment, this chapter is about place attachment and its implicature to environmental stewardship in peri-urban areas. The next paragraphs of this Section therefore narrows down to the conception of place attachment and its interlinkages to environmental stewardship behavior.

Place attachement is defined as an attitude of bonding (trust and belonging), mutual concerns and shared values with other members of one's group or locale' [5]. "…. Place attachment may contribute to the formation, maintanance and preservation of the identity of a person, group, or culture. And it may also be that place attachment plays a role in fostering individual, group and cultural self-esteem, self-worth, and self-pride …." [3]. Place attachment is also assumed to be of importance to the community since it facilitates engagement in local affairs [6]. Studies variously justify that place attachment motivates indiviaduals to contribute to civic activity on behalf of one's residential location, in the form of sustainable behavior [7, 8], ecological behaviors [9] or reactions to encroachment of one's territory [10, 11]. On this basis, place attachment serves both the individual and large community [6]. Place attachment has also been correlated to community resilience building processes [12] and fostering local friendship [13].

2.2 Peri-urban in the context of environmental stewardship

Situating the peri-urban concept is largely a daunting endeavor. This owes to its increasingly contested milieu by both academia and development practitioners [14, 15]. This Section reviews the conceptual contestation of peri-urban not in the sense of engaging in the debate of its contestations but for understanding its complex nature towards realizing environmental stewardship. The remainder of this Section therefore provide the contested environment and deployed discussion of the peri-urban concept, thereby its linkage to environmental stewardship and to the broad landscape planning discourse.

Broadly, the peri-urban concept is variously and increasingly viewed and conceptualized across both the geographical and disciplinary niches. Arguments behind the multiple conceptions of the peri-urban include lack of scientific definition Forsyth [15], diversity of engaged disciplinary orientations Thuo [14], the difficulties linked to the delimiting the spatial extent of this dynamic region Brook [16] and the equivocation of the concept itself. Notably, scholars increasingly argue that rural, peri-urban and urban environments operate as a system rather than independent entities [17, 18]. At the same time, peri-urban area is increasingly claimed to constitute the intersection point between urban and rural areas [19, 20]. In nutshell, there are growing yet converging understandings within academia on the lived reality of the diverse and context-laden definitions of the peri-urban concept [21]. Another converging understanding linked to peri-urban conception is the co-existence of urban and rural features within cities and beyond their limits [17, 21]. Notably, inspired by the aforehinted converging understandings, this chapter adopts the conception of peri-urban as defined by Mngumi [22], i.e. peri-urban is a city transitional zone, amalgamating both urban and rural landscape functions and features. In addition to the conceptual definition, the empirical materials discussed herein are drawn from the Pugu and Kazimzumbwi forest reserves at the peri-urban of Dar es Salaam city. Dar es Salaam is among the most rapidly growing cities in Africa and currently the business capital of Tanzania. Similar to other peri-urban areas in the Global South [23], natural resources and/ecosystem services in these peri-urban forest reserves are increasingly claimed to deteriorate [22].

Literature has increasingly established that the quantity and quality of natural resources are deteriorating in peri-urban areas, and particularly so in Sub-Saharan Africa [23, 24]. One explanation behind the decline of natural resources in peri-urban areas is the ongoing urbanization which has increasingly resulted in the expansion of the built environment on ecologically sensitive land, especially in peri-urban areas [25–27]. Onother explanation of the growing degradation of natural resources in peri-urban areas in Sub-Saharan Africa is the rapid urbanization due to poverty [28, 29]. The poverty aspect which is largely prevelant in the Global South context renders it difficult for the community members to inculcate the environmental stewardship habit. This is due to the fact that their livelihoods are intricately connected to the resources in their vicinity and so despite having strong natural attachment to the environment, this is compromised by livelihood hardships and as a result leading to low stewardship to the environment. Notably, there is a host of other contributing factors to the decline of natural resources at the peri-urban areas of Pugu and Kazimzumbwi forest reserves. These include climate change, the increase of anthropogenic activities and the expansion of the city's built area to the peri-urban areas [22].

3. Materials and methods

3.1 Case description

The Pugu and Kazimzumbwi Forest Reserves (PKFR), adjacent to the city of Dar es Salaam in Tanzania (**Figure 1**) forms the area where this study was carried

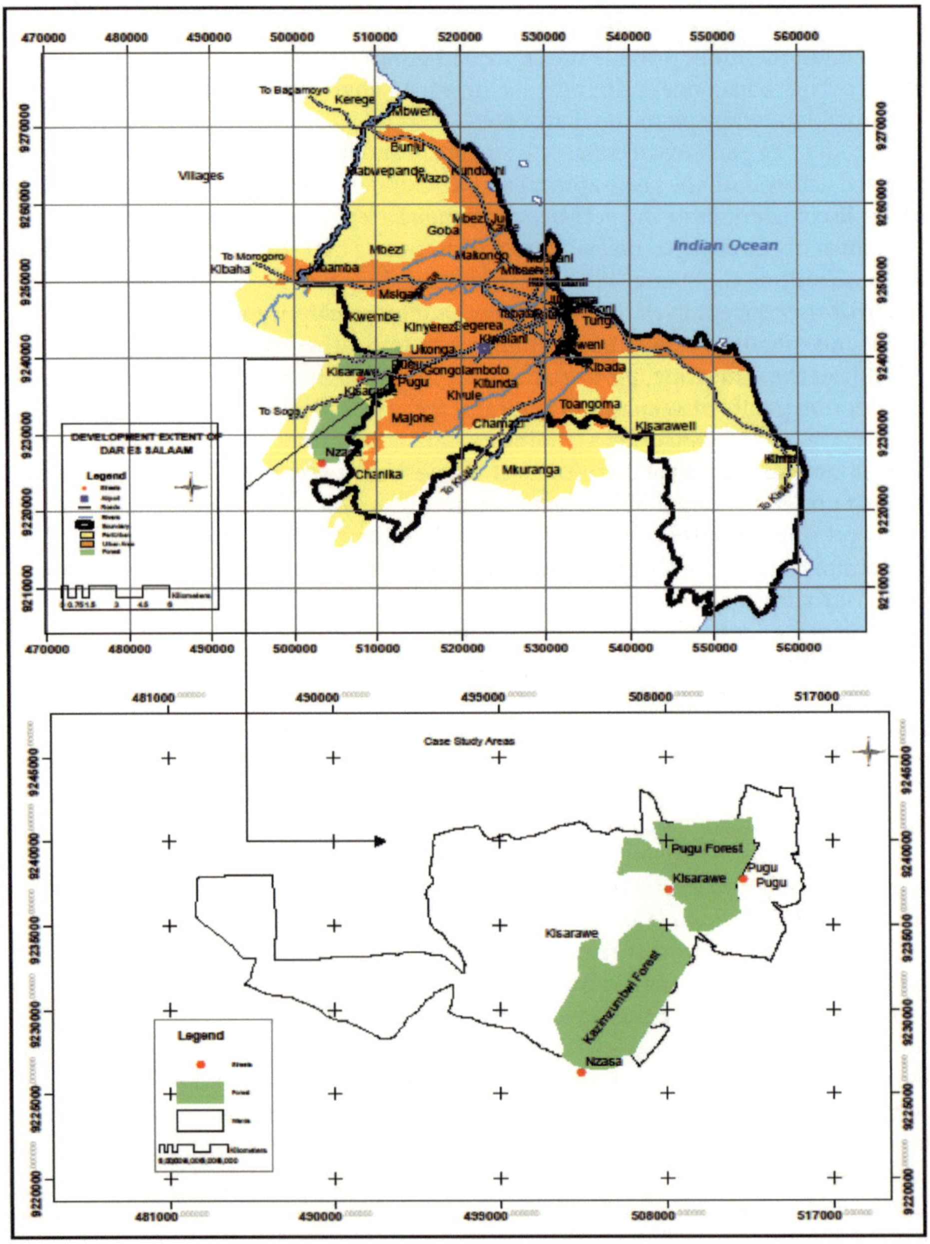

Figure 1.
Pugu and Kazimzumbwi forest reserves, Tanzania. Source: Erimina Massawe, Ardhi university 2019.

out. Administratively, the PKFRs are under the jurisdiction of both districts i.e. Kisarawe district found in Coast region and Ilala district found in Dar es Salaam region; both regions are located along the coastal belt of Tanzania. Historically, the two forest reserves are of global importance as they provide habitats to 37 endemic vertabrate species and about 554 endemic plants [30]. These forests are remnants of few ancient forests in the world, and form the catchment sites for a good number of rivers namely; Msimbazi, Mambizi, Mzumbwi, Vikongoro, Kimani, Nzasa and Nyeburu [1, 31]. Pugu forest gazetted as a forest reserve in the year 1954 lies in the northeastern part of the Pugu Hills, about 25 km southwest of Dar es Salaam and 20 km from the Indian Ocean, and it is positioned adjacent to the Kazimzumbwi Forest Reserve gazetted in the year 1936 [32].

The Indian Ocean tropical monsoon climate is influencing the rainfal and temperature characteristics of both forests. The two forests are characterized by bimodal rainfall pattern i.e. long rains between late March and early June typically known as Masika and short rains between October and December typically known as Vuli. The rainfall in the area i.e. (Pugu and Kazimzumbwi forests) is averaging at approximately 1,100 mm per year whereas Temperature is ranging from 24 to 31° proportional to elevation [32, 33]. In terms of topography, the two forests (Pugu and reserves are positioned between 100 and 305 and 120–280 metres above the sea level respectively [31].

The survey was done in three sub-wards, two of them are located adjacent to the Pugu forest reserve i.e. Pugu-Kibaoni and Kisarawe and the third sub-ward namely; Nzasa is located adjacent to the Kazimzumbwi forest reserve (**Figure 1**). The selection of the three study sub-wards was purposively done so as to capture the possible diversity on attachment discourse in the Pugu and Kazimzumbwi forest reserves.

3.2 Methods

Case study, an imperative approach in characterizing local state of the art on ecosystem services [34] was deployed. Household survey comprise the technique deployed in collecting field data. This was realized through household questionnaire which served as a tool in collecting data on various attributes regarding community attachment to the Pugu and Kazimzumbwi forest reserves and its associated ecosystem services. A standard closed-ended questionnaire for assessing the extent of attachment of the community was deployed. The questionnaire contained 5 score likert scale questions. Literature review on the other hand was used to map baseline information of the study area including the historical significance of the Pugu and Kazimzumbwi forest reserves. A total of fifty (50) questionnaires were randomly administered in each of the study sub-ward amounting to a total of one hundred and fifty (150) in the entire study area.

Interview consent was requested to all participants from the three study sub-wards, this was done so as to be able to proceed with the interview smoothly. This was done so as to comply with ethical requirements. The household interview through structured (close-ended) questionnaire took place after three days training enumerators. The training was purposed at making sure that the main themes of the questionnaire and the specific questions and likely response from the interviewees/respondents are aparently clear among the enumerators. This procedure was crucial as enumerator's work involved asking questions to respondents, providing clarification in case the question (s) becomes not so clear to respondents and thereafter fill the questionnaire after capturing response from interviewees/respondents. This was again important since it ensured maximum response as some respondents could not read and right. Furthermore, at every evening of the field work, a brief assessment with enumerators was held regarding the work of the day including cross checking the way questionnaires have been filled and planning for the day ahead. This was necessary to check the work progress and address the challenges if any for ensuring validity of the data collected. Data analysis was done through SPSS and excel software. The analysis was dominated by descriptive statistics whereby the findings in terms of barcharts and pie charts showing distribution of various attributes on attachment were produced.

4. Results and discussion

This section presents the findings on both respondents characteristics and the findings on community attachment on the Pugu and Kazimzumbwi forest reserves. The section starts by describing the demographic characteristics followed by attachments to the forests reserves of Pugu and Kazimzumbwi.

4.1 Selected participants' characteristics

Participants of this study included 150 randomly selected adults 18 through 82 of age with the mean age of 42.8 and (S.D = 17.9, 58 female); 83 were natives, 51 were migrants from the capital Dar es Salaam and 13 were migrants from upcountry. The participants were included from three sub-wards surrounding the Pugu and Kazimzumbwi forest reserve in order to capture hybrid and diverse attributes on attachment to the reserves. Each sub-ward had 50 participants in the survey. The main characteristics of participants are summarized in **Table 1**.

4.2 Community attachment to forest reserves in Pugu and Kazimzumbwi forest reserves

The aspect of community attachment to the forest reserve were assessed through household survey via a number of questions as detailed in the following section (s).

Respondent's characteristics		**Male**		**Female**		**Total**	
		N = 92	%	N = 58	%	N = 150	%
Education status	Primary or less	67	72.8	53	91.4	120	80
	Secondary	21	22.8	4	6.9	25	16.7
	High school	1	1.08	0	0	1	0.7
	Certificate and above	3	3.3	1	1.72	4	2.7
Residence status	Native	47	51.1	36	62.1	83	53.3
	Migrant from DSM	36	39.1	18	31.0	51	34
	Migrant from upcountry	9	9.7	4	6.8	13	8.6
Geographic location	Nzasa	22	23.9	28	50.9	50	33.3
	Pugu-Kibaoni	36	39	14	25.5	50	33.3
	Kisarawe	34	36.9	16	27.6	50	33.3

Source: Field work, 2018.

Table 1.
Selected characteristics of participants.

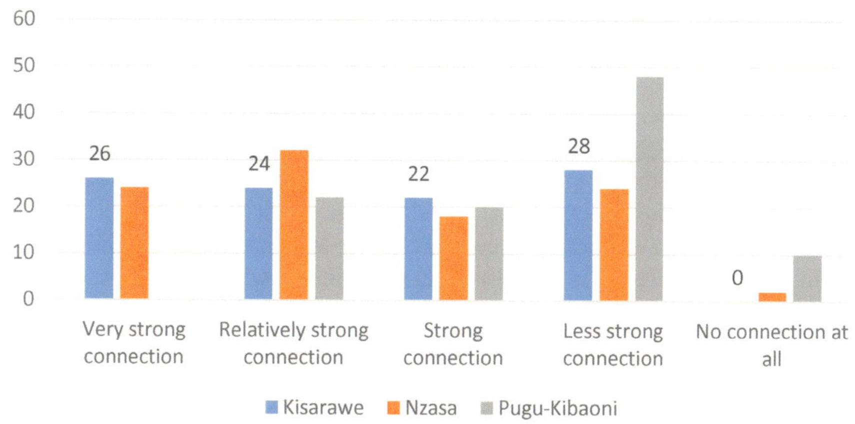

Figure 2.
Extent of connection to the forest reserves (%; N = 150). Source: field work 2018.

4.2.1 Extent of connection to the forest reserves

Figure 2 presents findings on the extent of community connection to the Pugu and Kazimzumbwi forest reserves. Generally; the extent of connection to the forest reserves are more or less similar between Nzasa and Kisarawe sub-wards. About a quarter (1/4) of respondents in these two sub-wards had shown relatively strong connection to the Pugu and Kazimzumbwi forest reserves. At the same time, Pugu-kibaoni sub-ward has outweighed the two sub-wards by having high percentage scores on less strong connection to the reserves. At Pugu-Kibaoni sub-ward, about half (1/2) of the respondents casted their scores on less-strong connection and a tenth (1/10) had their scores on no-connection at all to the forest reserves. The reason for Pugu-Kibaoni having considarable low level of connection to the reserves is due to the fact that the area is located near to the city as compared to the other two sub-wards. This being the case the livelihoods of the majority of its dwellers are not directly linked to the forest reserves as they engage mostly in trade and small business activities. This is backed up by the correlation analysis between the extent of connection to the forest reserves and the extent of one's life attachment to the forest reserves. This had a correlation score of 0.5 testifying that there is a considarable linkage between one's connection to the forest reserves and the connection of an individual's life to the forest reserves.

On the other hand the factors behind community attachment to the forest reserves were gathered during focus groug discussion (FGD) in all the three sub-wards. One of the autspoken reason for having strong connection to the Pugu and Kazimzumbwi forest reserve is the issue of livelihood connection with the forest reserves. During the FGDs it was underscored that a good proportion of the dwellers in areas at the vicinity of the forest reserves owes their living in the same. The activities in the forest reserves that derive a living to forest adjacent communities includes charcoal making and selling, timber business, fire wood fetching/cutting and selling, bee keeping and collecting other forest products such as medicene (herbs) and wild fruits. One participant in a FGD at Kisarawe had these to say ...

> *...I have been depending on the forest for earning a living for almost my entire life. This has made my life to be strongly connected to the forest reserves. This is the reason why I feel like I can not afford to live without these forest reserves. Nowardays there are some restrictions on how to access the reserves but I know how to find my way through the forest without ending up in the hands of those with authorities...*

4.2.2 Knowledge of the Pugu and Kazimzumbwi forest reserves

The question of wheather the community members were knowledgiable of the forest reserves was also asked in the course of establishing the level of community attachment. The findings on knowledgiability levels on Pugu and Kazimzumbwi forest reserves which partly tells on their attachment as as summarized **Figure 3**. The levels of knowledgiability was analyzed against sex of respondents.

The study revealed considerable contrast on forest knowledgiability levels among men and women in the Pugu and Kazimzumbwi forest reserves. Accordingly; males are generally revealed to be more knowledgiable of the Pugu and Kazimzumbwi forest reserves as compared to their female counterparts (**Figure 3**). Whereas a third (1/3) of female respondents were deduced to be less knowledgiable of the Pugu and Kazimzumbwi forest reserves, about a third (1/3) of males claimed to have complete knowledge of the reserves. The sex difference on knowledgiability on forest reserves also featured during the FGDs. It was explained that most men could get along the forest reserves with easy as compared to men. In a FGD at Nzasa, one female participant in a FGD expressed herself this way...

> *...In this forest reserve (Kzimzumbwi forest resrve) although both male and female access the reserves for gendered distinguised interests, males are more familier with the forest reserve. This is partly because men spends more time in the forest extracting resources for the living as compared to women....*

The variable of knowledgiability of the forest reserves was further correlated with the variable ' I know where all the paths leads to'. The correlation score for the two variables was 0.8 signifying that there were strong relationship between general knowledgiability of the forest reserves and the knowledge of where paths leads to in the forest reserve.

4.2.3 Thoughts about the present and the future of the forest reserves

Thoughts and feelings about the present and future of something tells the extent at which one is linked and/attached to the same. In this regard, the feelings and/thoughts about the present and the future of the Pugu and Kazimzumbwi forest reserves by the adjacent communities were examined. The findings on

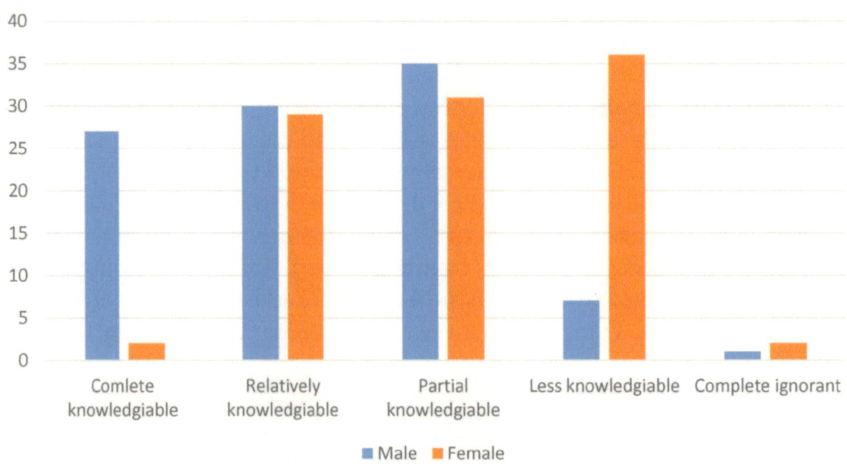

Figure 3.
Knowledge of the forest reserves by the community (%: N = 150). Source: Field work 2018.

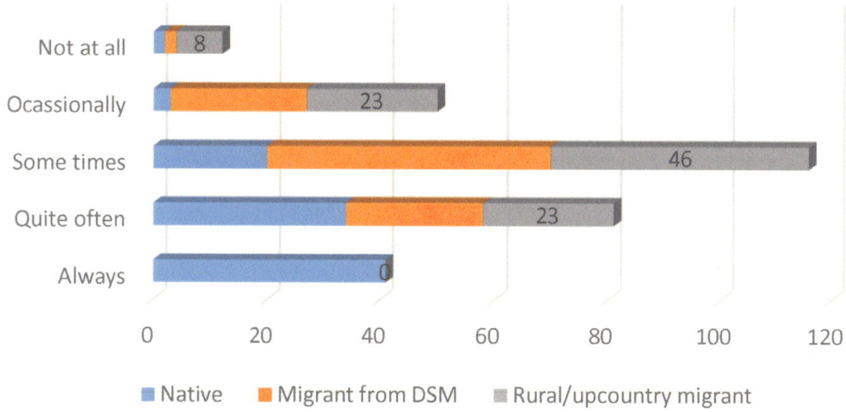

Figure 4.
Present and future thoughts about the forest reserves (%: N = 150). Source: field work 2018.

Figure 4 summarizes the stutus-quo on thoughts and feelings about the present and the future of the forest reserves.

The analysis on community thoughts and feelings about the present and the future of the forest reserves were done against the residence status.The study revealed that there was substantial relationship between residence status and the level of thoughts and feelings on the forest reserves. In nutshell, the study has shown that natives have more thoughts and feelings of the present and the future of the forest reserves as compared to those who migrated from other parts of the country. More than a third (1/3) of the natives claimed to *always* think about the present and the future of the forest reserves and there was non of the migrants who always had a thought about the present or the future of the forest reserves. On the other hand about half of the migrants from Dar es Salaam and those from upcountry had *some times* thought regarding the present and the future of the Pugu and Kazimzumbwi forest reserves. These findings on thoughts of the present and the future of the Pugu and Kazimzumbwi forest reserves suggests that the more one stays with a natural resource the more one becomes with feelings and thoughts about their present and their future prospects. This in turn tells on the extent at which one is attached to the particular resource and in this case the Pugu and Kazimzumbwi forest reserves.

5. Conclusion

This chapter questions the negligence of attachment scholarship in the context of environmental stewardship with a specific focus in peri-urban areas. Peri-urban being inhabited by a mixture of migrant populations and pressured by urban externalities poses unique attachment characteristic features in the context of environmental stewardship unlike the urban counterparts. This chapter has illuminated the imperatives of considering place attachement as an important factor in realizing environment stewardship in peri-urban areas. Whilist the empirical evidence are drawn from the peri-urban areas of Pugu and Kazimzumbwi forest reserves of Dar es Salaam city, the message thereoff is representing a broad reality in the peri-urban areas of the Global South. The inclusion of community attachment perspective in negotiating environmental stewardship is advocated for as it might contribute in addressing the growing degradation of natural resources in peri-urban areas which has been increasingly declining. Researchers and policy makers alike should take this up by including a focus on community attachment in the course of forging environmental stewardship in these transitional spaces in the Global South. This will in turn contribute to increase environmental welfare and reduce the growing degradation of natural resources in peri-urban areas.

Author details

Lazaro Eliyah Mngumi[1,2]* and Yonggui Chen[3]

1 Institute of Human Settlements Studies (IHSS), Ardhi University, Dar es Salaam, Tanzania

2 Faculty of Natural Resources and Agricultural Sciences, Swedish University of Agricultural Sciences (SLU), Uppsala, Sweden

3 College of Landscape and Arts, Northwest A&F University, Yangling, China

*Address all correspondence to: lazaromngumi@gmail.com

References

[1] Lupala, J.M., M.V. Mdemu, and S.P. Butungo, Effects of peri-urban land use changes on forest ecosystem services: the case of settlements surrounding Pugu and Kazimzumbwi Forest Reserves in Tanzania. Journal of Geography and Geology, 2014. 6(4): p. 231.

[2] Raymond, C.M., G. Brown, and D. Weber, The measurement of place attachment: Personal, community, and environmental connections. Journal of environmental psychology, 2010. 30(4): p. 422-434.

[3] Low, S.M. and I. Altman, Place attachment, in Place attachment. 1992, Springer. p. 1-12.

[4] Morgan, P., Towards a developmental theory of place attachment. Journal of environmental psychology, 2010. 30(1): p. 11-22.

[5] Perkins, D.D., J. Hughey, and P.W. Speer, Community psychology perspectives on social capital theory and community development practice. Community Development, 2002. 33(1): p. 33-52.

[6] Lewicka, M., Ways to make people active: The role of place attachment, cultural capital, and neighborhood ties. Journal of environmental psychology, 2005. 25(4): p. 381-395.

[7] Uzzell, D., E. Pol, and D. Badenas, Place identification, social cohesion, and enviornmental sustainability. Environment and behavior, 2002. 34(1): p. 26-53.

[8] Guàrdia, J. and E. Pol, A critical study of theoretical models of sustainability through structural equation systems. Environment and Behavior, 2002. 34(1): p. 137-149.

[9] Vorkinn, M. and H. Riese, Environmental concern in a local context: The significance of place attachment. Environment and behavior, 2001. 33(2): p. 249-263.

[10] Kyle, G., et al., Effects of place attachment on users' perceptions of social and environmental conditions in a natural setting. Journal of environmental psychology, 2004. 24(2): p. 213-225.

[11] Warin, M., et al., The power of place: space and time in women's and community health centres in South Australia. Social Science & Medicine, 2000. 50(12): p. 1863-1875.

[12] Garcĺa, I., F. Giuliani, and E. Wiesenfeld, Community and sense of community: The case of an urban barrio in Caracas. Journal of Community Psychology, 1999. 27(6): p. 727-740.

[13] Perkins, D.D., et al., Participation and the social and physical environment of residential blocks: Crime and community context. American journal of community psychology, 1990. 18(1): p. 83-115.

[14] Thuo, A.D.M., Unsettled Settled Spaces: Searching for a Theoretical 'Home'for Rural-Urban Fringes. International Journal of Scientific and Research Publications, 2013. . 3(7).

[15] Forsyth, A., Defining Suburbs. Journal of Planning Literature, 2012. 27(3): p. 270-281.

[16] Brook, R.M., The Peri-Urban Interface Hubli-Dharwad, India Changing Frontiers. Books for Change. 2001, Bangalore, India.

[17] Allen, A.D., and Hofmann P, Governance of Water and Sanitation Services for the Peri-urban Poor; A Framework for Understanding and Action in Metropolitan Regions, U. Development Planning Unit, United Kingdom, 2006

[18] Wandl, A. and M. Magoni, Sustainable Planning of Peri-Urban Areas: Introduction to the Special Issue. Planning Practice & Research, 2016. 32(1): p. 1-3.

[19] Olujimi, J. and K. Gbadamosi, Urbanisation of Peri-Urban Settlements: A Case Study of Aba-Oyo in Akure, Nigeria, The Social Sciences, 2007. 2(1): p. 60-69.

[20] Birkmann, J., et al., Adaptive urban governance: new challenges for the second generation of urban adaptation strategies to climate change. Sustainability Science, 2010. 5(2): p. 185-206.

[21] Salem, M., Peri-urban dynamics and land-use planning for the Greater Cairo Region in Egypt. Sustainable Development, 2015. 168: p. 1109.

[22] Mngumi, L., Climate change resilience: exploring socio-ecological system resilience for livelihood effects of climate change in peri-urban areas. 2021.

[23] Roy, M.K., et al., Climate change and declining levels of green structures: Life in informal settlements of Dar es Salaam, Tanzania. Landscape and Urban Planning, 2017.

[24] MEA, Ecosystem and human well-being: biodiversity synthesis. World Resources Institute, Washington, DC, 2005.

[25] Kombe, W.J., Land use dynamics in peri-urban areas and their implications on the urban growth and form: the case of Dar es Salaam, Tanzania, Habitat International, 2005. 29(1): p. 113-135.

[26] Marshall, F., Waldman, L., MacGregor, H., Mehta, L. and Randhawa, P., On the Edge of Sustainability: Perspectives on Peri-urban Dynamics, STEPS,. 2009.

[27] Radford, K.G. and P. James, Changes in the value of ecosystem services along a rural–urban gradient: A case study of Greater Manchester, UK, Landscape and Urban Planning, 2013. 109(1): p. 117-127.

[28] Kestemont, B., L. Frendo, and E. Zaccai, Indicators of the impacts of development on environment: A comparison of Africa and Europe. Ecological indicators, 2011. 11(3): p. 848-856.

[29] Niemelä, J., et al., Using the ecosystem services approach for better planning and conservation of urban green spaces: a Finland case study. Biodiversity and Conservation, 2010. 19(11): p. 3225-3243.

[30] Burgess, N.D., Global importance and patterns in the distribution of coastal forest species, in Coastal Forests of East Africa. 2000. p. 235-248.

[31] TFCG, Two surveys of the plants, birds and forest condition of Pugu and Kazimzumbwi Forest Reserves in 2011 and 2012, T.T.P. 36, 2013.

[32] Clarke, G. and A. Dickinson, Status Reports for 11 Coastal Forests in Coast Region, Tanzania Frontier-Tanzania Technical Report No. 17. The Society for, 1995.

[33] Lupala, J. and C. Maglan, Climate change and its effects on livelihood strategies of peri-urban coastal communities in Tanzania. 2015.

[34] Lamarque, P., et al., Stakeholder perceptions of grassland ecosystem services in relation to knowledge on soil fertility and biodiversity. Regional environmental change, 2011. 11(4): p. 791-804.

Chapter 5

Contributing to Healthy Landscapes by Sustainable Land Use Planning: A Vision for Restoring the Degraded Landscape of the Centre Region of Portugal

Selma B. Pena, Maria Luísa Franco and Manuela R. Magalhães

Abstract

The ecological-based methodologies are determinant to develop complete strategies in restoring the ecosystems at a landscape scale. Those methodologies start with comprehending ecological processes by mapping fundamental structures of the territory (water, soil, biodiversity), also called green infrastructures. The adequate land use planning and its forthcoming implementation will guarantee a multifunctional landscape, better ecosystem services provision, and a possibility of developing new economies. The intervention of Landscape Architecture at the landscape scale will also provide information about the place and the type of restoration actions to be implemented. The Centre Region was the most affected by rural fires from 2017, representing 15% of the total region area (416 thousand hectares). These events reflect the high importance of rethinking the territory with more suitable land uses, considering the concepts of sustainability, resilience, and ecological integrity. This work proposes a Landscape Transformation Plan for the Centre Region of Portugal, applying the FIRELAN model. The results show that about 35% of the Centre Region should have restoration action towards a more sustainable landscape.

Keywords: fire resilience, green infrastructure, ecosystem restoration, landscape transformation

1. Introduction

The first Portuguese Landscape Architect, Francisco Caldeira Cabral, reflected on how nature conservation should not be seen from a museology perspective, where Man is external to the object of protection. He defended that every person is an integral part of nature conservation by actively participating in protecting natural resources and constructing the landscape.

> *(...) a campaign of general mentalization began for the need as a condition of urban and rural life, to maintain in congruent form the essential elements of the*

IntechOpen

> *natural landscape, conserving or even reconstituting its continuity and functionality. Thus became aware of maintaining the " continuum natural " and the " continuum cultural [1].*

Since our existence, humankind has established interactive and empirical relationships with the Landscape [2], searching for defensive systems at higher elevations and safeguarding the fertile valleys as food producers essential for survival. But the interaction between Man and the ecosystems is bidirectional, where an action of Man on a particular ecosystem will imply a reaction and an adaptation of the ecosystem [3].

Landscape and land-use planning are intended to plan human interventions that maintain or promote the landscape's dynamic stability. The stability of the landscape is associated with slow landscape evolutions (pedogenesis), while instability is characterized by rapid changes (morphogenesis) [4]. The balance between morphogenetic and pedogenetic processes is a natural process of the landscape, which can be intensified towards instability by an incorrect action in the territory. The planning of human intervention in the territory in harmony with ecological systems has resulted in preserving natural resources and nature conservation.

The understanding of the functioning of natural systems, or ecosystems, as a support for decision making [5] emerged after the foundation of ecology as a science, in the mid-nineteenth century, by Ernest Haeckel. The evolution of knowledge about the ecological processes in a given territory allowed the development of ecology-based landscape planning methodologies.

Following the Convention on Biological Diversity [6], one of the decisions taken at the fifth Conference of the Parties to the Convention on Biological Diversity [7] was that biodiversity conservation objectives could only be achieved through an ecological-based approach:

> *(...) a strategy for the integrated management of land, water, and living resources that promotes conservation and sustainable use in an equitable way. It is based on the application of appropriate scientific methodologies focused on levels of biological organization that encompasses the processes, functions and interactions among organisms and their environment. It recognizes that humans, with their cultural diversity, are an integral part of ecosystems [7].*

Ecological-based methodologies start from the knowledge and spatialization of natural processes that occur in a given territory [8]. With this approach, the most significant areas for ecosystem functioning are identified in the landscape, with various potentialities, and actions are planned without compromising the stability and balance of the landscape. This approach is related to the concept of ecological suitability to the various human activities that do not compromise the proper functioning of ecological processes [9].

The concept of ecological suitability was used in the United States of America and the United Kingdom, using the manual overlay of transparent supports, whose methodologies were refined throughout the 1960s [10]. Other landscape architects followed, such as the work done by Philip Lewis, in 1964, to classify all the environmental resources of the State of Wisconsin, with the purpose of delimiting the areas where building should not be done [11]. The concept of ecological suitability was also considered in McHarg's [12] and Steiner [13] methodologies with the study of environmental processes and cultural integration in the choice of the best use, according to the intrinsic characteristics of the systems. In Portugal, contemporary with McHarg's methodology, the Algarve Plan was completed by landscape architects A. Barreto, A. Castelo-Branco and A. Dentinho, with methodologies based on ecological suitability of the Landscape [11].

The adequate planning of the landscape, according to the ecological processes that occur in it, impacts not only the ecological balance but also on the economic and social balance. This type of intervention has the ability, in a cost–benefit analysis, to be the best way to prevent future costs arising from natural disasters (flood prevention, fire risk reduction, mass movement), maintain water quality, ensure greater agricultural productivity, and contribute to the enhancement of urban areas [14]. Of this last point, the study done [15] in the case study of Cologne (Germany) concluded that increasing urban parks by 1%, about 500 meters from housing, leads to the growth of housing sales prices by 0.1%. With a priori protection of natural resources, engineering solutions are less, and the cost is lower [14].

The economic valuation of ecosystems was developed with the emergence of the ecosystem services concept [16] and spread at the beginning of the 21st century with the publications of the Millennium Ecosystem Assessment. Different local scientific and political communities started raising awareness of the importance and benefits of ecosystem services. Several initiatives and methodologies have emerged to quantify these services in monetary value. However, there are sometimes limitations in quantifying an ecosystem service [17].

Ecological-based planning has the added advantage of helping to increase the number or quality of services provided by ecosystems. Inherent in the concept of ecological-based planning is the continuity and ecological network [18]. In fact, landscape connectivity is a component of landscape structure that facilitates or impedes the flows of natural cycles [19]. Those authors believe that it is more critical to establish connectivity than the proximity of areas studied in "island biogeography" [20].

Like nature conservation, the ecological continuity of ecosystems is a broad concept that involves the need to promote continuity between plant and animal species and that of all ecological cycles: water, nutrients, carbon, etc. For this, the planning of these continuities (ecological structures or networks) cannot be disconnected from the function and its "congruent form", as defined by the landscape architect Francisco Caldeira Cabral [1]. The land-use planning approach advocated, plans the landscape from a biophysical and not only biological perspective [11, 21], where the continuous structures of the landscape are identified and planned according to their function and coherent form.

The importance of connecting nature protection areas, establishing a network or infrastructure has been reinforced [22]. According to this publication, this type of infrastructure, designated by green infrastructure, can mitigate fragmentation and promote the various benefits of maintaining and restoring ecosystems and their services, not only inside Natura 2000 areas.

However, as mentioned above, nature conservation should be understood as a comprehensive and integrative concept of Man and ecological processes, not limited to classified areas (RAMSAR Areas, Biogenetic Reserves, Natura 2000 Network, etc.). Nature conservation will emerge from the correct occupation and use of the Landscape by Man, both in terms of building, forest, woodland and/or agriculture. Therefore, the incorporation of ecological processes in the nature conservation strategy is necessary [23], not forgetting that: (1) ecosystems are spatially and temporally dynamic; (2) ecosystem components interact with each other contributing to biodiversity; (3) ecological processes act as species-selective forces; (4) in highly modified sites ecosystem restoration is a conservation priority.

2. Integration between green infrastructure, ecosystem services, and ecosystem restoration

The concept of structure in landscape architecture is necessarily related to spatialization. In this structure, relationships between ecosystems or elements are expressed [24]. The structure integrates the landscape system's objective and subjective components, articulating significant features that relate to each other [11]. Infrastructure is a structure that serves as the base for something to be developed. Thus, when we refer to green infrastructure, we are talking about a planned network of structural spaces rather than a network of spaces disconnected from the biophysical structure of the territory. The concept of green infrastructure is broad and varied, but it is considered the one summarized by Naumann [25], whose definition is frequently used:

> *Green infrastructure is the network of natural and semi-natural areas, features and green spaces in rural and urban, terrestrial, freshwater, coastal and marine areas, which together enhance ecosystem health and resilience, contribute to biodiversity conservation and benefit human populations through the maintenance and enhancement of ecosystem services. Green infrastructure can be strengthened through strategic and coordinated initiatives that focus on maintaining, restoring, improving and connecting existing areas and features as well as creating new areas and features [25].*

Green infrastructure planning involves an assessment of the types of natural and cultural resources available and a prioritization of the resources most important to present and future needs [14]. Therefore, a green infrastructure strategy includes the process of identifying, assessing, and prioritizing areas that are critical to preserving a healthy community. In addition to prioritizing areas, there is also a need to implement actions to ensure their conservation [14]. Mapping natural resources are thus the first step in building a green infrastructure map to inform which areas need conservation actions and which need restoration actions.

The European Biodiversity Strategy for 2020 [26] highlighted the importance of using green infrastructures in landscape planning since it can ensure the "best functional connectivity between ecosystems within and between Natura 2000 areas and in the wider countryside" ([27]: 6). The indication in a European document of the importance of ensuring functional connectivity between ecosystems, even outside protected areas, through landscape planning was a crucial step towards elevating nature conservation to a more comprehensive status than protecting particular species. Recently, the European Biodiversity Strategy for 2030 [28], as a core part of the European Green Deal, defines an action plan towards protecting nature and reversing the degradation of ecosystems.

Green infrastructure is also a tool to achieve economic and social benefits through natural solutions. This concept of natural solutions (or nature-based solutions) was further developed by the working groups of the European Commission [27] as a solution inspired, supported or copied from nature. The green infrastructure strategy itself states that "Green infrastructure can make a significant contribution to the effective implementation of all policies where some or all of the desired objectives can be achieved in whole or in part through nature-based solutions." ([29]: 3).

The scientific community widely refers to ecosystem services as benefits that a population acquires, directly or indirectly, from ecosystem functions [16]. Ecosystem services result from flows of materials, energy, and information from natural capital stocks capable of producing human well-being [16]. Their monetary quantification [30], with specific units [31] or measured through indicators [32] as well as a qualitative assessment [33] has been addressed in the last decades.

These services were categorized into several typologies by the Millennium Ecosystem Assessment [34]: supporting (services required to produce all other ecosystem services); provisioning (products obtained from ecosystems); regulating (benefits obtained through the regulation of ecosystem processes); and cultural (non-material benefits obtained from ecosystems). The landscape is intended to be one where ecosystem services are provided in balance with the physical structure that supports them, so the various actors dealing with land-use planning must understand the support structure of the landscape.

To contribute to a consistent definition of ecosystem services, Fisher [35] introduces the importance of ecosystems' structure and their processes and functions. In this approach, ecosystem services are characteristics of ecosystems used directly or indirectly by humans to produce well-being. Accordingly, ecosystem structure is itself a service because it provides a platform for ecosystem processes. Related to this idea, the same authors say that the configuration of ecosystem structure and processes is necessary for the healthy functioning of ecosystems and their services, relating them to the concept of (green) infrastructure. The spatial characteristics of ecosystems are also a way to classify their services, so it will be important in planning to know what services are available and how they flow through the landscape. In this way, it is important to understand the relationships between the production of the service and the place where the benefit occurs, recognizing the dynamic characteristics of ecosystems. In this regard, [35] propose a system for classifying ecosystem services into three categories: (1) in situ (services produced and benefits provided occur at the same location); (2) Omni-directional (where services are provided at a single location, but the benefit occurs in the landscape surrounding the service production with no defined direction); (3) directional - where service provision benefits from a specific location due to the direction of flow.

The Ecosystem services were also an integral part of the European Biodiversity Strategy for 2020 [26]. According to this strategy, ecosystems and their services would be maintained and enhanced by creating green infrastructure and restoring at least 15% of degraded ecosystems. Ecosystem services were then identified through indicators associated with each ecosystem, assessed according to the Common International Classification of Ecosystem Services [36]. The European methodologies followed for mapping ecosystems [36] use the interpretation of different land use and land cover classes and relate them to the European Habitat Classification (EUNIS). As a result, the analysis of ecosystem services while assessing land-use mapped ecosystems tends to present itself transformed into an "in situ" category, in the sense of [36], where the production of the service and the benefit are located in the same place. Another consequence of a methodology based on current land use mapping is to assess the use and service regardless of whether it is in an area of greater ecological suitability. Such is the example of forests that are all converted into ecosystems capable of producing the same services regardless of the type of forest species. This is an incorrect approach since it is well known that different forest trees or stands provide different ecosystem services. In Portugal, this is very relevant since most of the forest stands are not native (maritime pine, eucalyptus) providing poorer ecosystem services when compared with native stands (oak).

It is essential to consider that the landscape has different capabilities to provide specific ecosystem services [33], being of a more profound complexity than just an assessment by current land use. The structure of the ecosystem assumes a vital role in supporting the very functioning of the ecosystem. Interestingly, authors such as Burkhard [33] consider that the typology of supporting ecosystem services (defined in the Millennium Ecosystem Assessment), is understood as those that ensure ecological integrity. However, supporting services are considered difficult to map [37], and it is considered that the link between supporting services and human well-being occurs indirectly [34]. Therefore, supporting services have received much less attention among four types of ecosystem services. Despite European recommendations to map ecosystems and their services, recent publications have failed to include supporting services [38].

In this sense, it is considered that by mapping green infrastructure, the supportive services are provided in conjunction with other ecosystem services. Incorporating structural components of ecosystems also allows for a complete approach to mapping ecosystems and their services by encompassing the various relationships between the area of production of the service and the site of benefit from that service.

Including ecosystem services in landscape planning will need to go through defining the goal of achieving multiple ecosystem services [39]. However, the function of an ecosystem must be ensured in planning regardless of the benefit it may provide [40]. Maximizing one ecosystem service may jeopardize the balance of all other ecosystem services, as exemplified by Dosskey [39] in the US Green Belt region, where agricultural productivity was put as a priority at the expense of water quality and wildlife. It should be desirable for public policy to seek a degree of multifunctionality across cultural landscapes and to achieve the greatest degree of multifunctionality in green infrastructure [40]. The same author considers that monofunctional landscapes will require greater inputs to continue to provide values and functions and are likely to become unsustainable and require restoration.

Ecosystems are naturally multifunctional, making available services determined by landscape structure [39]. Modifying landscape structure can rebalance the services available. Thus, the landscape planning process will need to include an understanding of the current functioning of the landscape and an assessment of whether changing the landscape structure can affect the ecological functions and services provided [39]. Ecologically based planning aims to define the best use which implicitly includes the best use of natural resources without compromising their existence and their stability in the system [41]. This applies both to landscapes dominated by native vegetation, forming a dense woodland, and to agricultural or production forest areas properly integrated into the landscape and its structure with good management practices and appropriate design.

Ecosystem restoration is often focused on the recovery of a particular plant or animal community, often appearing related to landscape fragmentation [42]. The restoration of an ecosystem, which is itself a complex system, will involve the recovery of the ecological functions of the system. The functions that are easily altered by the degradation of an ecosystem are soil structure, nutrient flow, and water cycling [43]. The restoration of an ecosystem will mainly involve the recovery of the lost functions, because this loss has also contributed to its own degradation:

> *When a reintegrated landscape is achieved, it will be a landscape that is a mosaic of agricultural, natural, and semi natural systems, which together maximize biodiversity and economic returns by maintaining the landscape amenity function (minimizing the loss of landscape qualities through soil salinization of water and wind erosion) and so make possible a sustainable agriculture and a functional diverse natural system [43].*

The ecosystem restoration will involve the recovery of its ecological integrity. The integrity of an ecosystem includes the integrity of the system's structure and function, the maintenance of its components and its dynamic interactions [44]. From this perspective, any loss of system components leads to a loss of system integrity. According to Forman [45] and Thorn [46] a system with ecological integrity exhibits natural conditions of productivity, biodiversity, soil and water conservation, which are the goal of any sustainable environment [47]. The integrity of an ecosystem includes an adequacy of uses to the ecological characteristics of the system, meeting the dynamic stability of the landscape, making it resilient.

Ecosystem restoration is defined by the Society for Ecological Restoration as the process of supporting the recovery of an ecosystem that has been degraded, damaged, or destroyed [48]. These last three states can be equated to various states of morphogenesis, where an imbalance of landscape stability occurs. A morphogenesis ecosystem can be at different levels of imbalance from degradation to complete destruction. When ecosystems are being overexploited or degraded the health of the ecosystem goes into decline as well as its integrity and resilience [49]. This state of decline can be reversed with recovery actions, which in turn lead to ecosystem rebalance in Tricart's [3] interpretation of dynamic equilibrium. An ecosystem in equilibrium provides a greater number or a higher quality of services.

A Landscape-scale ecosystem restoration involves restoring a set of ecosystems to recover natural and cultural values and ecosystem service flows [49]. Hobbs [50] argues that the recovery of ecosystems should not focus on replicating the conditions prior to disturbance, but should be managed in a future perspective, not forgetting that the landscape is temporally and spatially dynamic. Besides, there are two types of recovery [51]: one in which the goal is the recovery of biotic continuity, and another, corresponding to more severe situations of degradation, will involve the physical recovery of the ecosystem. An example of the latter is the case of the obstruction of a river, where the goal will be to recreate the continuity of water flow. According to those authors, there are thus two thresholds that, if crossed, imply different interventions, the biotic threshold, and the abiotic threshold. The least severe situation of degradation of an ecosystem will thus imply changes in land use.

The ecosystem restoration can integrate cultural values, for example, an agricultural area located on productive soils and with good management practices (including compartmentalization hedges, contour farming) contributes to increased productivity, biodiversity, and soil and water conservation. Designing green infrastructure with planned land uses, consistent with its different ecological characteristics, guarantee the ecosystem's integrity and allow it to assess restoration needs.

3. The need for a restoration solution: centre region of Portugal

The absence of an ecological-based landscape and land-use plan can have severe consequences in the increase of soil degradation and floods, decrease of biodiversity, and increased fire risk. This landscape degradation is present in several areas of Portugal. Also, a consequence of the set of policies followed since the beginning of the 20th century. In the 1930s, Portugal went through a wheat campaign that destroyed the fertility of the soil. Later, the monoculture campaigns of maritime pine and eucalyptus continue the degradation of the soil and the destruction of the landscape, which we still see happening, especially in the north of the Tagus river, in the Centre and North regions.

The Centre Region (**Figure 1**) corresponds to a Statistical Territorial Unit (NUT) and comprises about 2,819,936 hectares. This region includes different landscape typologies, such as the southwest western zone with fruit productivity, the coastal zone with low and high coastlines, and an inland zone dominated by maritime pine and eucalyptus in formations of schist and granite. This area is characterized by a rugged relief in the most central location, such as Serra da Estrela, where Rio Mondego begins, the Serra da Lousã and the Serra do Açor. The Natura 2000 network includes 21 Special Areas of Conservation.

In an evaluation of land use from the 1990s to the present, it was possible to understand the evolution of the different land uses (**Figure 2** and **Table 1**).

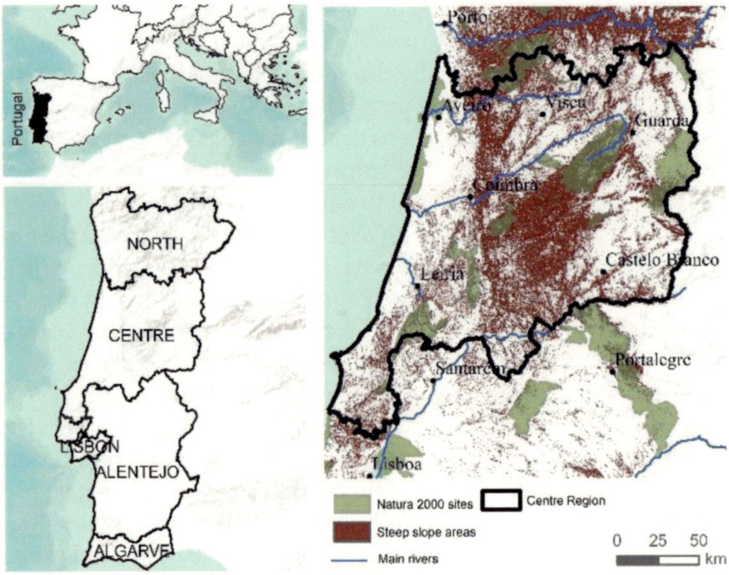

Figure 1.
Location of the Centre region in Portugal.

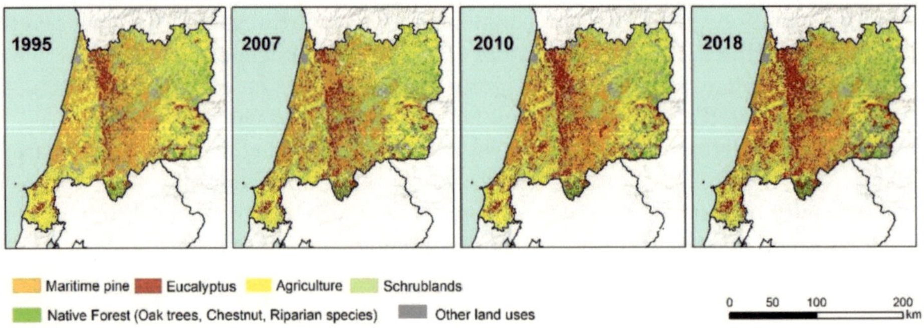

Figure 2.
Evolution of the main land use classes in the Centre region, between 1995 and 2018 (data base: DGT, land use and land cover map).

Main Land Uses in Centre Region	% in 1995	Variation	% in 2007	Variation	% in 2010	Variation	% in 2018
Maritime pine	26,7%	↘	20,0%	↗	22,5%	↘↔	22,3%
Eucalyptus	9,7%	↗	10,9%	↗	13,5%	↗	17,2%
Schrubland	11,7%	↗	16,4%	↘	12,3%	↗	13,3%
Agriculture	31,3%	↘	25,5%	↘	26,2%	↘	23,3%
Native	7,6%	↘	5,8%	↘	6,4%	↗	9,3%

Table 1.
Evolution of the main land uses in the Centre region, between 1995 and 2018.

This analysis was done with the interpretation and reclassification of the Land-Use and Land Cover maps produced by DGT (1995, 2007, 2010, 2018) [52]. In this period of 24 years, there is an oscillation of maritime pine, which tends to stabilize at 22%. The percentage of eucalyptus in the central zone has increased since 1995, from

9.7% of the total area, to 17% in 2018, i.e., it has practically doubled. On the other hand, the area occupied by agriculture has decreased since 1995. Native species include cork oak, holm oak, other oaks and also chestnut stands (archaeophyte), and oscillate between the years analyzed, with occupations between 6 and 9%.

There is in fact a very serious problem of inappropriate land uses that lead to the destruction of landscapes with negative consequences for those who live there, but also for those who live further away, for example, the impacts derived from water quality. Since the end of the 19th century, the oaks and chestnut trees and the traditional pastures, were replaced, first by maritime pine, which were planted on the community lands ("Baldios"), and then by eucalyptus. Land property fragmentation and, consequently, the landowners' increase also aggravated the land management problem.

A balanced landscape constituted by agriculture on the best soils, mixed woodland complementing agriculture and all its by-products, pastoralism in articulation with woodland and agriculture, villages, towns, and cities strategically located in situations of greater comfort and proximity to the food and materials produced was replaced by a landscape ecologically degraded, humanely depopulated and which burns extensively and repeatedly.

Fire frequency (number of times between 1990 and 2017)	Area (ha)	%
1 time burnt area between 1990 and 2017	698 775	24.8
2 times burnt area between 1990 and 2017	321 099	11.4
3 times burnt area between 1990 and 2017	93 965	3.3
4 times burnt area between 1990 and 2017	16 105	0.6
5 times burnt area between 1990 and 2017	2 882	0.1
6 times burnt area between 1990 and 2017	327	0.012
7 times burnt area between 1990 and 2017	224	0.008
No burned area between 1990 and 2017	1 686 559	59.8

Table 2.
Fire frequency between 1990 and 2017, area and percentage of case study.

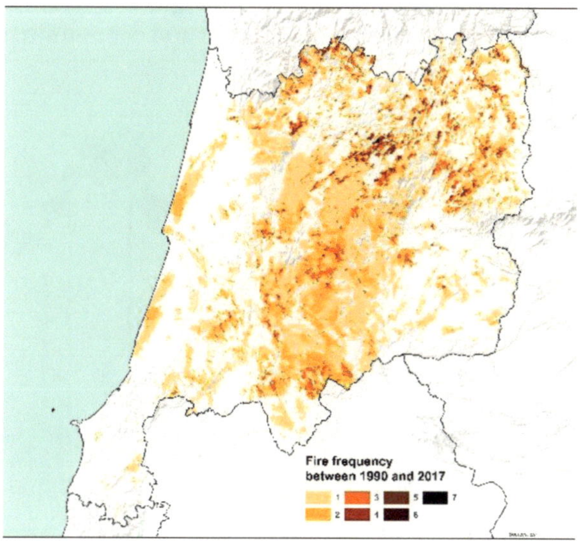

Figure 3.
Fire frequency between 1990 and 2017.

The policies disconnected from the ecological capacity of the land, led to the current situation of mega-fires, with loss of life and property and land abandonment. Analyzing the centre region in terms of burnet areas, 40% of the region was burned between 1995 and 2017. About 25% of the Centre Region burned once (**Table 2**), but 11% burned twice (**Figure 3**). In the megafires from 2017, the Centre Region was the most affected, representing 15% of total region area (416 thousand hectares). It is very urgent to develop adequate land-use plans for the rural areas.

The creation of a healthier landscape implies the conservation of natural resources, the creation of a balanced, multifunctional system with landscape recovery using native species. This will lead to creation of different economies, where ecosystem services payment can also take place.

4. The solution for healthier landscape – centre region of Portugal

The solutions developed to attain a healthier landscape involve the creation of a multifunctional landscape, with native or archaeophytes species, agriculture, and pastureland. The planned landscape will allow to create businesses, generate employment (landscape recovery companies, native species nurseries, forest management companies, reactivation of native wood business), and unique products (non-wood products, such as flour from oak acorns, chestnut, walnut, honey and mushroom production) capable of attracting nature tourism as well.

This landscape will then provide better ecosystem services, such as water quality, soil conservation, and biodiversity improvement, among others. Together, it will develop a fire resilient landscape, which is the landscape's capacity to absorb the disturbance caused by rural fires without losing its function, structure, and identity and ultimately weakening fire frequency and intensity or magnitude [53].

The vision for a healthier landscape of the Centre Region was developed by the application of FIRELAN [53], which is an ecologically based model that integrates different principles related to landscape fire resilience and ecological sustainability into a land-use plan, using the river basin as a landscape unit. The FIRELAN pretends to provide a multifunctional landscape (**Figure 4**) with benefits to the environment, but also developing economies and rural communities.

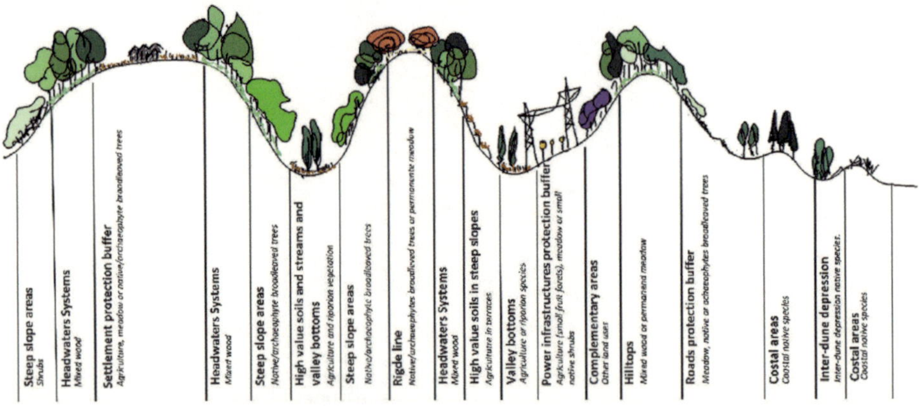

Figure 4.
Ecological and cultural system and land use potential from FIRELAN model.

The different components of the FIRELAN network for the Centre Region are mapped in **Figure 5**. The FIRELAN network is the main landscape structure with physical, biological, and cultural elements (**Table 3**). For each component there is a set of adequate land uses that should be promoted. Those land uses are identified in **Figure 6** and in **Table 4**. In the interstices of the FIRELAN network, also called Complementary Areas, the land use possibilities are wider.

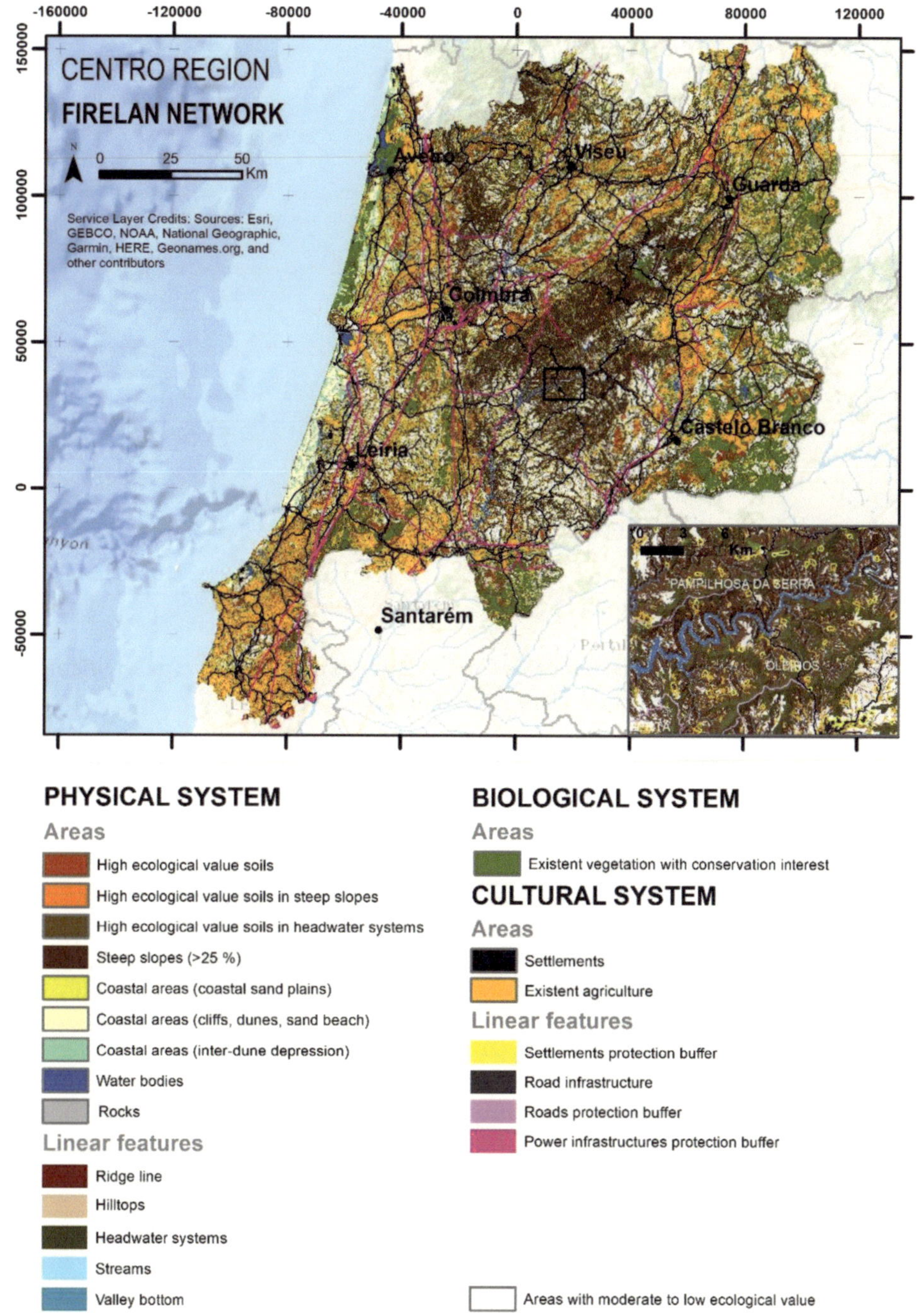

Figure 5.
FIRELAN network components in the Centre region of Portugal.

Firelan network component			Data Source
Physical System	Areas	Soils with high and very high ecological value	[54–56]
		Steep slopes (>25%)	[56, 57]
		Coastal area	[56, 58]
		Wetlands	[53]
		Water bodies	[53]
	Linear	Ridge line	[56, 57]
		Hilltops	[56, 57]
		Headwater systems	[56, 59]
		Streams and Valley bottoms	[56, 57]
Biological System	Areas	Existent vegetation with conservation interest	[52, 56, 60]
Cultural System	Areas	Settlements	[52]
		Existent agriculture	[52]
	Linear	Settlement protection buffer	—
		Roads	OpenStreetMap©
		Roads protection buffer	—
		Power infrastructures buffer	[61]

Table 3.
Components of FIRELAN and sources for the Centre region of Portugal.

The potential land uses plan (**Figure 6**, **Table 4**) highlights that:

- Native and archaeophytes species (including agroforestry systems), which represent 11% of the study area, can expand into further 31%.
- The agriculture can be increased in about 12% of the case study area, in addition to existing agriculture (23%).
- Existent vegetation with conservation interest is present in 14% of the Centre region.
- Complementary areas (areas with moderate to low ecological value) represent 16% of the study area.

Comparing potential land use and current land use map allows defining a Landscape Transformation Plan with conservation and restoration actions. According to the developed plan (**Figure 7**, **Table 5**): 35% of the Centre region should have restoration actions, and 57% should be maintained and conserved. Also, according to the results:

- The eucalyptus can be kept in 5% of the case study area, only in complementary areas, and with environmental measures.
- The maritime pine can be kept in 6% of the case study area, only in complementary areas, and with environmental measures.
- About 0,8% of the case study area should have recovery of the riparian vegetation.

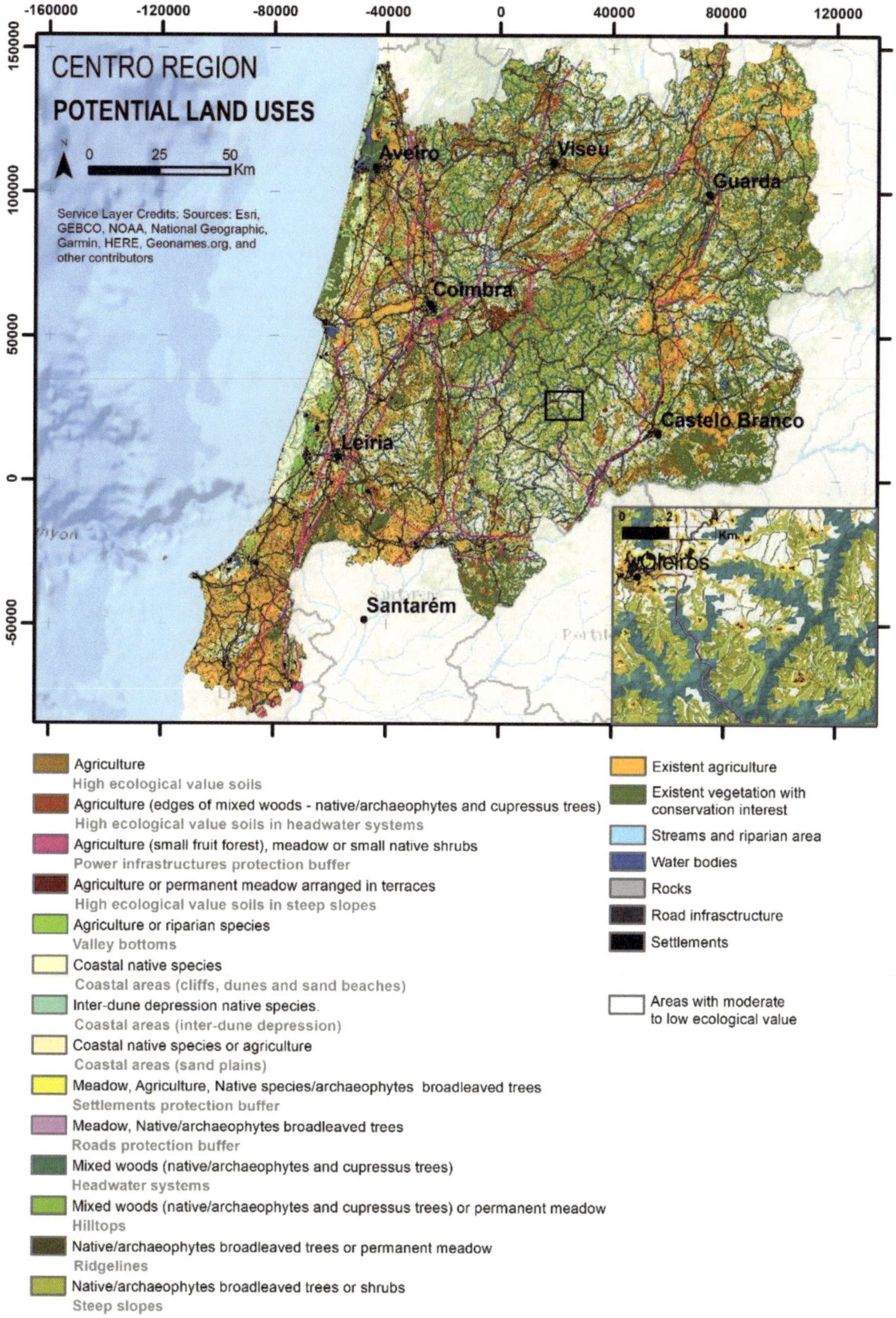

Figure 6.
Potential land uses in the Centre region of Portugal.

5. Conclusions

The United Nations General Assembly proclaimed the Ecosystem Restoration decade between 2021 to 2030, aiming to halt the degradation of ecosystems and restore them to achieve healthier landscapes. Landscape Architecture is an interdisciplinary discipline helpful in attaining those goals through planning and design

Firelan network components	Potential land uses	Area (ha)	%
High ecological value soils	Agriculture	131627	4.67
High ecological value soils in headwater systems	Agriculture (edges of mixed woods - native/ archaeophytes and cupressus trees)	52733	1.87
Power infrastructures protection buffer	Agriculture (small fruit forest), meadow or small native shrubs	5545	0.20
High ecological value soils in steep slopes	Agriculture or permanent meadow arranged in terraces	14408	0.51
Valley bottoms	Agriculture or riparian species	24395	0.87
Coastal areas (cliffs, dunes and sand beach)	Coastal native species	35653	1.26
Coastal areas (inter-dune depression)	Inter-dune depression native species	14178	0.50
Coastal areas (coastal sand plains)	Coastal native species or agriculture	5147	0.18
Settlements protection buffer	Meadow, Agriculture, Native species/ archaeophytes broadleaved trees	105883	3.76
Roads protection buffer	Meadow, Native/archaeophytes broadleaved trees	27454	0.97
Headwater systems	Mixed woods (native/archaeophytes and cupressus trees)	314469	11.15
Hilltops	Mixed woods (native/archaeophytes and cupressus trees) or permanent meadow	42221	1.50
Ridgelines	Native/archaeophytes broadleaved trees or permanent meadow	86392	3.06
Steep slopes	Native/archaeophytes broadleaved trees or shrubs	248223	8.80
Existent agriculture	Existent agriculture	646346	22.92
Existent vegetation with conservation interest	Existent vegetation with conservation interest	381180	13.52
Streams and riparian area	Streams and riparian area	22309	0.79
Water bodies	Water bodies	26709	0.95
Rocks	Rocks	3310	0.12
Road infrastructure	Road infrastructure	50764	1.80
Settlements	Settlements	132965	4.72
Areas with moderate to low ecological value	Areas with moderate to low ecological value	447731	15.88

Table 4.
Potential land uses for the Centre region of Portugal, area and percentage.

restoration in different contexts and scales, from rural to urban, from inland to coastal. The ecological-based planning methodologies contribute to better landscapes by starting from the knowledge and spatialization of natural processes. As part of an ecological-based planning methodology, working with green infrastructure allows the ecological integrity of the landscape, increasing the number or quality of services provided by the ecosystems and defining restoration actions and locations.

The Centre Region of Portugal has a severe problem of inappropriate land use. Since the end of the 19th century, the oaks and chestnut trees were replaced by

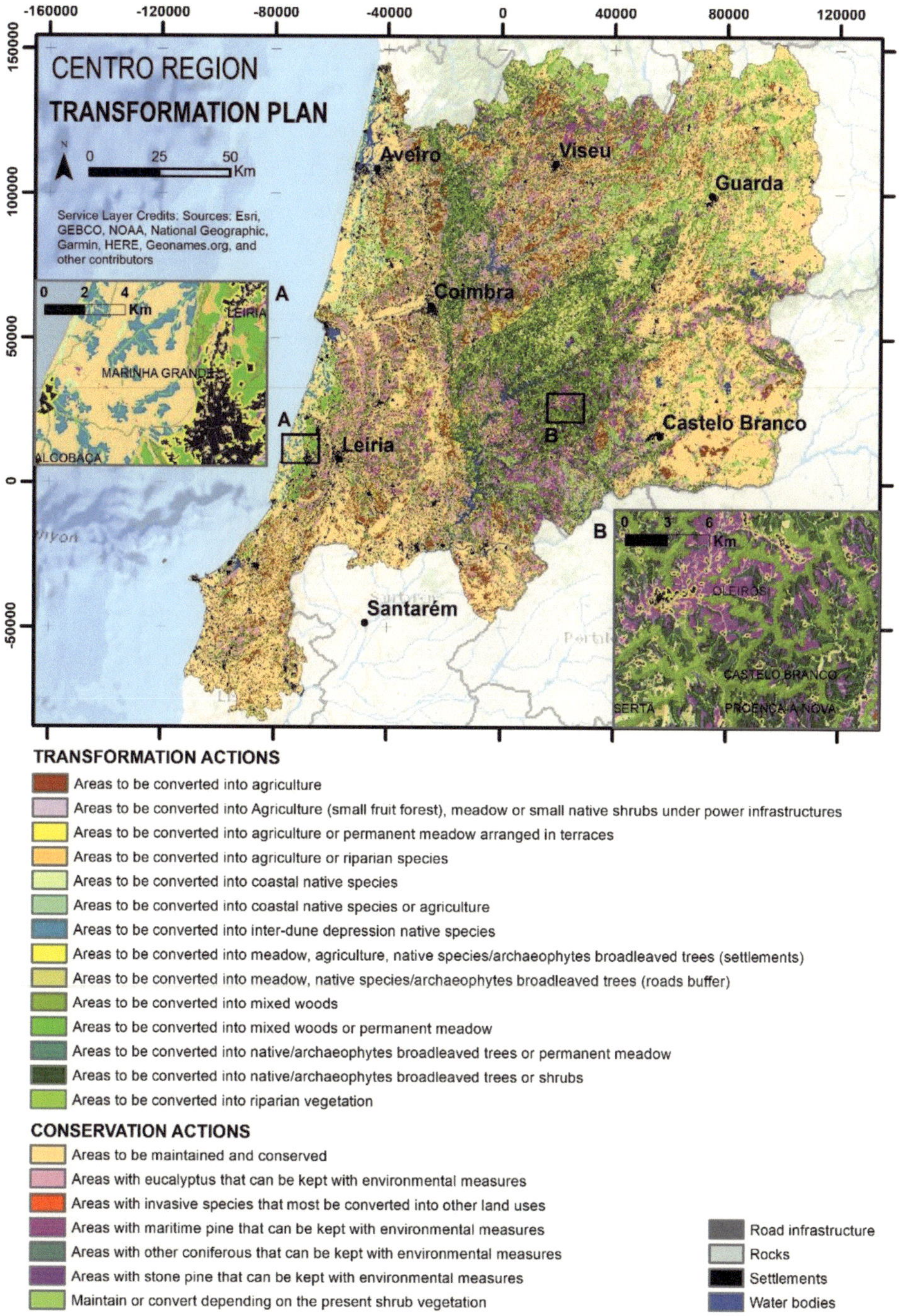

Figure 7.
Landscape transformation plan for the Centre region of Portugal.

maritime pine and eucalyptus. The number of fires and their severity increased, the land is continually degrading, and biodiversity loss increases. Applying an ecologically based landscape plan to prioritize restoration actions at the landscape scale is necessary to reverse this situation. The FIRELAN model was applied to the Centre Region to establish a vision for a healthier landscape. According to the results, 35% of the Centre Region should have restoration actions.

Landscape Transformation Actions		Area (ha)	%	Area (ha)	%
Restoration	Areas to be converted into agriculture	157670	5.59	980608	34.78
	Areas to be converted into Agriculture (small fruit forest), meadow or small native shrubs under power infrastructures	4332	0.15		
	Areas to be converted into agriculture or permanent meadow arranged in terraces	13938	0.49		
	Areas to be converted into agriculture or riparian species	18149	0.64		
	Areas to be converted into coastal native species	7840	0.28		
	Areas to be converted into coastal native species or agriculture	2786	0.10		
	Areas to be converted into inter-dune depression native species	14125	0.50		
	Areas to be converted into meadow, agriculture, native species/ archaeophytes broadleaved trees (settlements)	96844	3.43		
	Areas to be converted into meadow, native species/archaeophytes broadleaved trees (roads buffer)	24944	0.88		
	Areas to be converted into mixed woods	314469	11.15		
	Areas to be converted into mixed woods or permanent meadow	35461	1.26		
	Areas to be converted into native/ archaeophytes broadleaved trees or permanent meadow	80508	2.86		
	Areas to be converted into native/ archaeophytes broadleaved trees or shrubs	185648	6.58		
	Areas to be converted into riparian vegetation	22309	0.79		
	Areas with invasive species that most be converted into other land uses	1584	0.06		
Conservation	Areas to be maintained and conserved	1212952	43.02	1625284	57.64
	Areas with eucalyptus that can be kept with environmental measures	142909	5.07		
	Areas with maritime pine that can be kept with environmental measures	161113	5.71		
	Areas with other coniferous that can be kept with environmental measures	3283	0.12		
	Areas with stone pine that can be kept with environmental measures	2951	0.10		
	Maintain or convert depending on the present shrub vegetation	102077	3.62		
—	Road infrastructure	50764	1.80	—	—

Landscape Transformation Actions		Area (ha)	%	Area (ha)	%
—	Rocks	3310	0.12	—	—
—	Settlements	132965	4.72	—	—
—	Water bodies	26709	0.95	—	—

Table 5.
Transformation action, area and percentage of Centre region case study.

Subsequently, the eucalyptus area must drop from 17% of the Centre Region area to 5%, and the maritime pine from 22–6%. The results also show that agriculture could increase from 23–35% of the Centre Region. In the restoration actions, the native species should be used in more than 24% of the case study, especially in the headwaters systems and streams where mixed woods and riparian galleries should be developed.

The landscape transformation plan contributes to the definition of adequate policies to tackle ecosystem restoration through landscape and land use planning.

Acknowledgements

This work was supported and financed by the Portuguese Foundation for Science and Technology (FCT) under the project SCAPEFIRE PCIF/MOS/0046/2017. This work was also under the research unit Linking Landscape, Environment, Agriculture and Food Research Centre (LEAF) UID/AGR/04129/2020.

Author details

Selma B. Pena*, Maria Luísa Franco and Manuela R. Magalhães
Linking Landscape Environment, Agriculture and Food (LEAF), Instituto Superior Agronomia, Universidade de Lisboa, Lisboa, Portugal

*Address all correspondence to: selmapena@isa.ulisboa.pt

References

[1] Cabral FC. O "Continuum Naturale" e a Conservação da Natureza. (The "Continuum Naturale" and nature conservation.) Proceedings of the seminar Conservação da Natureza, Lisboa: Serviços de Estudos do Ambiente, 18 e 19 de Abril de 1980. p. 35-54

[2] Telles GR. 1992. A conservação das Paisagens Históricas e Rurais. Correio da Natureza 17. 1992 p. 53-55

[3] Tricart J. Ecodinâmica. Rio de Janeiro, IBGE, Diretoria Técnica, SUPREN. 1997. Available from: https://www.docdroid.net/D8HPCLd/jean-tricart-ecodinamica-pdf [Accessed: 2016-05-02]

[4] Tricart J. Morphogènese et pédogenèse, géomorphologie et pédologie. Science du Sol 1, 1965. P. 69-85

[5] Odum EP. Fundamentals of ecology. Saunders College Publishing, Philadelphia, PA (US), 1971. 574 pp.

[6] ONU. Convention on biological diversity. 1992. Available from: https://www.cbd.int/doc/legal/cbd-en.pdf [Accessed: 2015-09-01]

[7] COP 5. Fifth Ordinary Meeting of the Conference of the Parties to the Convention on Biological Diversity, 15 - 26 May 2000 - Nairobi, Kenya. Available from: https://www.cbd.int/decisions/cop/?m=cop-05 [Accessed: 2015-09-01]

[8] Magalhães MR, Abreu MM, Lousã M, Cortez N. Estrutura Ecológica da Paisagem. Conceitos e delimitação – escalas regional e municipal. ISAPress, Lisboa, 2007. 361pp.

[9] Magalhães MR, Cunha N, Pena S. The Ecological Land Suitability in the Land-Use Plan. In Proceedings of the 7th International Conference on Virtual Cities and Territories, Nova University, TULisbon, October 11-13, 2011, Portugal. 2011. p. 307-312

[10] Collins MG, Steiner FR, Rushman MJ. Land-use suitability analysis in the United States: historical development and promising technological achievements. Environ. Manage. 2001; 28 (5), 611-621. DOI: 10.1007/s002670010247

[11] Magalhães MR,. A Arquitectura Paisagista. Morfologia e Complexidade. Editorial Estampa, Lisboa, 2001. p.525.

[12] McHarg IL. Design with Nature. NHP, New York (US), 1971. 197 pp.

[13] Steiner FR. The Living Landscape, Second Edition: An Ecological Approach to Landscape Planning, Island Press. 2008.

[14] Firehock K. Strategic Green Infrastructure Planning. A multi-scale approach. Island Press. 2015. 138pp.

[15] Kolbe J, WUstemann H. Estimating the Value of Urban Green Space: A hedonic Pricing Analysis of the Housing Market in Cologne, Germany. J. Chem. Inf. Model. 2014. 5, 45-61. DOI:10.1017/CBO9781107415324.004

[16] Costanza R, d'Arge R, De Groot R, Farber S, Grasso M, Hannon B, Limburg K, Naeem S, O'Neill RV, Paruelo J, Raskin RG, Sutton P, Van Den Belt M, The value of the world's ecosystem services and natural capital. Nature 1997. 387, 253-260. DOI:10.1038/387253a0

[17] Hermann A, Schleifer S, Wrbka T. The Concept of Ecosystem Services Regarding Landscape Research: A Review. Living Rev. Landsc. Res. 2011. 5, 1-37. DOI:10.1177/0170840609104565

[18] Bennett G, Mulongoy KJ. Review of experience with ecological networks, corridors and buffer zones. 2006. Secretariat of the Convention on Biological Diversity, Montreal, Technical Series No. 23.

[19] Taylor PD, Fahrig L, Henein K, Merriam G. Connectivity is a vital element of landscape structure. Oikos. 1993. 68, 571-573. DOI:10.2307/3544927

[20] MacArthur Rh, Wilson EO. The theory of island biogeography. Princeton Univ. Press. Ed., Princeton. 1967.

[21] Cunha NS, Magalhães MR. Methodology for mapping the national ecological network to mainland Portugal: A planning tool towards a green infrastructure. Ecological Indicators 2019. 104C: 802-818. DOI: 10.1016/j.ecolind.2019.04.050

[22] EEA. Green infrastructure and territorial cohesion. 2011. Technical report No 18/2011

[23] Bennett AF, Haslem A, Cheal DC, Clarke MF, Jones RN, Koehn JD, Lake PS, Lumsden LF, Lunt ID, Mackey BG, Nally RM, Menkhorst PW, New TR, Newell GR, O'Hara T, Quinn GP, Radford JQ, Robinson D, Watson JEM, Yen AL, Ecological processes: A key element in strategies for nature conservation. Ecological Management & Restoration, 2009. 10: 192-199. DOI: 10.1111/j.1442-8903.2009.00489.x

[24] Forman RTT, Godron M. Landscape ecology. John Wiley, New York. 1986. 619p.

[25] Naumann S, McKenna D, Timo K, Mav P, Matt R. Design, implementation and cost elements of Green Infrastructure projects. Final report to the European Commission, DG Environment, 2011. Contract no. 070307/2010/577182/ETU/F.1, Ecologic institute and GHK Consulting.

[26] EC. Our life insurance, our natural capital: an EU biodiversity strategy to 2020. COM(2011) 244 final. 2011.

[27] EC. Towards an EU Research and Innovation policy agenda for Nature-Based Solutions & Re-Naturing Cities. 2015. DOI:10.2777/765301

[28] EC. EU Biodiversity Strategy for 2030 Bringing nature back into our lives COM/2020/380 final. 2020.

[29] EC. Green infrastructure (GI): enhancing Europe's natural capital COM(2013) 249 final. 2013.

[30] De Groot R, Brander L, van der Ploeg S, Costanza R, Bernard F., Braat L, Christie M, Crossman N, Ghermandi A, Hein L, Hussain S, Kumar P, McVittie A, Portela R, Rodriguez LC, ten Brink P, van Beukering P. Global estimates of the value of ecosystems and their services in monetary units. Ecosyst. Serv. 2012. 1, 50-61. DOI:10.1016/j.ecoser.2012.07.005

[31] Boyd J, Banzhaf S. What are ecosystem services? The need for standardized environmental accounting units. Ecol. Econ. 2007. 63, 616-626. DOI:10.1016/j.ecolecon.2007.01.002

[32] Müller F, Burkhard B. The indicator side of ecosystem services. Ecosyst. Serv. 2012. 1, 26-30. DOI:10.1016/j.ecoser.2012.06.001

[33] Burkhard B, Kroll F, Müller F. Landscapes' Capacities to Provide Ecosystem Services – a Concept for Land-Cover Based Assessments. Landsc. Online 2010, 1-22. doi:10.3097/LO.200915

[34] Alcamo J. Ecosystems and human well-being, Ecosystems. ISLAND PRESS, Washington, D.C., USA. 2003. p. 245.

[35] Fisher B, Turner RK, Morling P. Defining and classifying ecosystem

services for decision making. Ecol. Econ. 2009. 68, 643-653. doi:10.1016/j.ecolecon.2008.09.014

[36] EU. Mapping and assessment of ecosystems and their services in the EU. Indicators for ecosystem assessments under Action 5 of the EU Biodiversity Strategy to 2020, 2nd Report. 2014. DOI:10.2779/75203

[37] Benis E, Drakou E, Dunbar MB, Maes J, Willeman L. Indicators for mapping ecosystem services: a review. 2012. DOI:10.2788/41823

[38] EEA. European ecosystem assessment— concept, data, and implementation Contribution to Target 2 Action 5 Mapping and Assessment of Ecosystems and their Services (MAES) of the EU Biodiversity Strategy to 2020. 2015. DOI:10.2800/629258.1.

[39] Dosskey MG, Wells G, Bentrup DW. Enhancing ecosystem services: Designing for multifunctionality. J. Soil Water Conserv. 2012. 67, 37A – 41A. DOI:10.2489/jswc.67.2.37A

[40] Selman P. Planning for landscape multifunctionality. Sustain. Sci. Pract. Policy, 2009. 5, 45-52. DOI: 10.1080/15487733.2009.11908035

[41] Ndubisi F. Managing change in the landscape: a synthesis of approaches for ecological planning. Landscape Journal. 2002. University of Wisconsin Press. EBSCO Publishing. p. 138-155

[42] Jaeger JAE, Soukup T, Madriñán LF, Schwick Chr, Kienast F. Landscape fragmentation in Europe. EEA-FOEN Report. 2011. ISBN 978-92-9213-215-6

[43] Main AR. Landscape Reintegration: Problem Definition. In: Richard J. Hobbs, Denis A. Saunders (editors) Reintegrating Fragmented Landscapes. 1993. Springer New York. p 189-208.

[44] Regier HA,. The Notion of Natural and Cultural Integrity. In: Woodley, S., Key, J., Francis, G. (editors) Ecological Integrity and the Management of Ecosystems. USA. St. Lucie Press. 1993. p3-18

[45] Forman RTT. Land Mosaics. Cambridge University Press, Cambridge, 1995. p. 632.

[46] Thorne JF. Landscape ecology. A Foundation for greenway design. In Paul S.Smith & Paul Cawood Hellmund (editors) "Ecology of greenways" University of Minnesota press. 1993. p 23-42

[47] Cook EA. Landscape structure indices for assessing urban ecological networks. Landsc. Urban Plan. 2002. 58, 269-280. DOI: 10.1016/S0169-2046(01)00226-2

[48] SER, 2004. The SER International Primer on Ecological Restoration, in: Www.ser.org & Tucson: Society for Ecological Restoration International.

[49] Clewell AF, Aronson J. Ecological Restoration. Principles, Values, and Structure of an Emerging Profession. 2ª edition. USA. ISLANDPress. 2013

[50] Hobbs RJ, Hallett LM, Ehrlich PR, Mooney H a,. Intervention Ecology: Applying Ecological Science in the Twenty-first Century. Bioscience, 2011. 61, 442-450. DOI:10.1525/bio.2011.61.6.6

[51] Hobbs RJ, Harris JA. Restoration Ecology : Repairing the Earth ' s Ecosystems in the New Millennium. Restor. Ecol. 2001. 9, 239-246. DOI:10.1046/j.1526-100x.2001.009002239.x

[52] DGT, Land Use and Land Cover Maps from different years: 1995, 2007, 2010, 2018. Available from: https://snig.dgterritorio.gov.pt/ [Accessed: 2020-09-01]

[53] Magalhães MR, Cunha NS, Pena SB, Müller A. FIRELAN—An Ecologically

Based Planning Model towards a Fire Resilient and Sustainable Landscape. A Case Study in Center Region of Portugal. Sustainability 2021. 13, 7055. DOI:10.3390/su13137055

[54] Leitão M, Cortez N, Pena SB. Solo (Soil). In Magalhães, M.R.M. (editor) Estrutura Ecológica Nacional. Uma Proposta de Delimitação e Regulamentação (National Ecological Network. A Delimitation and Regulation Proposal). ISAPress, Lisboa. 2013. p. 83-104, ISBN 978-972-8669-53-9.

[55] Pena SB, Abreu MM, Magalhães MR, Cortez N. Water erosion aspects of land degradation neutrality to landscape planning tools at national scale. Geoderma, 2020. 363, 114093. DOI: 10.1016/j.geoderma.2019.114093.

[56] EPICWebGIS (http://epic-webgis-portugal.isa.ulisboa.pt/)

[57] Cunha NS, Magalhães MR, Domingos T, Abreu MM, Withing K. The land morphology concept and mapping method and its application to mainland Portugal. Geoderma, 2018. 325: 72-89. DOI: 10.1016/j.geoderma.2018.03.018.

[58] Silva JF, Cunha N, Lopes AM, Abreu MM, Magalhães MR. Litoral (Coastal area). In: Magalhães, MRM (editor), Estrutura Ecológica Nacional. Uma Proposta de Delimitação e Regulamentação (National Ecological Network. A Delimitation and Regulation Proposal), ISAPress, Lisboa. 2013. p. 67-82ISBN 978-972-8669-53-9.

[59] Pena SB, Magalhães MR., Abreu MM. Mapping headwater systems using a HS-GIS model. An application to landscape structure and land use planning in Portugal. Land Use Policy 2018. 71, 543-553. DOI: 10.1016/j.landusepol.2017.11.009

[60] Mesquita, S. (2013). Vegetação (Vegetation). In: Magalhães MRM (editor), Estrutura Ecológica Nacional. Uma Proposta de Delimitação e Regulamentação (National Ecological Network. A Delimitation and Regulation Proposal), ISAPress, Lisboa. 2013. p. 105-120. ISBN 978-972-8669-53-9

[61] REN. Rede nacional de transporte de electricidade (National electricity transport network), 2012, Portugal. Escala 1:1.000.000. REN - Rede Elétrica Nacional, S.A. Lisboa

Chapter 6

The American Landscape Architecture Research Universe and a Higher Education Ordination: Descriptive Insights into the Discipline and Profession of Landscape Architecture

Chunqing Liu, Xiaowen Jin, Zhi Yue, Zhen Wu and Jon Bryan Burley

Abstract

Landscape scholars, educators, and academics are interested in the structure and nature of the knowledgebase that comprises both the discipline of landscape architecture and the profession of landscape architecture. In this study, the latent nature of the landscape architecture discipline was revealed by constructing a principal component citation analysis representation (the landscape architecture research universe) concerning several decades of literature (1982–2017) in *Landscape Journal*, a preeminent American journal addressing landscape architecture research. In addition, an ordination was developed describing the curriculum relationships between fifteen top American universities teaching landscape architecture as identified by 'DesignIntelligence,' preparing students for practicing in the profession of landscape architecture. The results revealed that in the discipline, the research activity is highly diverse along many dimensions, constantly evolving as new topics are explored. The pattern in landscape architecture research is broad, as the discipline integrates knowledge and ideas in many fields. In contrast, landscape architecture curriculums, teaching the fundamentals of the profession, are fairly closely clustered together and quite similar, with small differences reflecting emphasis in either landscape history or the visual arts, and mathematics or course electives. This dual identity is both a source of conflict and a unique opportunity.

Keywords: information science, multivariate analysis, environmental design, higher education, citation analysis

1. Introduction

Over the past few decades, information scientists, plus others, have been interested in the structure and nature of the knowledgebases that comprises both disciplines/professionals in many academic areas and subjects, including landscape architecture. A discipline is the body of information collected, studied, analyzed, and reported by a group of individuals who collectively are affiliated with a subject [1]. For the most part, a discipline is usually associated with being a science—describing the way of the universe as best as it can be deciphered, interpreted, and explained, usually with the scientific method. On the other hand, a profession is an activity where a group of individuals practice the art of the profession—making decisions about what to do and how to accomplish the task. For example, in the area of medicine, researchers study the body, conduct experiments, and report results in the discipline of medicine. In contrast, medical doctors give advice to patients and perform operations, deciding what to do and what should be done, often without perfect information, practicing the art of medicine. Usually those who study the discipline are found at research institutions and organizations. Those who practice the profession are typically in business applying their art. Doctors, lawyers, architects, planners, musicians, and athletes are all examples of practitioners applying their skill, deciding what to do and what should be done; thus, it is called the 'art of practice.'

In landscape architecture, dominance has been expressed through the activities of the profession, where individuals practice the art of decision making for planning, design, construction, and maintenance of the exterior environment [1]. It was only relatively recently (1980s) that any attention was given to the discipline of landscape architecture, although some may claim landscape research extends back at least to the a thesis by Frank Waugh concerning campus planning and design at Oklahoma State University for a master's degree at Kansas State University in 1894 [2–6]. The debate concerning the difference between professional practice and the need for the accreditation of schools offering professional practice degrees and the role of research in graduate education is illustrated in *Graduate Education in Landscape Architecture: a Compendium* [7]. Much has changed since the 1980s in the discipline of landscape architecture. An undated report by the American Society of Landscape Architects (ASLA) illustrates how little activity in landscape research was undertaken by American landscape architecture programs in the United States in the early 1970s [8]. A report titled: *Metrics Evaluating Multivariate Design Alternatives: Application of the Friedman's Two-way Analysis of Variance by Ranks: A Personal Reflection*, provides some insight into the progression and develop of landscape research over the last 50 years from the viewpoint of one American scholar [9].

Research in landscape architecture can be divided into two aspects. The first is the development of predictions (models) [1]. Models can be equations, graphs, or even 3-dimensional representations. The other aspect is the development of theories (explanations) [1]. In addition, theories can be further divided into scientific theories (explanations about the universe that if shown to be false are discarded) and normative theories (explanations about reasons and ideas forming a a foundation for decision making, such as a set of ideas about why a designer created a design in a particular manner—exceptions can always be found and all of these normative theories are false, but they are not scientific theories and are simply guides or principles to make decisions in an imperfect world of knowledge—for a designer this is very useful) [1]. There is very little in the way of scientific theory in landscape architecture as most of the theory is normative, useful for practitioners. Most books on landscape theory are about normative theory, ideas and approaches for creating and managing landscape. For example, the deployment of a concept in a design is a normative theory [10, 11]. In contrast, landscape scholars often focus their energy upon developing predictive models accepting the models as evidence but rarely focusing upon scientific theory. Examples of predictive models developed by landscape architects are in human perception research related to assessing visual quality [12–15] and in natural resources to develop soil reclamation Eqs. [16–21]. 'Human intrusion theory' in explaining visual quality equations [22] and 'mesic preference theory' for reclaiming surface mines [23] are explanations that are scientific theories developed by landscape scholars. Within this context/framework of models and theories, landscape research has evolved.

As the volume of landscape research accumulated. Research about research was of interest to some. One approach to study this research was to derive a structure to examine citations in articles written and published in journals [24]. This general approach was reported in a study by Dr. Burley and his spouse Cheryl (an information scientist) concerning the landscape architecture literature for a journal titled: *Landscape Journal* [25]. The co-authors of this book chapter queried Dr. Burley about the reception of this effort. "Well, for the conference, they gave us a premiere setting at the beginning of the conference. With the exception of a few conference people in the room who were required to attend the session, the room was empty. There were a lot more people in the hotel bar. At the time, I really do not think anyone went to these conferences to learn about research, but rather to escape their academic institution, converse with friends and colleagues, and unwind. No one was interested. A better venue would have been an information science setting. Still, I kept looking for opportunities to expand the research endeavor. I was undaunted, very independent; I still am." reflected and commented Dr. Burley. In 2009, a similar expanded study was reported examining the landscape research literature in transportation [26]. Surprisingly, this study was noticed and featured in a seminal book about landscape research [27]. And the study earned an ASLA state award for research. An interesting finding in the study was that the results indicated a fractured, weakly linked research universe where investigators were deep into their line of research and not tied or integrated into other areas. In contrast, in the landscape architecture discipline, there were many connections and interrelationships. "The blending and borrowing across different subject areas was something that landscape architects have claimed for a long time. The study supported those claims. Often in academia, other disciplines tout their depth and wonder why landscape architects do not do the same? Again, here was evidence that in one area, environmental transportation, they were deep but unconnected. I believe both approaches are beneficial, but the differences illustrate where conflicts from those who believe in one approach over the other can be generated. Because landscape architects borrow and integrate, it can go unappreciated by other academics." assessed Dr. Burley.

The foundation of the research is to employ multivariate principal component analysis (PCA), something that landscape architects rarely study. "During my time as a graduate student, my professors at the University of Manitoba urged me to take as many advanced statistic courses as possible and I took even more at the University of Michigan for my PhD. It was like learning the analytic tools for conducting research. If one does not know the tools, it can be difficult to understand the possibilities. Similarly in landscape architecture, if one does not know the design process, it is difficult to generate a design. Many landscape programs around the world have research programs, but seem to emphasize learning more about the environment and less about the tools of research." noted Dr. Burley. In ecology and other fields, multivariate analysis was essential to study and compare settings and ideas. Curtis studied vegetation communities in Wisconsin and ordinated the communities by recording the frequency, density, and dominance of each plant type in a stand [28]. An ordination of research activity can be accomplished by treating the category of literature cited in an article (like a vegetation type) and the article itself as a stand of vegetation. "When it was first proposed to me about studying research structure of literature with citations, it only took me about 15 seconds to develop the experimental design, but it had taken half a lifetime to be prepared for those 15 seconds." stated Dr. Burley. With this basic analytic tool (PCA), other kinds of studies related to garden design, cemeteries, cultural heritage landscapes, and paintings have been examined by those working with Dr. Burley [29–34].

The intent of the study reported in this book chapter, an expanded investigation of the landscape research literature to visualize the changes across time for *Landscape Journal* were initiated. The study provides insight into how topics studied change and evolved.

2. Methodology

For this investigation, the analysis reveals a latent underlying structure for the landscape architecture discipline (the landscape architecture research universe) concerning the citation literature of *Landscape Journal* from several decades of articles (1982–2018). *Landscape Journal*, is a preeminent American journal addressing landscape architecture research and is affiliated with the Council of Educators in Landscape Architecture (CELA).

For each issue, the study team collected all of the peer reviewed published articles, 'source articles' for the study years. Each source article comprised one observation set. For one observation set there would usually be journal articles cited in the bibliography. These cited journal articles contained within the bibliography of a source article are called 'citation articles'. To classify a citation article, the Library of Congress classification number for the journal title of each citation article was recorded. If the same journal is cited more than once, the tally will be greater than one. Within an observation set, the total number of citation articles for a particular category was tallied. For example, if the subject category 'architecture' had 6 cited architecture citation articles in a source article, the architecture tally for the observation set would be six. The Library of Congress classification was chosen as it was an existing, broad, and easy to use system, recognized by many major state research universities. The Library of Congress system is non-hierarchical, meaning that the new bodies of knowledge that emerge are relatively easily incorporated into the classification system and thus as the system grows over time, it can accommodate modifications and development in the knowledge base. Flexibility over time was an essential component since historical research may span across a wide time frame.

In this study there were 38 subject variables. Thus each observation sets had 38 scores, each representing the tally of each subject from the source article. With the subject areas for all of the journals identified, one could then sort the citation articles from each source article into a subject category. Citations to literature such as monographs, technical reports, and books were not included in this study. I addition, proceedings were included only if they appeared to be published at least annually, meaning it was a serial. Once the subject areas for each source article were tabulated, the dataset could then be entered into a computer for statistical analysis.

Multivariate data analysis was performed using SAS 9.1 [35]. To conduct a PCA, the subject categories were each standardized to a mean of 0.0, standard deviation of 1.0. The standardization is important to the analysis [36]. Otherwise, the results will be dominated by categories with large scores. After standardization principal component analysis can be conducted upon the observation sets (an observation set is comprised of the scores in 38 subject category variables for a source article). The analysis produces a numerical table present eigenvalues which represent independent dimensions, from the largest value to the smallest. For interpretation, eigenvalues for standardized data with values over 1.0 were considered significant dimensions. The significant dimensions represent bodies of knowledge in the landscape research university. Significant dimensions were selected for further analysis by examining the eigenvector coefficients of each dimension which indicate the level of association that a subject category had with the dimension. In other words, eigenvector coefficients numerically illustrate the correlation between a variable (the subject category) and the dimension. The eigenvector coefficients are arranged in a table, sorted by the eigenvalue and would range in score between −1.0 and 1.0. Values near 0 indicate low correlation with the eigenvalue dimension while values near −1.0 or 1.0 indicate a strong association with the dimension. In this study, eigenvector coefficients with a value greater than or equal to 0.400 or less than −0.400 were considered to be affiliated with a particular dimension. Subject categories with more than one significant eigenvector coefficient meant that the subject was significant across more than one dimension, suggesting a dimension connecting subject category. Subject categories with only one significant eigenvector were considered primary to the associated eigenvalue. Primary categories were employed to label (name of identify) a dimension. Weak associations with the dimensions were considered to be eigenvector coefficients ranging from −0.4 to −0.20 and 0.20 to 0.4. The results of the PCA were plotted creating a structural map (universe) of the dimensions, associated subject categories, and connecting subjects. In other words, this map could graphically describe the latent properties of the data. The map would be a graphical depiction of the research universe in a given time frame. Several time frames were examined: the complete time frame from 1982 to 2018, 1997 to 2007, 1999 to 2009, 2001 to 2011, 2003 to 2013, 2005 50 2015, and 2007 to 2017.

In contrast to the research universe, an ordination was also developed describing the curriculum relationships between fifteen top American universities teaching landscape architecture as identified by 'DesignIntelligence,' preparing students for practicing in the profession of landscape architecture [37]. Each school was an observation set and the subjects taught in the curriculum were the categories in each observation set. The categories were standardized, and PCA invoked. The results of the latent dimension can be plotted to illustrate the relative position of one school to another. The intent is not to show which is better, but rather to identify similarities and differences. The plots can depict an educational univers in a manner similar to other types of plots [29–34].

3. Results

The results for the complete set of source articles studied (1982–2017) indicated that at least 16 dimensions were significant, meaning they had eigenvalue scores greater than 1.0 (**Table 1**). The eigenvector coefficients for the first four dimensions are included in **Table 2** to illustrate eignevectors from the tables. The complete tables are too extensive to print in this book chapter; however, they are available from the corresponding author. Across different time frames, the number of dimensions ranged for 14 to 17significant dimensions. The large number of dimensions suggest a fair number of topics are being studied within the profession. There is great diversity in what landscape scholars study and what comprises the breadth of the landscape discipline.

Twenty-two subjects were found in the study of the curriculums for the fifteen top 2016 undergraduate school in the United States, PCA analysis revealed that the subjects could be compacted into fourteen dimensions (**Table 3**). **Table 4** illustrates the first two eigenvector coefficients for the first two eigenvalues from **Table 3**.

Dimension	Eigenvalue	Dimension	Eigenvalue
1	3.37416400	20	0.86118014
2	2.37773394	21	0.82850031
3	2.30449507	22	0.79618238
4	1.72251904	23	0.76786013
5	1.57488980	24	0.73251054
6	1.48665358	25	0.70882083
7	1.35165038	26	0.70309916
8	1.28649532	27	0.60824656
9	1.24473535	28	0.60049263
10	1.19781171	29	0.58875709
11	1.14476595	30	0.56176491
12	1.11569426	31	0.53288311
13	1.08994852	32	0.49631484
14	1.05809264	33	0.46083954
15	1.03103935	34	0.42602992
16	1.01171969	35	0.38829465
17	0.98221308	36	0.36183629
18	0.96307548	37	0.32252224
19	0.93616755	38	0.00000000

Table 1.
Eigenvalue scores for the set of source articles from 1982 to 2017.

Categories	Prin1	Prin2	Prin3	Prin4
General works, newspapers, college publications	–.007198	–.002037	*–.310990*	*0.367060*
Psychology/ environmental psychology/ esthetics/ethics	0.042329	*0.302969*	0.021038	–.070908
History	–.075144	–.129145	*0.307017*	*0.367516*
Geography	0.031008	0.149171	0.119874	0.142929
Human ecology, anthropogeography	0.081835	0.098209	–.011167	0.057008
Anthropology/ folklore	–.036767	–.055084	0.141117	0.176610
Recreation/ leisure	–.043616	0.257909	0.133548	–.074602
Social sciences, general	–.056506	*0.333947*	0.004893	0.074859
Economics	0.038984	*0.375711*	0.101212	0.127883
Economics/business	–.028212	0.095794	0.085080	–.043298
Sociology	0.014627	**0.439660**	0.065118	0.061568
Law	0.011909	0.215752	–.020283	0.174770
Education	–.028176	–.037113	0.231629	0.006814
Visual arts	–.031139	–.024711	–.210592	–.046823
Architecture	–.044855	0.106333	–.183968	0.013958
Planning:	–.050149	0.270738	–.000570	–.082755

Categories	Prin1	Prin2	Prin3	Prin4
Science	*0.399670*	–.059300	0.082008	–.044951
Mathematics/computer science	–.028045	–.036028	0.003420	–.138301
Physics/meteorology	0.087573	–.057693	–.042900	–.106514
Geology	0.190450	–.077314	*0.321455*	0.002699
Natural history/ecology	**0.410378**	0.045850	0.015143	–.006438
General biology, zoology, botany	*0.389709*	0.001750	0.142236	0.031558
Agriculture	*0.340821*	0.044523	0.024557	0.049860
Plant sciences	–.017463	–.054205	–.132844	–.088730
Landscape architecture	–.050347	0.225171	–.044795	–.172863
Forestry	0.095814	0.209337	–.070056	0.203556
Technology/engineering	0.068859	0.061201	–.222159	0.284676
Library science	–.014933	–.029205	0.024091	–.037774
Language and literature	–.058329	–.019355	–.122869	–.030357
Military	–.023128	–.080632	–.163174	*0.352283*
Community health, medicine, nursing	*0.376630*	–.028026	–.250960	–.012459
Political sciences	*0.376630*	–.028026	–.250960	–.012459
Africa ecology	0.139597	–.071225	*0.363935*	0.017311
Religion, theology	0.032129	0.151920	0.117238	0.139279
Archeology, genealogy, civilization	–.042456	–.146986	0.195003	**0.412185**
Decorative arts	–.027927	0.047571	–.166330	0.110450
Physiology	–.010245	0.132418	0.042379	–.047434
Environmental sciences	–.026603	–.059235	–.116184	0.265003

*Note: *Bold* coefficients in red indicate categories with a *strong* association for a particular principal component (dimension); *Italic* blude coefficients indicate categories with *a modest* association for a particular principal component (dimension); *Underlined* coefficients indicate categories associated with more than one dimension.*

Table 2.
Eigenvector coefficients for the first four dimensions from source articles 1982 to 2017.

4. Discussion

4.1 Landscape research universe

The plotting and description concerning all the time frames examined would be longer than allowed for the space allotted to this book chapter. However three time periods from and the universe of research the educational program universe are of particular interest (**Figures 1–4**). For the decade from 1992 to 2002, the research universe had expanded to many more dimensions from 10 in as first reported by Burley and Burley to 16 with environmental science as the largest dimensions giving way to agriculture [25]. Yet by the decade from 1998 to 2008, agriculture gave way to a more amorphous environmental science dimension and a total of 17 dimensions within the universe. The trend for amorphous categories continued until the dominant dimension in 2006–2016 was an amorphous unlabeled dimension. This suggests that some of landscape research was clustered in undefined and uncategorized set, defying description. For some this may be refreshing and or others this may be disturbing. While the categories change and the size of them varies, the complexity remains

Eigenvalues of the Covariance Matrix				
	Eigenvalue	**Difference**	**Proportion**	**Cumulative**
1	4.488223	1.336905	0.1951	0.1951
2	3.151318	0.154661	0.137	0.3322
3	2.996657	0.213096	0.1303	0.4624
4	2.783561	0.550254	0.121	0.5835
5	2.233306	0.506218	0.0971	0.6806
6	1.727088	0.097973	0.0751	0.7557
7	1.629115	0.473776	0.0708	0.8265
8	1.155339	0.28917	0.0502	0.8767
9	0.86617	0.29367	0.0377	0.9144
10	0.5725	0.072126	0.0249	0.9393
11	0.500374	0.029942	0.0218	0.961
12	0.470432	0.233565	0.0205	0.9815
13	0.236867	0.047816	0.0103	0.9918
14	0.189051	0.189051	0.0082	1

Table 3.
Eigenvalue scores for the set of subjects studied at the 15 schools.

	Prin1	**Prin2**
Psychology	0.321304	0.194173
History	0.403998	0.070797
Geography	−0.02829	−0.00676
Anthropology	0.003392	−0.38156
Social sciences	0.123925	0.021618
Economics	−0.15176	0.295427
Sociology	−0.02066	−0.12461
Visual arts	−0.23055	0.162509
Architecture	0.011708	0.154676
Planning	0.271037	0.243036
Mathematics/computer science	−0.09343	0.446772
Physics	0.399808	0.055758
Geology	−0.04178	−0.00577
Natural history/ecology	0.39242	−0.00639
General biology, zoology, botany	−0.24822	0.074506
Agriculture	−0.0826	−0.1556
Plant sciences	−0.02562	−0.06083
Landscape architecture	−0.24405	0.145394
Language and literature	−0.08476	−0.04105
Political Sciences	0.266506	0.152063

DOI: http://dx.doi.org/10.5772/intechopen.99119

	Prin1	Prin2
Electives (dark matter)	0.046689	−0.51369
Humanities	−0.16304	0.149757
Chemistry	−0.08505	0.174573

*Note: *Bold* coefficients in red indicate categories with a *strong* association for a particular principal component (dimension);*

Table 4.
Eigenvalue scores for the subjects in the first wo dimensions. Studied at the 15 schools.

across the time frames. In any one time frame, much of the remaining research not placed in a dimension, representing the proportion of research not placed in an significant dimension is 35.84 percent (the sum of eignevalues in **Table 1** that are less than one and then divided by 38) of the research activity. This means that about 1/3rd of the research activity is not in a cluster and not categorized. There is a fair amount independent exploration.

Are **Figures 1–3**, what one expects to see? or desires to see? Some may call for a more unified focus and other may call for even more anarchy and diffusion in landscape research.

4.2 Landscape education universe

In comparison to **Figures 1**, **2**, and **3**, **Figure 4** presents a very different universe. Landscape architecture dominates with 54.9 percent of the subject material taught and in second place it the amorphous dark matter of electives which defy categorization. **Table 5** illustrates the average percent of academic categories taught at the five schools.

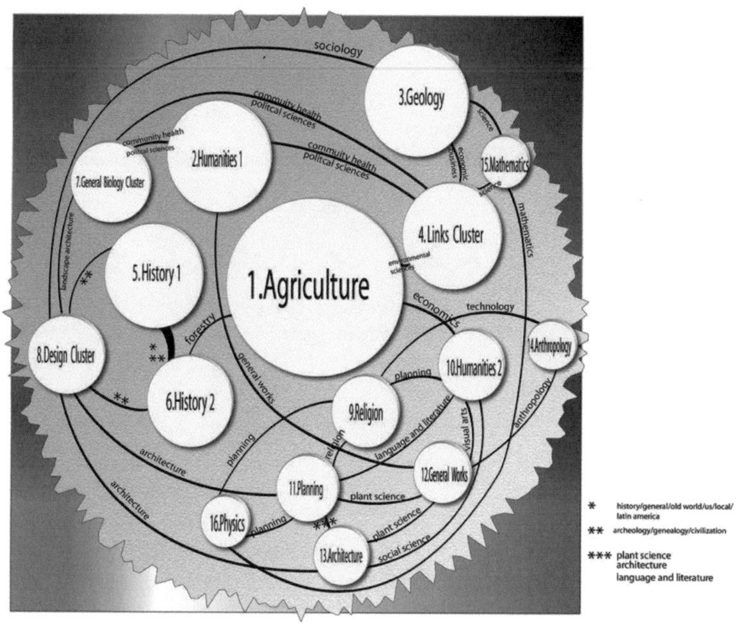

Figure 1.
A drawing of the landscape research universe from 1992 to 2002.

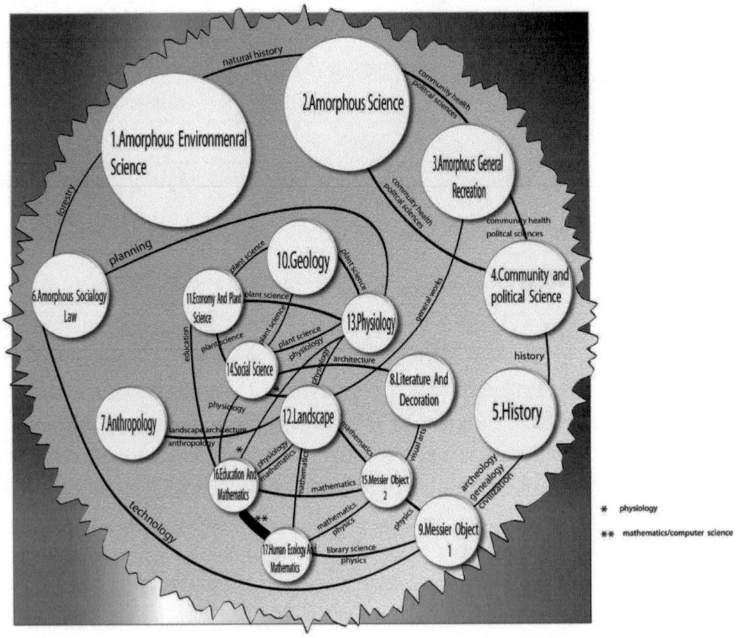

Figure 2.
A drawing of the landscape research universe from 1998 to 2008.

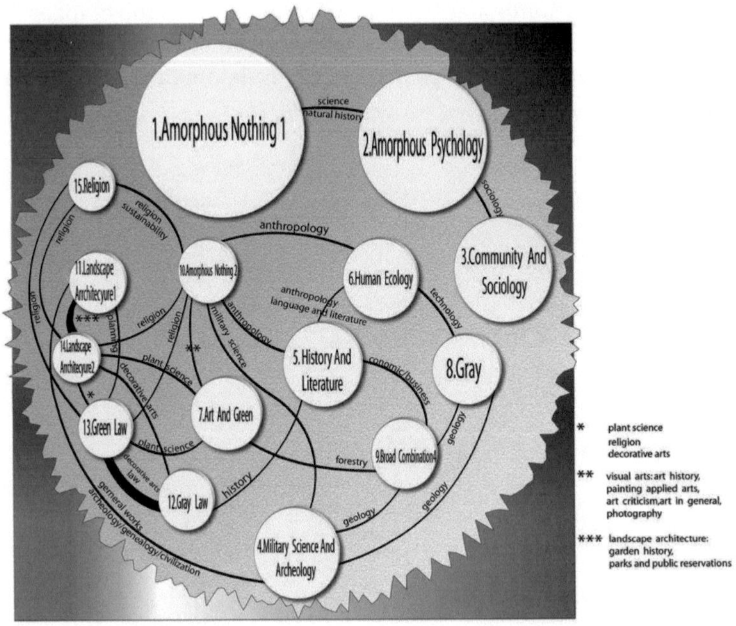

Figure 3.
A drawing of the landscape research universe from 2006 to 2016.

Figure 5 is an ordination plot of the fifteen schools based upon the first two dimensions. All fifteen schools are very good schools and share much in common. However, some schools emphasize one area over another. While the MSU landscape architectural program is not highly mathematical or visual in emphasis, it is relatively more than its peer institutions. If the schools were drastically different, the scale on the dimensions would be in the tens not the single digits.

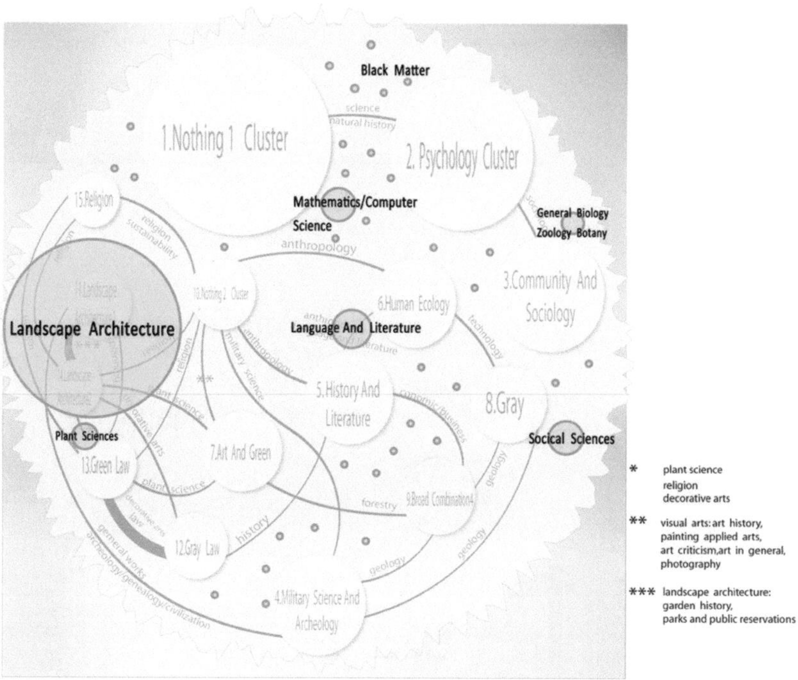

Figure 4.
A drawing of the landscape architectural program education universe 2016 overlaid upon ***Figure 3****.*

Category	Average
Psychology	0.16%
History	1.2%
Geography	0.26%
Anthropology	0.41%
Social sciences	2.42%
Economics	0.57%
Sociology	0.91%
Visual arts	1.67%
Architecture	1.28%
Planning	1.25%
Mathematics/computer science	3.81%
Physics	0.92%
Geology	0.75%
Natural history/ecology	0.79%
General biology, zoology, botany	2.33%
Agriculture	0.47%
Plant sciences	2.59%
Landscape architecture	54.91%
Language and literature	5.66%
Political Sciences	0.56%

Category	Average
Electives	15.04%
Humanities	1.78%
Chemistry	0.26%

Table 5.
Percentage of subject categories taught at the top 15 American undergraduate schools.

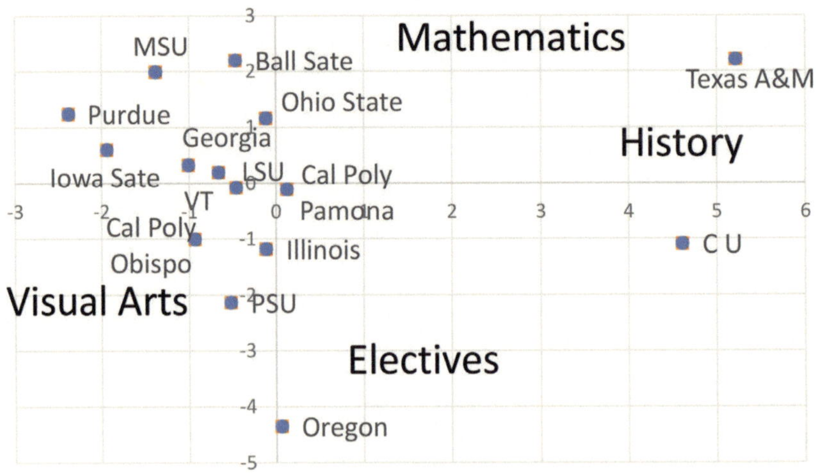

Figure 5.
An ordination plot of the fifteen school based upon the first two dimensions (dimension one horizontal, dimension 2 vertical).

4.3 Blending the two universes

It becomes apparent, that to teach landscape architecture and to do research in landscape architecture occupy two different realms. This understanding is not new to those who work in the academic treadmill. However, the results revealed in this study supports this belief. A tenure stream academic in landscape architecture may have to balance two worlds: the more narrow focus of training landscape architects and the extremely broad and diverse world of landscape research. And it is not surprising that many landscape architecture undergraduates would have little connection or interest in research. And it is not surprising that many newly hired professors coming from their professional training would be unprepared to tackle a research endeavor.

The co-authors asked Dr. Burley about his observations concerning this change, as he has observed, witnessed, and participated in this transition. "I believe much of the change began in the 1970s. The push for research has to do with money and university ranking. Schools around the world are now competing with each other for status and position. Administrators compete for a ranking, as the ranking is based upon publications, citations, and money. Therefore, administrators need to coerce/urge their faculty to obtain grants, publish, and be cited. I was told by someone who had been a faculty member in the MSU department of geography that in the mid-1970s, it was very rare for anyone to have a grant, although in their department many published. In the 1970s very few published, if ever in landscape architecture at MSU. This was frustrating for MSU administration. The merits of landscape architecture in service to society are admirable; however, these merits do not contribute to university ranking and comparatively, teaching landscape architecture as a major is expensive (small classrooms, dedicated studio space). The landscape program at Michigan State University was slated for closure in the 1980s because the faculty were slow to adopt the mission of publication and grant writing. Then Dr. Jo Westphal was hired in the landscape program and the transition began. The hiring of myself and Dr. Mary Ann Kniseley was the second phase of that transition. To explain further, university priorities change based upon where the money is to be found. In the past the state often funded many public schools, but that money has long disappeared at many institutions. The money has been found by raising tuition quickly and by seeking eternal funding for research. Majors in the humanities and general education may give way to majors in medicine and physics. Schools change their identity. It is in this academic environment landscape architecture educators and students find themselves placed within. It is easy to imagine where conflict can arise and also where opportunities may exist. In the 1970s, the mindset of landscape architecture academics was purely in a setting similar to **Figure 4**. But now on the research side of things it has evolved in the last 40 years to something similar to **Figure 3**. I was a part of that change. I am not saying it is for the good and the better, nor suggesting that it is negative either. That is for others to decide. I am too close to the middle of it to make a judgment. But indeed, it has been fun to discover and uncover measures and analytic approaches to understand what has happened and to work with fine colleagues from around the world." stated Dr. Burley.

The co-authors also asked Dr. Burley about his interpretation of what this change means for landscape architecture faculty. "Well, first it is a source of conflict at many levels. I have witnessed it many times around the world at many landscape architecture programs and in discussions with many faculty. To illustrate how successful this has been, not one hired landscape architecture assistant professor has made it to full professor at MSU in over the last 40 years. That is a tragic track record. But it really does not matter from the university's perspective because very few know this track record--administrators and most faculty have a very short time frame in their positions. The two deans who were recently hired to oversee our department/school did not even finish one full term—they left. When I was hired, there were four of us as new assistant professors in a multi-disciplinary department, but after 8 years, I was the only one who remained, the rest had left. Of the last nine landscape faculty to leave the landscape program over the last 30 years, I can say all left somewhat disgruntled, jaded and often disillusioned. I am sure it will be no different for me. Yet the university can present a positive perspective to the outside world. From the thousands of professors it hires, it only needs to show possibly 20 or so success stories each year to market the university in a very positive manner (that is 4/10ths of one percent of the instructor population at MSU). In the 30 years I have been at MSU, rarely are individuals in my department/school ever featured. It has happened; however planning and design scholars are not a priority (remember in the 1980s they tried to dispose of this group) and not what the university may wish to project as an image. Often, I see publications featuring laboratories and medicine. There is often an optimistic attitude about the future. While past events may have resulted in dismal failure, the belief that the next person hired will bring a bright and happy future is a consistent theme. Then reality sets in, problems occur, people leave, and the bright and happy future of the next forthcoming hire is all that is discussed. Over 40 years ago at another institution, I would listen to a certain dean's yearly report to stakeholders. He would always paint a bright and beautiful future. But after several years of this, I would recall the new initiatives he had promoted the years before, most ending in an unpleasant manner. But it did not matter, no one remembered them (but I did). All that mattered was that the forthcoming year was going to be marvelous. Universities struggle with this all the time. The quest for money, publications, and citations at an ever increasing level generate many internal problems. And because most universities accept this incremental race for recognition and prestige, in many ways they generate the problems and issues that arise at the institution. As has been said before, 'We have met the enemy and it is us.' While universities may claim to be bastions upholding diversity and equity; often instead, they are halls of elitism, intolerance, insecurity, and arrogance with no chance for true inclusion and diversity – in fact it can be quite brutal. This is often what I have observed for landscape architecture faculty at many institutions around the world. Still many try, and some do succeed; but one will rarely hear about the many who did not succeed. I am not attempting to present a dreary image, but rather I have been in academic for over 45 years and at one institution for nearly 30 years, plus have lectured at around 35 universities world-wide, and at many more conferences, so eventually one gets an understanding of what is occurring. **Figures 3** and **4**, make a lot of sense to me. They help to explain the setting and the situation." noted Dr. Burley.

To cope with this duality, one approach that universities have been employing is something known by some as the 'Stanford Academic Educational Model.' The model establishes two classes of instructors and researchers. In the Stanford Academic Educational Model, one academic class of employees, the researchers, are highly paid, in the tenure stream, teaching only advanced graduate student courses, focus upon producing research papers in the most highly respected journals possible, usually seeking research grants to support their efforts. The other academic class of employees, the instructors, are paid at a lower level, not in the tenure stream, teaching the masses of undergraduate students, have no research responsibilities, and are not required to produce journal articles. The researchers may have a very high opinion of themselves and the instructors will wonder why the researchers are not as engaged with the students. It is the difference between **Figures 3** and **4**. At some institutions and within departments this causes great internal strife and battles. The differences are reflected in the expectations of those serving educational professional practice and those serving the search for new knowledge. Universities attempt to be entrepreneurial with their research faculty and still serve the needs of the student body, searching for relevance, contributions, and meaning for the public [38–40].

"At MSU, it used to be that most of the faculty were a blend between the two types, one in the tenure stream, conducting research, writing papers, and teaching all levels of students. There were very few employees in the purely instructional model. But MSU has drifted towards the Stanford Academic Educational Model where now about half of the faculty are instructors. This approach saves the university substantial salary money. Since the instructors are not in the tenure stream, it brings administrators more flexibility to hire, fire, and change academic offerings/majors. It used to be that the instructors were not even considered faculty, but with about half of the academics now being non-tenured stream, universities have found means to label them as faculty. Titles are easy to give/anoint and cost almost nothing. And it would not help the university's cause for it to be known that the number of what had been known as faculty staffing had been reduced in half. Universities struggle to find approaches that still serve their student body clientele bringing in tuition dollars and striving to maintain their academic ranking and position with journal articles, citations, and research dollars. I find neither fault nor praise for what has transpired, but rather based upon the differences between **Figures 3** and **4**, I understand why this has happened. At one time there was an interesting documentary film shown on American Public Television, describing the struggles and challenges of one part-time instructor at Stanford, but I have been unable to find a citation for this film. It is very revealing. Stanford has a very well respected Department of Art and Art History which produces many excellent documentaries." reflected Dr. Burley.

"The push to maintain university rank and standing continues. In the past most educators in landscape architecture had master's degrees. I recall one European nation that urged landscape faculty to each earn a PhD., using termination as a stick and the promise of increased pay as a carrot. Eventually most earned a PhD. Upon conclusion, there no terminations but also was no money to support pay increases, but the faculty had earned PhD.s and started writing and publishing as part of their duties—mission accomplished. Faculty need to recognize that the goals and requirements for faculty by universities are going to constantly change, universities are going to expect more, not less and it will be driven by the need to sustain ranking and status above all else. For landscape architecture faculty, they need to understand that their existence is based both on the expectations of planning and design professionals to produce students illustrated in **Figure 4**, and to conduct research illustrated in **Figure 3**. And they need to understand that they are in competition with other departments and professional schools in their university. According to recent metrics in GoogleScholar, the top landscape architectural citated author in the world, William Sullivan at the University of Illinois, he has over 13,000 citations as of July 2021; but for example as my institution, the top cited authority was Joey Hudson a physicist, with over 336,000 citations. It was not until one reached down to about the 140th cited researcher that one was at the 13,000 metric. Approaching 1,000 citations, I am usually in the top 40 of cited landscape architecture researchers in the world, but I am not even close to the top 500 at my university. Universities look at these standings. It is not easy for landscape architecture faculty to compete in such an environment with the other departments and professional schools. When universities make decisions about where to invest, it is easy to understand their priorities. **Figures 3** and **4** offer insight into those challenges." observed Dr. Burley.

5. Conclusion

This study revealed that the landscape research universe has become complex with many dimensions growing and diminishing but always remaining complex. In contrast the landscape educational universe is more simplified and not congruent with the organization of the research universe. For landscape architecture academics, many reside in both of these dual worlds. Interest in this topic continues with recent published articles by Ozdil and by Newman (et al.) [41, 42].

Conflict of interest

The authors declare no conflict of interest."

Author details

Chunqing Liu[1], Xiaowen Jin[2], Zhi Yue[2], Zhen Wu[3] and Jon Bryan Burley[4*]

1 Jiangxi Agricultural University, Nanchang, China

2 Nanjing Forestry University, Nanjing, China

3 College of Architecture, Nanjing Tech University, Nanjing, China

4 Michigan State University, E. Lansing, Michigan, USA

*Address all correspondence to: burleyj@msu.edu

References

[1] Burley, J.B., Machemer, T. From Eye to Heart: Exterior Spaces Explored and Explained. 1st ed. San Diego, California, USA: Cognella Academic Publishing; 2016.

[2] Zube, E.H. Research and design: prospects for the 1980s. In: Alanen, A. R. editor. Conference Proceedings: Research in Landscape Architecture: Selected Papers Presented at the Annual Meeting of The Council of Educators in Landscape Architecture, August 20-23, 1980. Madison, Wisconsin: Department of Landscape Architecture, School of Natural Resources, College of Agriculture and Life Sciences & The council of Educators in Landscape Architecture, 1981. p. 1-11.

[3] Miller, S.C. Threshold to perspective: Developing research in landscape architecture. In: Alanen, A. R. editor. Conference Proceedings: Research in Landscape Architecture: Selected Papers Presented at the Annual Meeting of The Council of Educators in Landscape Architecture, August 20-23, 1980. Madison, Wisconsin: Department of Landscape Architecture, School of Natural Resources, College of Agriculture and Life Sciences & The council of Educators in Landscape Architecture, 1981. p. 12-21.

[4] Chenoweth, R.E. Research in Landscape architecture: an extended definition. In: Alanen, A. R. editor. Conference Proceedings: Research in Landscape Architecture: Selected Papers Presented at the Annual Meeting of The Council of Educators in Landscape Architecture, August 20-23, 1980. Madison, Wisconsin: Department of Landscape Architecture, School of Natural Resources, College of Agriculture and Life Sciences & The council of Educators in Landscape Architecture, 1981. p. 22-30.

[5] Zube, E.H. The roots of future innovations: research and theory. Landscape Architecture; 1980. p. 614-617.

[6] Burley, J.B. Charles Parker Halligan's impact on the MSU Landscape Architecture program. LandTEXTURE. E. Lansing, Michigan: Michigan State University Extension; 2020.

[7] Vanlandingham, M.C., editor. Graduate Education in Landscape Architecture: a Compendium. Washington, D.C.: American Society of Landscape Architects Council on Education; 1987.

[8] American Society of Landscape Architects, Council on Education. Research in Schools of Landscape Architecture. Washington, D.C.: American Society of Landscape Architects, Council on Education; undated.

[9] Burley, J.B., Li, X., He, S. Metrics Evaluating Multivariate Design Alternatives: Application of the Friedman's Two-way Analysis of Variance by Ranks: A Personal Reflection. Perrinton, Michigan: Whitemud Academics; 2020.

[10] Loures, L., Burley, J. B. Conceptual Precedent: Seven Historic Sites Revisited. WSEAS Transactions on Environment and Development. 2009; 5(1):55-64.

[11] Burley, J.B. The design concept: intellectual landscapes in Michigan. The Michigan Landscape. 2009; 49(12):33-40.

[12] Jin, Y., Burley, J.B., Machemer, P., Crawford, P., Xu, H., Wu, Z., Loures, L. 2018. The Corbusier dream and Frank Lloyd Wright vision: cliff detritus vs. urban savanna. In: Ergen, M. editor. Urban Agglomeration. Rijeka, Croatia: Intech; 2018. p. 211-230.

[13] Burley, J. B., Yilmaz, R. Visual quality preference: the Smyser index variables. International Journal of Energy and Environment. 2014;8:147-153.

[14] Mo, F., Le Cléach, G., Sales, M., Deyoung, G., Burley, J. B. Visual and environmental quality perception and preference in the People's Republic of China, France, and Portugal. International Journal of Energy and Environment. 2011;4(5): 549-556.

[15] Burley, J.B. Visual and ecological environmental quality model for transportation planning and design. Transportation Research Record. 1997;1549:54-60.

[16] Yue, Z., Burley, J.B.. Predictive Models for Reforestation and Agricultural Reclamation: A Clearfield County, Pennsylvania Case Study. Vegetation Index and Dynamics. London, United Kingdom: Intech Open; 2021. p. 1-16.

[17] Wen, B., Burley, J.B. Soil-based vegetation productivity model for Coryell County, Texas. Sustainability. 2020;12(5240):1-14.

[18] Burley, J.B., Wu, Z., He, S., Li, X. Soil-based vegetation productivity models for disturbed lands along the northern and central, western Great Plains, USA. Journal of Advanced Agricultural Technologies. 2020;7(1):1-7

[19] Corr, D.L, Burley, J. B., Cregg, B., Schutzki. R. 2020. Soil based vegetation productivity models for reclaiming northern Michigan landscapes: a monograph. Journal of the American Society for Mining and Reclamation. 2020;9(4):1-43. DOI: http://dx.doi.org/10.21000/JASMR20040001

[20] Bai, Y., Chang, Q., Guo, C., Burley, J. B., Partin, S. Neo-sol productivity models for disturbed lands in Wisconsin and Georgia, USA. International Journal of Energy and Environment. 2016;10:52-60.

[21] Burley, J.B., Thomsen, C., Kenkel, N. Development of an agricultural productivity model to reclaim surface mines in Clay County, Minnesota. Environmental Management. 1989;13(5):631-638.

[22] Burley, J.B. The science of design: green vegetation and flowering plants do make a difference: quantifying visual quality. The Michigan Landscape. 2006;49(8):27-30.

[23] Burley, J.B., editor. Environmental Design for Reclaiming Surface Mines. Lewiston, New York: Edwin Mellen Press; 2001.

[24] Small, H., Griffith, B. C. The structure of scientific literature. I: identifying and graphing specialties. Science Studies. 1974;4(1):17-41.

[25] Burley, J. B, Burley, C. J. (1998) Citation analysis of landscape architecture research literature: the generation of a discipline's structural domain typology. In: Conference Abstracts, The University of Texas at Arlington, Council of Educators in Landscape Architecture. Arlington, Texas: University of Texas Arlington; 1998. p. 51-55.

[26] Burley, J.B., Singhal, V.B.P., Burley, C.J., Fasser, D., Churchward, C., Hellekson, D,. Raharizafy, I. Citation analysis of transportation research literature: a multi-dimensional map of the roadside universe. Landscape Research. 2009;34(4):481-495.

[27] Deming, M.E., Swaffield, S. Landscape Architectural Research: Inquiry, Strategy, Design. Hoboken, New Jersey: Wiley; 2010.

[28] Curtis, J.T. Vegetation of Wisconsin: an Ordination of Plant Communities. Madison, Wisconsin: University of Wisconsin Press; 1959.

[29] Chen, D., Burley, J. B., Machemer, R. Schutzki, R. Ordination of selected traditional Japanese gardens, traditional Chinese gardens, and modern Chinese gardens. International Journal of Culture and History. 2021;8(1):14-51. DOI: https://doi.org/10.5296/ijch.v8i1.18250

[30] Li, N., Wang, L., Jin, X., Yue, Z., Machemer, T., Zhou, J., Burley, J. B. An ordination of selected artists, painters, and designers: line, composition, color. Journal of Architecture and Construction. 2020;3(1):37-51.

[31] Loures, L., Burley, J. B., Panagopoulos, T., Zhou, J. Dimensions in post-industrial land transformation on planning and design: a Portuguese case study concerning public perception. Journal American Society of Mining and Reclamation. 2020;9(3): 14-43. DOI: http://dx.doi.org/10.21000/JASMR20030014

[32] Wen, B., Burley, J. B. Expert Opinion Dimensions of Rural Landscape Quality in Xiangxi, Hunan, P.R. of China: Principal Component Analysis and Factor Analysis. Sustainability. 2020;12(1316).

[33] Xu, H., Burley,J.B., Crawford, P., Schutzki, R. Cross-cultural ordination of burial sites. International Journal of Cultural Heritage. 2017;2:92-104.

[34] Xu, Y., Burley, J. B., Machemer, P., Allen, A. A dimensional comparison between classical Chinese gardens and modern Chinese gardens. WSEAS Transactions on Environment and Development. 2016;12(2016):200-213.

[35] SAS. SAS 9.1 Cary, NC: SAS Institute Inc.; 2003.

[36] Johnson, R. A., Wichern, D. W. Applied Multivariate Statistical Analysis, 2nd edition. Englewood Cliffs, New Jersey: Prentice Hall; 1988.

[37] Cramer, J. P., editor. Landscape Architecture: Top 15 Programs. Design Intelligence: America's Best Architecture and Design Schools. 2026;21(6):43.

[38] Etzkowitz, H. StartX and the 'Paradox of Success': Filling the gap in Stanford's entrepreneurial culture. Social Science Information. 2013;52(4),605-627.

[39] Etzkowitz, H. Innovation in innovation: The triple helix of university-industry-government relations. Social science information. 2003:42(3),293-337.

[40] Etzkowitz, H. Entrepreneurial scientists and entrepreneurial universities in American academic science. Minerva. 1983;21(2-3),198-233.

[41] Ozdil, T. R. Who will teach the next generation of landscape architects? Ten-year review of academic position descriptions in landscape architecture in North America. Landscape Journal. 2020;39(1):55-69.

[42] Newman, G., Li, D., Tao, Z., Zhu, R. Recent trends in LA-based research: a topic analysis of CELA abstract content. Landscape Journal. 2020;39(2):51-73.

POLITICS AND STEEL IN BRITAIN, 1967–1988

For Frank Dudley

Politics and Steel in Britain, 1967–1988

The Life and Times of the British Steel Corporation

G. F. Dudley
and
J. J. Richardson

Dartmouth

Aldershot • Brookfield USA • Hong Kong • Singapore • Sydney

Published by
Dartmouth Publishing Company Limited
Gower House
Croft Road
Aldershot
Hants GU11 3HR
England

Dartmouth Publishing Company
Old Post Road
Brookfield
Vermont 05036
USA

British Library Cataloguing in Publication Data
Dudley, G. F. (Geoffrey F.)
Politics and steel in Britain, 1967-1988 : the life and times of the British Steel Corporation.
1. Great Britain. Iron & steel industries : British Steel Corporation, history
I. Title II. Richardson, J. J. (Jeremy John) 1942-
338.766910941

ISBN 1-85521-072-X

ISBN 1 85521 072 X

Printed in Great Britain by
Billing & Sons Ltd, Worcester

Contents

Preface

The initial impetus for this book arose from a collaborative research project sponsored by the European University Institute in Florence. We provided the chapter on UK steel policy for the book (*The Politics of Steel: Western Europe and the Steel Industry in the Crisis Years (1974–84)*, De Gruyter, 1986) which resulted from this project. We took as one of the two principal themes the disastrous results of a totally unrealistic development programme instituted by the state owned British Steel Corporation (BSC) in the early 1970s.

Although the UK government had serious doubts about the BSC's plans, Ministers placed their faith in the management of the Corporation, but with calamitous results. The other main theme of the chapter was concerned, therefore, with the inability of successive governments to gain either strategic or financial control over the BSC. We concluded that, if the government had been directly responsible for the steel industry, then it might have been more likely to ensure that decisions were economically and commercially sound.

This book now seeks to expand considerably on the contents of that earlier work, introduce several new themes, and, at least to some extent, modify our previous conclusions. These developments can be classified into three categories.

First, we examine and analyse in much greater detail the development of the BSC in the middle to late 1970s and early 1980s. For example, complete chapters are devoted to the work of the Shelton Steelworks Action Committee and the events of the 1980 steel strike.

Second, the book continues the BSC story up to its death in 1988, and organisational rebirth as the privatised British Steel plc.

Third, despite the importance of the BSC–government relationship, we have become increasingly aware that the Corporation's relationships with a whole range of interests played a crucial role in determining its fortunes. The book thus deals in depth not only with the greater authority exercised over the UK steel industry by the Commission of the European Economic Community during the 1980s, but also the crucial importance of the BSC's relationships with its trade unions, suppliers (including iron ore, scrap, coal and energy) and customers (with particular emphasis on the motor industry). Our analysis leads us to the conclusion that, to a large extent, government may be able to exercise greater power as an arbiter and regulator between interests rather than attempting to exercise direct control over an organisation such as the BSC.

We thank Alison Robinson, Grace Hunter, Alison Bennett and Margaret Leckie for typing the various drafts of this book. We are also grateful to Flo Dudley for her assistance with the preparation of the manuscript.

1 Introduction: Steel as a Political Issue

What must above all be avoided are solutions which are politically convenient in the short term, but industrially damaging for a generation. We are still suffering the effects of a political decision of the 1950s to have two sub-optimal size steel plants in Scotland and Wales (J. F. Safford, Director, British Iron and Steel Consumers' Council, 19 February 1981).

The political salience of steel

As can be seen from the above quotation, the steel industry in the UK has long been and remains an issue of high political salience. The salience of steel as a political issue has taken two principal forms:

1. The *ownership* of the industry has been at times the subject of dramatic and bitter party conflict. Thus between 1945 and 1967 steel policy may be characterised as a classic example of the adversary politics which some writers see as typical of the British policy process as a whole, (see Finer, 1975, 19) with each of the two main political parties seemingly determined to reverse each other's policies. As Ovenden suggests, by the mid 1960s steel had come to occupy a very special place in British party politics and signified: 'perhaps more than anything else the policy distance facing the

two major parties' (Ovenden, 1978, 16). In fact, as we have argued elsewhere, this was probably untypical of the British policy process as a whole (see Jordan and Richardson, 1982).

2. After nationalisation of the fourteen largest companies was completed in 1967, a large element of the adversarial approach went out of steel policy, and by the early 1970s it could be described as typifying the more normal British style of consensual politics. Ironically, it was this general consensus which agreed on a massive development programme for the state-owned British Steel Corporation (BSC). The ramifications of what proved to be a disastrous decision maintained the political salience of steel throughout the 1970s, albeit for different reasons than those which placed it in such prominence during earlier decades. By the 1980s, its own ideological base, together with the problems of the BSC, had led the Conservative government to once more favour a process of denationalisation. The policy was of an incremental nature, but nevertheless suggested that steel might once more become politically salient on adversarial grounds. In the event the privatisation process was completed after a rather low-key political debate, particularly when compared with the controversy aroused by earlier changes in ownership.

This book is thus chiefly concerned with the policy process surrounding the second form of political salience, for it is a policy style which shaped the development of the BSC from 1967–88. The size, shape and location of the industry, irrespective of ownership, was the principal subject of political debate on steel after 1967, despite the policy of the Thatcher government which led to privatisation in 1988. Moreover, decisions made at the political level, within the relevant government departments and latterly the European Coal and Steel Community, have had ultimately important effects on the nature of the industry itself. Rather than an inter-Party debate, the chief interest centred on the relationship between government and the steel industry itself. After nationalisation the private sector was left with mainly the special steels sector, together with a number of relatively small electric arc steelmaking plants. Until the 1980s this private sector attracted little political attention, for it was generally considered to occupy a minor, if comparatively prosperous, niche within the industry as a whole. In contrast, the BSC was given a statutory responsibility to satisfy the reasonable needs of industry consistent with its statutory financial duties, and for both Labour and Conservative governments the Corporation became a vital symbol of their intent to foster general industrial growth.

The built-in propensity for optimism of the political leaders might alone have been enough to ensure that the BSC would embark on a massive capital development programme, but in the event the management of the Corporation was even more susceptible to the grandiose vision. British steel policy and its implementation thus was highly dependent on an alliance between BSC and government.

The old alliance

Throughout its history, the UK steel industry has been a prominent actor on the political stage. The industry itself has more often than not conveyed the impression that its presence in this arena is an involuntary act, rather like someone who meanders mistakenly into a theatre, and suddenly finds himself 'volunteering' to assist the star conjuror in the performance of his latest trick. At times, the involuntary appearance may be the correct interpretation of events, for example in the twenty years after World War Two when the nationalisation/denationalisation debate was at the centre of attention. On the majority of occasions, however, the 'volunteer' has, in reality, been an integral part of the act. The strategic importance of the industry has led successive governments to take a close interest in steel, but the industry itself has not been averse to seeking political assistance when the situation demanded such action.

Indeed, if the analogy can be stretched slightly further, it is the conjuror's assistant who may in reality have devised the trick! Although the BSC itself was officially accountable to the Minister, and through him to Parliament, the power balance of the relationship undoubtedly favoured the management of the Corporation. The most obvious facet of the BSC's power was its specialised expertise, which both Ministers and civil servants challenged only occasionally and tentatively. As a result of this official passivity, the development of steel policy during the Corporation's life is largely a story of BSC autonomy interspersed with spasmodic bouts of relatively ineffectual intervention by government. The ubiquity of this expertise imbalance is illustrated by Hannah's description of how the British Electricity Authority (the forerunner of the Central Electricity Generating Board) manipulated government in the early days of nationalisation:

> Senior engineers and managers in the industry were often contemptuous of the amateurishness and lack of detailed knowledge of the civil service. They sometimes excelled in the art of misleading Whitehall with what at best were plausible half-truths, and expressed astonishment at the mixture of the trivial and the monumental which the political origins of some queries forced on civil servants. They soon learned that an internally consistent paper with logical relationships to what had gone before stood a better chance of acceptance than something which was right but new: civil servants

> after all knew nothing about the industry and had little basis on which to judge proposals except consistency with what had gone before. (Hannah, 1982, 44)

Complementing expertise as a tool of power for the BSC was the government's inability to devise policy instruments which could enable it to impose its will on the Corporation. As with most British public corporations, the BSC was intended to have an arm's-length relationship with government. In reality, the nature of this relationship prevented the government from closely monitoring the BSC's progress in implementing major policy decisions, and also inhibited valid assessments of the Corporation's true financial position.

Two specific areas of government authority over the BSC were the appointment of the Chairman and his main Board, and the control of finance. The government would signal clearly its intentions for future Corporation policy by its choice of a Chairman – the most obvious example being the appointment of Ian MacGregor in 1980 at a time when the government was anxious to convey the message that the BSC should cease to be a drain on public resources. At the same time, once installed, a Chairman was not obliged to go along with Ministerial policy, and could also dictate the shape of his Board. MacGregor himself had a large say in the restructuring of the Board during the early 1980s, when worker directors and civil servants were progressively removed. Confrontations between Ministers and BSC Chairmen of course were not unknown, most notably in the 1975 dispute between Tony Benn and Monty Finniston, but were not the norm. Sir William Barlow, Chairman of the Post Office from 1977–80, states:

> I can say with sincerity that despite many arguments with Ministers, all my dealings were friendly and amicable. In three years I never had a row with a Minister, although the press reported that I did (Barlow, 1981, 35).

In the case of the BSC it could be said that, because Ministers and civil servants were reluctant to challenge the authority and expertise of a BSC Chairman, then direct confrontation was kept to a minimum.

On finance, from the mid-1970s the government operated annual External Financing Limits (EFLs) which ostensibly restricted the size of BSC borrowing. In reality, whenever the BSC required to borrow more, then the limit was raised. In addition, a large part of this borrowing was from government in the form of Public Dividend Capital (PDC) or, in later years, what became known as 'Section 18' capital (from the 1975 Iron and Steel Act). In both cases, money was injected on the basis that interest and capital would only be repaid if and when the Corporation was in a position to do so. In the event, much of this money was written off by government with a shrug of resignation to the inevitable.

The passivity of government can thus be partly explained by its lack of expertise and its inability to impose control over the BSC, but there is another factor, seemingly more autonomous, which significantly shaped the development of the BSC. Steel became a political football during the 1940s, 50s and 60s because it was considered to be of great strategic importance in the development of British industry as a whole. After nationalisation the industry was in a more settled position, and it was hoped by governments of both Parties that the development of the BSC would demonstrate to manufacturing industry the extent of official confidence in the future.

In the event, the forecasts for future industrial growth proved to be hopelessly over-optimistic, and by the 1980s many of the industries which provided the main markets for the steel industry's products (e.g. construction, engineering and vehicle manufacture) were in a state of great depression. World-wide over-capacity in steel and tendencies towards protectionism left little scope for increased exports, while cheap imports created difficulties for the British steel industry in maintaining its share of home markets. The crisis had a great effect in concentrating steelmakers' minds on improvements in productivity, but the lack of demand threatened further contractions within the steel industry.

The harsh reality of the 1980s was a far cry from the optimism of the early 1970s, yet the political constraints on government actions were much the same. Ministers could encourage the steel industry to rationalise and hence improve productivity, but also needed to retain faith in a future for British steel. To do otherwise would be tantamount to admitting that steel, and indeed British manufacturing industry in general, had little or no viable future. The prohibitive political cost of such an admission compelled government to maintain its alliance with the BSC, for Ministers (even more than BSC managers) had a vested interest in maintaining optimism for the future. In Chapter Eleven we will argue that greater political responsibility for strategic BSC decisions (in the form of a Minister acting as BSC Chairman), together with an alternative source of expertise, would have clarified the discontinuities of accountability and also potential weaknesses in a decision-making process which may have been too enclosed and too subject to organisational self-interest for its own good. In dominating its sphere politically, the BSC was able to use the system in order to further its own organisational ends. Greater Ministerial responsibility and accountability, together with an alternative source of expertise, would of course be no guarantee in themselves of good decision-making, but they would at least have concentrated the minds of Ministers on a subject where vast amounts of public money had been required just to keep the industry afloat.

The BSC thus had great power within its own sector (i.e. British steel policy), yet the Corporation often perceived itself to be exploited as a victim of political circumstance. In the view of McEachern, British steel policy represents an example of state capital serving the interests of private capital (McEachern, 1980, 177). BSC management might not have agreed with McEachern's Marxist approach, but they would have been sympathetic towards an analysis which portrayed the Corporation as on the one hand being forced to pay exorbitant prices to its fuel suppliers (e.g. coal, electricity, gas) while on the other hand being placed in a position where its need to compete compelled it to supply 'cheap' steel to industry. Over the years the BSC conducted several running battles with the National Coal Board (now British Coal) and the Central Electricity Generating Board, while at the same time it attempted to prevent the collapse of steel prices (in this respect the EEC has become increasingly important). It is these cross-sectoral weaknesses which give a clue to a potential power of government. In the final chapter we will be discussing how government, by initiating intra and cross-sectoral co-operation; playing one interest off against another within a regulatory framework; imposing purchasing agreements on a manufacturer wishing to commence a new operation etc, can wield power in a constructive way in order to achieve a higher degree of 'co-ordinated rationality' (see Gustafsson and Richardson, 1979, 415–436).

Thus although this book is concerned with British steel policy, the fortunes of the BSC could be taken as symbolic of the general decline in British manufacturing industry. Although significant adverse criticism can be levelled at the quality of decision-making and policy implementation within the steel sector, by the early 1980s the steelmakers could also claim with justification that they had been let down by the poor performance of British manufacturing industry. The collapse in demand, and the consequent failure to match expectation with reality, at least assisted in a process which by the early 1980s had brought the British steel industry close to the point of complete collapse.

Rationalisation and the unions

From 1977 onwards, both Labour and Conservative governments recognised that the hoped-for expansion in BSC capacity was no longer a viable proposition. In this respect, at least, both the Callaghan and Thatcher governments became prisoners of circumstance. In the mid-1970s the Labour government had held a review of proposed steelworks closures, as a result of which a temporary reprieve had been granted to some of the older works. From 1977–79 (when the Labour government left office) the closures were implemented at a hectic pace. The title of the White Paper produced to herald this fundamental change: *The Road to Viability* (H.M. Government, 1978, Cmnd 7149) signified that henceforth the policy priority was to restore the BSC to a state of profit (a feat finally achieved in 1986).

The new Conservative government came into office with an aggressive policy commitment to quickly force 'lame duck' Nationalised Industries into a position where large-scale subsidies were no longer necessary and privatisation could take place. After a three month national steel strike and increasing BSC losses, the government conceded in 1980 that viability lay some distance in the future, and in effect handed the Corporation a blank cheque. Closures and cutbacks continued apace, to such an extent that the BSC workforce declined by over three quarters between 1975 and 1986 (from 228,300 to 54,200).

Generous redundancy terms ensured the minimum of opposition to closures from the BSC workforce. Indeed, despite the 1980 national strike (and a number of damaging local disputes), it could be argued that the steel unions have been remarkably compliant in the face of unprecedented change. The main steel union, the Iron and Steel Trades Confederation (ISTC), has a long tradition of moderation, and unlike the National Union of Mineworkers (NUM) has never taken militant industrial action in order to become part of the strategic planning process for the industry. Worker directors sat on BSC Divisional Boards from 1968, and on the Main Board from 1978–83 (they were removed at the instigation of Ian MacGregor), but played a passive role in shaping BSC policy. In contrast with the year long miners' strike of 1984–85, the 1980 national steel strike was about pay rather than the closures programme. The lack of fellow feeling between the ISTC and NUM reached a climax in 1984, when the steel workers refused to shut down their works in support of the miners. The action of the steelmen led to some of the most violent scenes of the coal strike (particularly at the BSC's Orgreave coke depot in Yorkshire), but the determination of steel union members to continue working undoubtedly acted as a serious political and psychological blow to the miners' cause.

Ironically, the steel unions have often reserved their most vitriolic quarrels for inter-union rivalry. There is a mutual suspicion between craft and general unions, and pay differentials and membership levels are guarded zealously.

Integration with Europe

In 1980 'a state of manifest crisis' for steel was declared within the European Coal and Steel Community (ECSC). This gave the European Economic Community (EEC) Commission unprecedented powers to control steel output and prices within the Community. In addition, negotiations were concluded to restrict imports into the EEC from increasingly aggressive external competitors such as Japan. The Industry Commissioner from 1976–84, Count Etienne Davignon, became particularly associated with the new controlled and regulated market. After his departure the controls were loosened somewhat, but there was still considerable political pressure from many EEC governments to maintain a substantially regulated market. Despite the free market principles of the Thatcher government, both Ministers and BSC management came to recognise that the Davignon Plan (as it became known) had given the Corporation valuable breathing space in its quest for viability.

In addition to output and prices, in 1983 EEC governments reached a joint agreement on a co-ordinated rationalisation plan. Implementation of the UK closures programme had been in operation since the late 1970s, but countries such as France and Italy had been reluctant to reduce the size of their own industries. At this time there was considerable pressure on the government from the BSC itself to close down the Ravenscraig plant in Scotland (one of the five major steelmaking plants). In the event the government decided that the plant should remain open (a decision repeated in 1985), partly on the grounds that other EEC countries had yet to close plants more inefficient than any operated by the BSC at that time.

The degree to which the UK steel industry has become integrated with the steel policy-making processes within the EEC as a whole is a political fact still not fully comprehended in the domestic political arena. The EEC cartel was strengthened in 1976 by the formation of a club for major steelmakers known as Eurofer. The implementation of the regulatory framework set up by Davignon was considerably assisted by the work of Eurofer, particularly with regard to the allocation of output quotas for particular products. After the departure of Davignon from the EEC Commission his regulatory system was gradually dismantled. In reality, however, major steelmaking companies in EEC countries such as France, Italy and Belgium continued to make heavy losses, and even in West Germany, traditionally the largest and most efficient of EEC steelmaking countries, there was considerable political opposition to the complete removal of controls. Davignon's great strength was his untiring ability to push national governments towards some sort of consensus and co-ordinated approach. His successors appeared to lack this political acumen with the result that a combined approach to regulation and rationalisation became difficult to achieve. On the other hand, if co-operation were to break down, then a new price war such as that which brought about the 'state of manifest crisis', would be likely to ensue.

The BSC itself enjoyed little success in penetrating the EEC market. The controls of the 1980s tended to ossify trade, but after the UK's accession to the EEC in 1973 the BSC's share of the domestic market quickly fell from 75 per cent to around 54 per cent, while the importers' share rose from 5 to 21 per cent. Private sector share held steady at around 25 per cent. By far the greater part of this increased import penetration came from the EEC countries themselves (in particular West Germany). In contrast, BSC exports to the EEC remained virtually static throughout this period (although in the late 1980s there were at last signs of improvement). The Davignon Plan undoubtedly gave the BSC a useful breathing space while it improved its efficiency, but the Corporation would not recapture the share of the UK market it held prior to 1973.

Two further difficulties for the BSC were, firstly, the burgeoning steel industries of what were previously high import countries. Nations such as South Korea, Brazil and India were no longer content to obtain their steel supplies from the US, Japan and Europe, but instead became largely self-sufficient in a wide range of products. Secondly, the US adopted an increasingly protectionist approach during the 1980s, and a number of trade agreements were struck with the EEC. Negotiations on these agreements were conducted by the EEC Commission on behalf of the Community as a whole, and these heightened the sense in which the UK steel industry had become a sub-unit of a wider European industry. The ostensibly free market Thatcher government had been as zealous as any in negotiating politically with the EEC, and also supported the Commission in its trade battles with Japan, the US etc. In general, there is still often a reluctance within the UK to see a particular issue in European dimensions, but in the case of steel, force of circumstance led the British government towards a Community-wide politically negotiated approach.

The spread of UK steel policy debate from an intra-national to an international focus highlights the extent to which steel has become a politically managed industry worldwide. Despite its relative decline within the EEC, the steel industry continues to occupy a position of strategic importance which politicians ignore at their peril. UK membership of the EEC could thus be said to have brought one large adverse effect and one large benefit to the BSC. On the adverse side, the opening up of the UK domestic market led to a large increase in steel imports which was not counterbalanced by a similar increase in exports to the EEC. On the positive side, the Davignon Plan gave the Corporation an unmatched opportunity to hold markets and revenues in a steady state while productivity and efficiency were improved.

Explanations for the BSC's decline

To ask the fundamental question: What went wrong with the BSC?, will inevitably provoke a number of answers. Some events, at least with hindsight, speak for themselves:

- The over-optimism of the early 1970s;
- the reluctance of all parties in the UK steel community, including the Labour Government, to recognise during the mid 1970s that the domestic market they had once taken for granted no longer existed;
- the naivety of the 1979 Conservative government in assuming that the BSC could quickly achieve viability.

These three factors can be seen as the most obvious and striking answers to the question. Others include:

- The severe shortage of steel industry expertise to be found within Whitehall;
- the BSC's inability to control raw material and energy prices;
- the Corporation's failure to form an export strategy within the EEC;
- a general lack of co-ordination between the steel industry and its customers within the UK itself.

These latter, less obvious, answers to the key question do not usually receive the degree of political attention which they deserve, but they are of great importance in understanding how the BSC was pushed to the limit of economic (and, probably more important, political) viability.

Fundamentally, the position of steel as what might be termed an intermediate process industry (lying between the mining of raw materials such as iron ore and coal, and the production of finished goods such as cars, washing machines and houses) gives it both a particular vulnerability and a particular strength. Its vulnerability stems from its reliance on suppliers and customers who have their own interests, and are not primarily concerned with the prosperity of the steel industry. Its strength stems from its very position at the crossroads of the manufacturing process. The decline of steel as an employer of manpower has reduced its prestige somewhat, but politically nearly all governments recognise that to abandon their steel industries would be seen as tantamount to forfeiting any long-term commitment to manufacturing industry.

Thus in the 1980s the Thatcher government supported a tough regulatory framework for output and prices while fighting to keep open the vulnerable Ravenscraig plant; a Conservative West German government used its economic and political muscle to safeguard the interests of its predominantly privately owned steel companies; while the 'liberal' Reagan Administration introduced protectionist measures at the behest of its private sector steel companies. The truth appears to be that, despite official rhetoric, steel retains its capacity to influence governments of all colours. There is a real sense in which the steel companies of the world continue to represent a powerful interest which, at a moment of crisis, can still rely on government to come to its aid.

It could be reasonably argued, in the mid to late 1980s, that it was time for governments to call the bluff of the steel industry, and that indeed government weakness is the basic answer to the question: What went wrong with the BSC? The problem for those who advocate this solution is that it brings them face to face with an even more fundamental question: Whither the state itself?

All Western governments place great emphasis on economic growth, not only as a means of creating wealth directly, but also because it provides resources for such essential public services as education, health, social benefits etc. In the UK, the apparently inexorable decline of manufacturing industry during the twentieth century has been an unceasing source of concern to successive governments. The Thatcher government has placed great emphasis on the service sector as both an employer of manpower and a wealth creator, but a powerful counter argument was put by the House of Lords Select Committee on Overseas Trade in its widely publicised 1985 Report. The Committee claimed that while 20 per cent of Gross Domestic Product (GDP) was derived from manufacturing, and over 60 per cent from services, growth would continue to depend primarily on manufacturing. In support of that claim, they argued that the service sector is very dependent on manufacturing, that the value added in manufacturing exports is some three times higher than in services, and that services only slightly dependent on manufacturing, such as tourism, would not be able to make up for a loss of manufacturing. As the Managing Director of the General Electric Company (GEC), Lord Weinstock, put it in giving evidence to the Committee:

> What will the service industries be servicing when there is no hardware, when no wealth is actually being produced? We will be servicing, presumably, the production of wealth by others. We will supply the Changing of the Guard, we will supply Beefeaters around the Tower of London. We will become a curiosity. I do not think that is what Britain is about; I think that is rubbish (House of Lords Select Committee on Overseas Trade, 1984–85, Q. 1381).

In 1983 the UK balance of trade in manufacturing moved into deficit for the first time, and by 1988 stood at a record £20.34 billion. A further increase in steel imports could only increase this balance of payments deficit, while lack of domestic steel demand would indicate a continuing decline in manufacturing industry. As North Sea oil revenues fall, this manufacturing decline will become still more obvious. Sir Charles Villiers, Chairman of the BSC during the particularly traumatic period of 1976–80, summarises his view of the position thus:

> The products of British industry have not been good enough in quality, design, reliability, delivery, service and price to meet the existing demand at home and abroad; this demand has therefore been met by goods made in other countries. So, Britain has been in chronic trouble and has had heavily to restrain government and industry spending; industry and public services had to cut back and make people redundant. This is the downward spiral in which this country is caught, and we will not escape unless we make a drastic effort to break out of it on a scale not so far contemplated (Villiers, 1984, 23).

Successive governments have sought to reverse this decline with a variety of remedies, but only limited success. Politically, it is a problem which cannot be by-passed permanently, for its consequences are too far-reaching to be ignored. In these circumstances, it is understandable that no government would be a willing accomplice in the ultimate fall of its domestic steel industry. Indeed, to abandon it would be to lose a fundamental *raison d'être* by which the government governs. It is unsurprising that, when governments face the potential consequences of jumping into this abyss, they draw back. In examining the work of the BSC, therefore, we should perpetually bear in mind the enormously powerful bonds, both political and economic, which bind together the steel industry and successive governments. In this sense, questions of public or private ownership are of secondary importance to the inherent alliance of interests between steelmen and politicians.

The development of the BSC

In the final two chapters, we will be summarising how the BSC's relationships with a number of actors, including government, unions, suppliers, customers and EEC, together with the culture of the BSC itself, shaped UK steel policy. In the intervening chapters we will describe and analyse the frequently controversial and turbulent events which characterised the development of the BSC (See Appendix A for a summary of key BSC statistics). Chapter Three will examine the early years of the BSC, from 1967–73. After a quiet start, an organisational change from geographical to product Divisions broke up the old private sector company structure and heralded the beginning of a new era. The period of the early 1970s was characterised by a determination on the part of the BSC management rapidly to modernise and expand steel production. This plan was based on highly optimistic forecasts concerning future demand, and its implementation clearly required considerable financial help from the government. The Heath government was initially sceptical of BSC market projections, but after a lengthy review process, Industry and Trade Secretary Peter Walker ratified the development strategy. In the event, BSC forecasts of 40 million tonnes of steel required by the 1980s were hopelessly over-optimistic, when the actual manned capacity figure proved to be 14.4 million tonnes.

The development strategy of 1973 implied that BSC steelmaking would eventually be restricted to five large coastal plants. Closure of the older and smaller plants became a sensitive political issue, and when a Labour government was returned in 1974 it immediately instituted a review of all proposed BSC closures. In Chapter Four we will see how this review progressed, and in particular concentrate on the campaign mounted by the Shelton steelworks in Stoke-on-Trent to retain steelmaking at the site. The Beswick review (named after the Labour Minister who led it) came out in favour of new investment at the works, although this prize was to be taken from Shelton at a later date. At other works, closure dates were delayed, to the clear annoyance of BSC management.

From 1975 onwards the financial position of the BSC quickly worsened. Demand slumped and import penetration began to creep up. As the losses increased, it became clear that the ever accelerating level of investment would have to be curtailed. By 1977, the Labour government and BSC management agreed on the abandonment of the ambitious development strategy, in favour of an incremental step-by-step approach. At the same time the closures programme was considerably accelerated (one of its victims being Shelton). Chapter Five will examine this period, when the facade of at least moderate success, which had fronted the UK steel industry since the 1940s, was finally stripped away to reveal a highly depressed and ragged Corporation.

The Conservatives returned to power in 1979 with a fervent zeal to improve rapidly the efficiency and hence profitability of the Nationalised Industries. As a prime loss-maker, the BSC was inevitably one of its principal targets. The long-term aim was that the BSC should eventually be returned to the private sector. It was intended initially that the Corporation would be in profit during 1980, but the government was quickly brought down to earth by a national steel strike over pay which dragged on from January until April 1980. Chapter Six will examine this period in some detail, not only for the manner in which it illustrates how political misunderstanding and miscalculation may aggravate a delicate industry-union relationship, but also because it demonstrates how the Thatcher government was compelled to recognise that an industry such as steel could not be abandoned to its fate.

The government having acknowledged this reality, Chapters Seven, Eight and Nine describe and analyse subsequent events during the 1980s on the long road to viability (including the increasing involvement of the EEC in UK steel policy and the role of the Corporation's suppliers) while in Chapter Ten we will examine the BSC's relationship with its customers.

Earlier, we referred both to the BSC's advantages in expertise over the government, and also the need to introduce a greater degree of 'co-ordinated rationality' into the steel policy process. In the remainder of this chapter (and in Chapters Eleven and Twelve) we will expand on these points, for they are vitally important elements in understanding the shape of UK steel policy. Firstly, we outline briefly the nature of the BSC interest and the organisation's development in terms of the character of its six Chairmen. Secondly, we look at the UK steel community in terms of its tendency (with one or two exceptions) towards fragmentation.

BSC Chairmen

From its inception, the BSC itself was the most powerful organisation within the UK steel policy community. The factors which determine the character of such a large organisation are many, but the temperament and approach adopted by each of its Chairmen reflects fairly the political and economic mood of the Corporation during the relevant period in its history. Until his sudden and premature death in 1973, the BSC's first Chairman, merchant banker Lord Melchett, projected an image of quiet authority and optimism for the future, together with a conciliatory approach towards the Corporation's workforce. The optimism continued under Melchett's successor, metallurgist Monty Finniston, but was mixed with a Glaswegian pugnacity which did not make for the smoothest relationship with government. After the merchant banker Melchett and the metallurgist Finniston, the BSC turned to another merchant banker in 1976 with the appointment as Chairman of Sir Charles Villiers. Undoubtedly well-intentioned, it was Sir Charles' lot to find himself running an organisation in a state of headlong decline. The U-turn of 1977–78 was inevitably a difficult feat to accomplish, and was bound to have political costs. Sir Charles became increasingly sandwiched between Labour and Conservative governments desperate to reduce losses, and unions understandably anxious about the BSC's accelerating job cuts programme. Villiers hoped to introduce a 'Steel Contract', which would have devolved authority within the BSC, but the plan aroused such opposition from his own management that he was forced to drop it. Matters came to a head during the 1980 steel strike, when Villiers lost the support of key Ministers, and his departure later that year appeared to be mourned by few.

Sir Charles had presided over a revolution of sorts at the BSC, yet most of the public credit and blame for this revolution fell on his successor, Ian MacGregor. Like Finniston, a Scottish metallurgist (they were indeed students together in Glasgow), Ian MacGregor arrived at the BSC with a background of working in the US for around forty years. The potential transfer fee of £1,150,000 to be paid by the government to MacGregor's former employers, the New York merchant bankers Lazard Freres, immediately created a storm of political controversy (the actual sum paid, in 1985, was £875,000). MacGregor's image was that of a ruthlessly tough businessman (an image which he himself appeared to cultivate), who would quickly lick into shape this inefficient state-owned loss maker. As such, MacGregor's philosophy appeared to be perfectly in tune with that of the Prime Minister, Mrs. Thatcher.

Nevertheless, when the three years of MacGregor's Chairmanship (1980–83) are analysed, it cannot be said that he was a particularly innovative Chairman. The workforce *was* reduced by more than half during his Chairmanship, but this course had already been set by Villiers. Similarly, the devolution of management authority and pay bargaining pushed through by MacGregor had also begun under Villiers who, ironically, had encountered a great deal of internal opposition towards his 'Steel Contract' (although MacGregor's changes did not include the moves toward greater worker participation envisaged by Villiers). MacGregor did help to initiate a campaign for lower energy prices, and also designed an imaginative plan to ship plate steel from Ravenscraig to be processed in the US (a scheme later abandoned). At the same time, he was quite happy to go along with the Davignon Plan, and reluctantly accepted the government's decision to keep open Ravenscraig. It could therefore be argued that, like the Thatcher government itself, the image of MacGregor as a radical instigator of change at the BSC owes more to popular mythology and rhetoric than reality. On the other hand, having set upon a course of action, he was always a man determined to see it through.

In 1983 MacGregor left the BSC for the even more politically treacherous job of leading the National Coal Board (at one stage he had proposed doing both jobs). His successor at the BSC (and later the NCB), Robert Haslam, was the BSC's first part-time Chairman, and also the first non-merchant banker or metallurgist! A trained mining engineer, Haslam kept his job as Chairman of the sugar company Tate and Lyle. Under Haslam, the BSC became a less politically prominent organisation. This could be attributed partly to having a part-time Chairman, together with Haslam's own propensity for adopting a low public profile, and also to the fact that many of the steel closures had already been implemented.

Nevertheless, Haslam's Chairmanship did include the year long miners' strike of 1984–85. During this period, the National Union of Mineworkers placed great pressure on the steelworkers to come out in sympathy. The steelmen refused, with the result that violent picketing took place in a number of areas. The steelmen felt that during their own strike of 1980 they had not been given sufficient support by the miners, and were not prepared to risk losing further jobs in the steel industry in order to support the miners. The events of the miners' strike (described in Chapter Nine) demonstrated that the vision of the NUM's President, Arthur Scargill, for a Triple Alliance of steel, coal and rail unions was a chimera. The steelmen's lack of support for the miners was also of great political value for the steel unions in their dealings with government, for Ministers could hardly fail to support an industry which came down on their side in that cataclysmic industrial dispute.

When Haslam left the BSC in 1986, the Corporation had reported its first profit for a decade. The long time number two at Corporation headquarters, Bob Scholey, became the new Chairman. He was the first career steelman to head the organisation. During the 1980s the government's long-term aim had been to privatise the BSC (privatisation objectives were built into the contracts of MacGregor, Haslam, and Scholey). In the event, only relatively small parts of the BSC had rejoined the private sector by 1987, and in any case the Corporation usually retained such large shareholdings in these companies that they could not truly be described as pure denationalisation. With the return to profitability, the government could revive its hopes for full privatisation, but the long-term viability of the BSC appeared to rest on a comparatively fragile base of policy co-ordination within the ECSC. Scholey was still at the helm when British Steel was privatised in 1988. His survival epitomised the manner in which the BSC had come through so many crises into a period of relative financial and political stability.

The UK steel community

The general reluctance of government to challenge the judgement of BSC management inevitably placed the principal responsibility for decision-making on the Corporation itself. In Chapter Eleven we will examine the implications of the Corporation's performance in steel policy-making (although policy implementation is a different matter altogether). This chapter will also summarise and analyse the development of the BSC as an organisation, together with its place within the UK and EEC steel communities. Despite privatisation, there is no indication that the conjuror and his assistant will go their separate ways. In the highly political theatre of the EEC decision-making process they need each other too badly for the act to split up.

The BSC and the government held by far the most powerful positions with regard to strategic policy-making decisions on UK steel policy. Yet the vulnerability of the UK steel policy-making community is demonstrated by a large number of factors, including: (a) the economic strength of other EEC steel industries; (b) the political power of the EEC Commission; (c) the development of national steel industries by many developing countries; (d) the price of raw materials such as iron ore, scrap, coal and energy prices; (e) the state of steel consuming industries such as vehicle manufacture, construction and engineering; (f) the propensity of steel consuming industries to import; (g) the willingness of the trade unions to accept (or at least not actively resist) BSC output, price and productivity goals; (h) the relationship between the BSC and the remainder of the UK steel industry.

Each of these factors has a strong political dimension which can decisively influence a particular policy outcome (as opposed to a goal). What can be termed the steel policy community is thus a complex network of relationships, in which the size and power balance of the parties involved varies from one stage of the policy process to another. The dynamic nature of the process, together with its inherent characteristics, ensures that no single interest is dominant throughout. Thus the power of the BSC itself rested chiefly on an ability to instigate major strategic policy decisions e.g. building a new works or closing an old one, and also on its possession of knowledge and information which no other agency could authoritatively challenge. The power of the government over the BSC resided in its ability to decide the ultimate status of the steel industry (private or public sector), the authority to appoint BSC Chairmen and Board members, and control over Corporation finances (e.g. borrowing and the External Financing Limit). Elsewhere, we have described the bargaining between the government and a Nationalised Industry as being rather like a duel between someone wielding a broadsword (the government) and an opponent holding a rapier (the BSC) with the blood usually being spilled behind closed doors (or, to sustain the earlier analogy, off-stage!) (Dudley and Richardson, 1984, 130).

Implementation and steel communities

The obvious power of the BSC with regard to strategic policy-making should not obscure the fact that policy implementation could place it in a much weaker position. The over-optimistic investment plan of the early 1970s is a good example of this wish fulfilment, in that although it was apparently based on a rational appraisal of market prospects, there was a large element of self-delusion that by the early 1980s the BSC could sell about 40 million tonnes of steel per annum. Although the massive rise in oil prices during 1974 (which led to worldwide industrial depression), could not have been forseen, there is no doubt that, even without this cataclysmic event, the demand for BSC steel would have fallen far short of the hoped-for targets.

The increased steel import penetration after the UK entered the EEC (not matched by equivalent exports), combined with the tendency of UK consumers to 'second source' steel supplies, threw the strategic investment plan hopelessly off course. Yet these were events which both the government and the BSC were apparently powerless to change. In a King Canute-like exercise in 1976 the Labour government paid for the stockpiling of BSC steel, but in 1977 had to admit defeat and abandon the ambitious investment plan.

Similarly, as we outlined above, when the Conservative government came to power in 1979 it was determined that 'lame duck' Nationalised Industries, such as the BSC, should quickly be made to see the error of their ways. In the event, it was the government which was forced to perform a U-turn. As previous governments had found, where the BSC was concerned policy decisions usually became the inevitable result of the BSC's inability to control events. Thus although the UK steel policy-making community was a tightly knit and enclosed circle, the community required to implement the policy was less united and compliant. The BSC discovered that it was possible to implement the plant closures programme by means of by-passing the unions and offering generous redundancy terms to the workers, but suppliers and customers were a more difficult matter.

From the BSC's point of view, a primary policy aim was to secure some sort of stability in its environment. In order to achieve this goal, the Corporation needed to obtain greater influence over events, not only in its own steelmaking sector, but also in the more peripheral and overlapping policy areas occupied by suppliers and consumers. Some suppliers, such as iron ore mining and scrap companies (we discuss the relationship between the BSC and its suppliers in Chapter Seven), are integrally involved with the steelmakers, while others, such as energy undertakings and transport operators, have many other interests.

Similarly, with regard to customers (the subject of Chapter Ten), private sector steelmakers and steel stockholders buy their product exclusively from the steel industry, while car manufacturers, construction companies etc, look at steel (and particularly the BSC) as only one part of their concerns, and are principally involved in other policy areas and communities.

In countries such as West Germany and Japan, privately owned steel giants such as Thyssen and Nippon Steel have diversified into other (more profitable) lines of business, but the public corporation BSC was basically restricted to its core activities. Nevertheless, in two particular areas the BSC succeeded during the 1980s in winning greater influence within its own sphere of the environment.

Firstly, greater integration with the private sector, although portrayed politically as privatisation, indirectly gave the BSC greater control over the entire UK steelmaking industry. Secondly, with regard to consumers, the

Davignon Plan imposed production quotas and minimum prices on the steel industry, thus preventing steel consumers from creating a cut-throat price war amongst the steel companies.

These successes for the BSC can be counterbalanced by a particularly notable failure to secure stable supplies of iron ore, and the now established tendency of both stockholders and consumers to 'second source' their steel supplies from abroad. A brief outline of how these developments have occurred will indicate the trends towards consolidation in some areas, and fragmentation in others.

Fragmentation and the politicisation of BSC suppliers

As we will see in Chapter Seven, suppliers of raw materials and energy to the BSC were themselves powerful interests which were determined to maintain their own position. As the BSC came to rely more on imported iron ore and coal for its coastal works, it inevitably became dependent on world markets. In an attempt to stabilise its supply of iron ore, the BSC made a large investment in a Canadian mine, which turned out to be a disastrous venture.

The price and supply of scrap also became a more problematical affair for the BSC. Although nearly all the BSC's scrap was supplied domestically, scrap merchants discovered new and lucrative export markets (in particular Spain). Thus the British Scrap Federation found a more powerful political voice, as the value of scrap exports cannot be ignored by government, despite pleas by the steel industry for lower prices.

The issue of coal supplies to the BSC became more openly politicised in the 1980s, particularly under the Chairmanship of Ian MacGregor. The Corporation adopted an aggressive policy in its negotiations with the National Coal Board (NCB), with the result that the government intervened to ensure that the BSC did not radically increase its level of coal imports. Notwithstanding the Thatcher government's free market philosophy, it could not afford the economic and political cost of watching one public corporation trying to severely undermine another.

With regard to energy, during the 1980s the BSC fought with increasing political intensity for lower prices from its chief suppliers, British Gas and the Central Electricity Generating Board. Although the Corporation enjoyed some success in its campaign for lower energy prices (in which it formed political alliances with other high energy using industries), the political and economic power of its opponents was an inevitable bar to radical progress.

The BSC's increasing integration with the private sector

The BSC's relationship with the private sector steel industry in the UK changed during the 1980s. For most of the 1970s, despite the BSC's problems, the private sector was apparently still in a relatively stable and prosperous state. After 1980 it became increasingly obvious that several of the private sector companies could no longer stand on their own feet. Consequently, the government was compelled to use public money in order to encourage rationalisation in the private sector. At the same time, the government was anxious to privatise at least some of the more viable BSC operations. Thus a number of joint BSC/private sector companies were founded. As we will see in Chapter Seven, these so-called Phoenix projects were launched with large sums of public money, and had a sizeable BSC shareholding. They were private sector companies, but could only have been founded, and in one case (Sheffield Forgemasters) operated, through public sector action. Indeed, the new companies indirectly increased the BSC's empire, for with directors on the boards of Phoenix companies the corporation had a say in areas which had previously been the preserve of the private sector. On the other hand, private sector steelmen continued to have no say in the affairs of the BSC. This BSC involvement in the private sector inevitably caused tensions within the latter's representative group, the British Independent Steel Producers' Association (BISPA).

Non-compliant customers

While the BSC has tended towards a more integrated relationship with the private steel sector, in the case of other customers the trend has been towards greater fragmentation of the steel community. The BSC's loss of its domestic market share during the 1970s (from 75 per cent to around 50 per cent) was an undoubted catastrophe for the Corporation. Many customers became highly dissatisfied with both the quality of BSC steel and the standard of service, and consequently either switched entirely to imported steel or at least 'second sourced' supplies. The Japanese began to invade the UK market during the early 1970s, but after the UK joined the EEC in 1973 other member countries, and in particular West Germany, became the chief beneficiaries of the BSC's vulnerability. At the same time, the BSC was singularly unsuccessful in increasing its exports to the EEC, with the result that the UK's trade balance in steel sharply deteriorated. US owned vehicle companies such as Ford and Vauxhall were reluctant to use large amounts of BSC steel in their UK manufactured products, while even the Rover Group (formerly BL) periodically reduced its UK component content.

A further encouragement towards fragmentation of the UK steel community itself was the increased use of stockholders. By the late 1980s, about 35 per cent of UK finished steel deliveries were sent to stockholders, who also dealt with a substantial amount of imported steel (NEDO, 1986, 49). For their part, customers were more concerned with the price and quality of the steel rather than its point of origin. Although the BSC itself operated one of the largest steel stockholding companies, it only controlled a small part of the total market.

This breakdown of a community of interests between the BSC and its UK consumers was exacerbated by the general decline of UK manufacturing industry. The problems of BSC's principal customers, such as the coal, shipbuilding, vehicle manufacture, engineering and construction industries, have only increased the inherent tensions. The area of policy co-ordination and implementation is one where government has a potential power and a particular role to play, and in Chapter Twelve we will examine briefly the nature of this power and the variety of policy instruments it may wield.

The political salience of steel over most of its history means that events earlier in the twentieth century are relevant to a discussion of later crises. Before going on to consider the development of the BSC, it is therefore important to outline briefly the rather long history of governmental involvement in the industry.

2 The Development of the UK Steel Industry

The early years and twentieth century crisis

The modern history of the steel industry dates from the middle of the nineteenth century, when technical developments by Bessemer and Siemens revolutionised the steelmaking process. It was now possible to blow hot air into a furnace containing molten pig iron, and thereby remove impurities such as carbon and silicon in order to tap molten steel. This so-called open hearth process thus became the most widely accepted method of making steel for the next hundred years (Vaizey, 1974, 5). As Vaizey notes, it is highly significant that, even in 1875, UK steel production, unlike that of iron, was already subject to intense competition from the United States and Germany (Vaizey, 6). By the end of the century, technical advances in the US and Germany had placed them ahead of the UK in annual steel production, a lead they have held ever since:

> By 1906 world output of steel was 50.40 million tons with the UK producing only 6.46 million tons, while the US produced 23.40 million tons, and Germany 10.53 million tons. The authoritative historians of the industry write that '1896 inaugurated a new era of world expansion... The British Steel industry, its leadership of the world already gone, faced fundamental problems of production and trade and found no answers' (Vaizey, 9).

It should be emphasised that there has never been a 'golden age' for the UK steel industry. Even in the early twentieth century, UK steel companies were demanding protection from government. Although some amalgamations took place during the First World War, by the early 1920s the industry was still fragmented. Many of the larger firms also owned coal, shipbuilding and engineering companies. This integrated structure made them powerful, but also rendered them vulnerable to slumps in heavy industry.

During the 1920s trade in heavy industry was generally poor, with the result that many of the steel companies became heavily in debt. The slump in 1929 was the last straw, and brought about a fundamental reappraisal amongst the UK steel companies themselves, and also in the political and business world generally. Most of the companies were highly conservative, with little attention being given to rationalisation, investment or marketing (Vaisey, 23). In 1928 one of the largest firms, the Scottish based company Stewarts and Lloyds, called in American consultant engineers, Brasserts, to advise them on investment in the tube-making business (calling in the American consultants McKinseys was to become standard practice for the government during the 1960s and 1970s). Brasserts advised Stewarts and Lloyds to build a new 'greenfield' works based on home ore at Corby in Northamptonshire (Corby was to be the one major UK steel development of the 1930s, although in its turn the works was to be cast adrift by the BSC when the new coastal works, using imported ore, came on stream in the 1970s). The Brasserts expertise was picked up by Montague Norman, Governor of the Bank of England, who commissioned a special study from the American firm on how to tackle the problems of the UK steel industry. The government itself was not to be left out, and in 1929 appointed a committee under Lord Sankey to inquire into the conditions and prospects of the iron and steel industry.

Both Brasserts and the Sankey Committee found that the steel companies needed to achieve greater integration and consequent rationalisation. Brasserts recommended six main plants, some based on home ores, but others on imported ore (the BSC plan of the early 1970s was based on five major coastal plants supplied with imported ore, but there are nevertheless important similarities between the two plans, separated as they were by forty years!)

The outcome of the Brassert and Sankey Reports was an arrangement involving the steel companies, the Bank of England and the City, and the government. The companies were to implement the rationalisation and investment plans, while in return the City (coaxed by the Bank of England) would help to finance these changes. Finally, the government would grant the companies the trade protection they had been seeking since the late 1920s. By the early 1930s countries such as Germany and Belgium had already introduced protective measures for their steel companies, and a European cartel was emerging. The 1932 Import Duties Act placed a tariff of 33.3 per cent on most imported steel products (Vaizey, 67).

It was now assumed by the government and Montague Norman that the industry would set about implementing the development and rationalisation plans, but in the event the companies dragged their feet. In 1934, the British Iron and Steel Federation (BISF) was founded to represent the interests of the industry. Until its disbandment at the time of nationalisation in 1967, the BISF was an important component of the relationship between government and the steel industry, but its creation did not speed change. In 1935, the BISF negotiated a trade agreement with members of the European cartel, an arrangement somewhat reminiscent of the Davignon Plan fifty years later (although with its integrated policy on rationalisation the Davignon Plan was a more comprehensive and balanced structure than its 1930s predecessor). The cartel held until the beginning of the Second World War, by which time the BISF had fostered a close relationship with the banks and the government. In addition to the tariffs, the BISF had also evolved a complex system for fixing prices in order to prevent a price war between the companies.

As a result of the protectionist policies, the UK steel companies ended the 1930s in much sounder shape than they had been during the 1920s. The industry also derived obvious benefit from the gathering pace of rearmament. During the War itself the industry was placed largely under the control of the government, with BISF officials being integrated into the Ministry of Supply (Vaizey, 97). From 1939-1945 the steel plants were worked flat out, with little thought being given to new investment. Thus by 1945 much of the plant was in a poor state, with the developments hoped for in 1930 still unfulfilled.

British steel policy 1945–1967

The Labour government, elected in 1945, came to office with a commitment to a massive programme of nationalisation. Thus by the time it had left office in 1951 it had succeeded in nationalising a number of key industries including the railways, parts of road haulage, electricity, gas, coal and, at the end of its period of office, steel. Steel nationalisation appears to have presented considerable problems to the new government, not the least of which was that they arrived in office with no detailed policy. As Ross notes, the sum of Labour's thinking on steel when it came to office in 1945 '...was contained in the 1934 TUC scheme, of little practical use, and in two sentences of "Let Us Face the Future"' [Labour's election manifesto] (Ross, 1965, 47). Not until October 1948 did the government manage to introduce its Bill to nationalise the industry. Even then it was hardly a revolutionary measure, proposing:

> the creation of a central Iron and Steel Corporation of Great Britain, whose principal task would be to act as a holding company for a number of iron and steel firms to be compulsorily purchased by the State (Ross, 87).

Indeed, as Burn points out, the Bill was presented 'as one likely to cause as little as possible unneeded disturbance in the industry' (Burn, 1961, 295). In the event the Iron and Steel Corporation of Great Britain was set up in September 1950 (for a detailed account of this period see Burn, Chapter VI).

The fact that steel came late in the nationalisation programme may be seen, with hindsight, as having had profound implications for the future of the industry. Thus the Labour government was defeated in the 1951 election and was succeeded by a Conservative government pledged to return the industry to private ownership. This they did, though interestingly this was not the fate of other publicly owned industries (parts of road haulage were sold off, with difficulty) (see Richardson, 1971, 385–402). In 1951 the new government suspended the 1949 Nationalisation Act and by 1953 had introduced its own Bill to denationalise the industry. The return to private ownership (excluding one company, Richard Thomas and Baldwins, which remained in public ownership) was conducted through the Iron and Steel Holdings and Realisation Agency. At the same time another public body was set up – the Iron and Steel Board. The Board was given the function of keeping the industry under review. In doing so it was charged with the promotion of the efficient, economic and adequate supply under competitive conditions of iron and steel products (see Iron and Steel Act 1953).

In principle the Board was given powers in two very important areas:

- it had the power to set maximum prices;
- it could veto development proposals and capital expenditure if it considered that these would seriously prejudice the efficient and economic development of production facilities in the UK (for a detailed discussion see Burn, Chapter IX).

In practice the Board did not prove to be an effective body for controlling and directing the industry. Thus Burn suggested that at best the Board was associated with an industry trying to 'catch up':

> When members of the Board made up their minds they had to follow a course of persuasion: if they met with resistance they would usually have to persuade the government – whether they thought more should be done, or if they wanted to prevent what was proposed by a firm. These are not procedures which promise quick reactions to changing circumstances (Burn, 671–8).

In so far as major adjustment in the industry was needed, the Board made little contribution. Apart from some pressure on the question of the greater use of home ores:

> there were no overt signs that the Board made any effort to promote the major adjustment, but the Government were clearly not favourable to it, and there are no signs that the Government departments concerned thought it significant. Had the Board wished to assist the kind of reorganisation into fewer bigger units with more specialisation...the Act of 1953 hardly provided adequate power (Burn, 679).

The Board was also unable to prevent a crucial investment decision being made on the grounds of political expediency. By the late 1950s it was generally considered that a large new strip mill was needed, with the two main candidates being the Welsh based Richard Thomas and Baldwin's (RTB), and Colville's in Scotland. The Board argued that sufficient demand existed for only one of these mills to be built, but for reasons of regional employment and electoral advantage (a General Election was held in 1959) the government convinced itself otherwise. After a decision made at Cabinet level it was announced in November 1958 that two new strip mills would be built – one each for Wales and Scotland. The Companies themselves were extremely doubtful about the viability of this plan, and were in any case unable to provide the necessary finance. The government thus brought in the 1960 Iron and Steel (Financial Provisions) Act, which supplied £70 million to RTB and £50 million to Colville's. In the event both projects proved far more expensive than originally envisaged, while the demand for steel produced from these two mills did not exist to the extent anticipated. As a result, RTB and Colville's were both brought to the verge of bankruptcy. Government intervention not only resulted in a bad business decision, but also committed the industry to large-scale development at two particular sites (Llanwern in Wales, and Ravenscraig in Scotland) (Vaizey, 169–177).

This example of a major investment decision being made on purely political grounds has, perhaps surprisingly, not been repeated. Subsequent strategic decisions on plant configuration have been based more on the goal of achieving overall efficiency and competitiveness, although as we will see the options selected contained their own particular traps. Nevertheless, this decision of the 1950s inevitably circumscribed the development options available in the 1970s. Investment in the steel industry has a long 'lead' time, and there is also a need to match the production facilities found on a particular site. It could also be said that the decision to develop a major steelworks in Scotland represented an attempt by the government at a kind of political 'co-ordinated rationality'. Ministers were anxious to develop a major car production plant in Scotland, with the Ravenscraig plant acting as its principal steel supplier. The Linwood factory opened in the early 1960s, but despite several changes in ownership did not prove viable and eventually closed in 1981. Over the years, doubts have recurred about the future existence of the Ravenscraig works, and emphasise that, regardless of its efficiency, a steelworks without a market is something of an industrial dinosaur.

According to Ovenden, the Board itself 'became increasingly frustrated and, although it did not campaign for renationalisation, its demands for reform – for an Iron and Steel Board "with teeth" – were well-known in establishment circles, although not to the general public' (Ovenden, 17). Moreover, after the rejection of Britain's application to join the EEC in 1963, the Board made 'proposals for a public controlling body with powers to force nationalisation on the industry' (Ovenden, 17). The proposals were never discussed for fear that they would provide the Labour Party with a convenient exit for its electorally unpopular commitment to renationalisation (Ovenden, 17).

By 1964 and the election of a new Labour government, the industry was by common consent in throughly bad shape in terms of profitability, levels of investment, size of plant and technological advance (for a brief review see Ovenden, 18–74). As the First Report of the newly formed British Steel Corporation noted, the profitability of the fourteen main companies which formed the Corporation had been declining very rapidly since 1962. In many senses the steel industry was thus not untypical of Britain's industrial decline after the Second World War. For a while it had managed to defend its world market share but by the mid 1960s it was clear that the industry had, like the vast majority of British industry, slipped seriously behind its competitors in terms of productivity. Thus Pryke noted that:

> The industry's productivity performance was also disappointing. It is true that output per equivalent worker rose by 42 per cent over the period 1958–68, which was not far short of the increase in manufacturing. However, when allowance is made for the use of capital, productivity increased much less than in manufacturing, viz. by 21 per cent as against 39 per cent. What was also worrying was

> that, although labour productivity was increasing at a moderate rate by British standards, our foreign competitors were making much faster progress. Between 1955 and 1965 output per man hour grew by around 150 per cent in Italy and Japan, by about two-thirds in France, Holland, Canada and Belgium, and by 58 per cent in Germany. In Britain there was a rise of only 39 per cent and the US was the sole major steel producing country whose increase was smaller, viz. 25 per cent. The British steel industry was labouring under the handicap that its output was growing less rapidly than in any of the other countries except the United States. Nevertheless the fact remains that whereas in 1955 productivity in Britain was only slightly lower than in the Common Market, by 1965 output per man hour in the Six was about 25 per cent higher. Moreover it appears that Poland and Czechoslovakia were level or ahead, that in Sweden and Russia output per man hour was around 50 per cent greater, and that in the US it was well over twice as great!
>
> Productivity was low in Britain because plant was over-manned and because the utilisation of equipment was poor. For instance blast furnaces, in comparison with good foreign practice, were operated at an average efficiency of only 43 per cent, partly due to the use of low grade British coking coal, much of which was produced at a high cost. In addition the great bulk of our steelmaking capacity was obsolescent (Pryke, 1981, 183–4).

The explanations of why this situation was allowed to develop are no doubt complex and almost certainly not peculiar to the steel industry. In very general terms it is reasonable to see the 'steel problem' as resulting from an amalgam of the following – an industry incapable of restructuring itself, even though the seriousness of the problem was recognised in the industry (Ovenden, 24); trade unions resistant to technological change and determined to maintain restrictive labour practices; a civil service unused to (and often opposed to) an interventionist style of governmental/industry relations; and politicians who were prepared to influence strategic decisions for political ends.

In fact the 1964 Labour government did not immediately renationalise the industry as the party manifesto had promised. The reason for not doing so was perfectly simple – the government had a very slim majority in Parliament and the issue caused considerable dissention within Labour's own ranks, let alone opposition from the Conservatives and from the industry itself. The internal machinations of the Labour Party need not concern us here (see Ovenden, Chapter 5). By 1966, however, the Labour government had been re-elected with a comfortable working majority and Harold Wilson, as Prime

Minister, was able to sanction the introduction of legislation to renationalise the industry in July 1966. By September of that year an Organising Committee had been appointed to plan for the implementation of nationalisation, notwithstanding the fact that the legislation was still going through Parliament (Ovenden, 139).

During 1966 the steel companies had produced one last card in their attempts to persuade the government that nationalisation was not necessary. The BISF appointed a committee, under the Chairmanship of Sir Henry Benson, to examine the technical requirements for the period up to 1975, and to consider the organisational changes that would be required (Heal, 137). In the event, only the technical survey emerged, and the detailed recommendations were never published. Nevertheless, Lord Melchett, the first Chairman of the BSC, did see some of the unpublished parts of the Benson Report, and Ovenden suggests that its contents were influential in determining the initial Corporation report on its own proposed structure (Ovenden, 147). In addition, Heal claims that the BSC's development plan of the early 1970s bore a remarkable resemblance to the broad-brush plan suggested by the Benson Committee (Heal, 181). Thus although the Benson Committee was overtaken by events, its apparent influence over the BSC represents a degree of policy continuity across the private/nationalised divide. As Heal also comments, the chief difference was that the BSC found itself in a position to implement proposals (see also Aylen, 1980, 200–34), whereas the BISF and the Iron and Steel Board had no powers of compliance (Heal, 181).

The Organising Committee was dominated by representatives from the steel companies, several of whom had been amongst the most vocal opponents of nationalisation. Ovenden notes the extent to which the Organising Committee sought to distance itself from the government. He considers that some key members of the Committee would have resigned if the Minister had attempted to impose his own plans on the industry (Ovenden, 137). The work of the Organising Committee thus indicated that, despite nationalisation, the management of the industry was determined to minimise the extent of government control.

The Bill received its Royal Assent on 22 March 1967 with 'vesting day' on 28 July 1967. As Ovenden notes, the task of organising the nationalisation of the bulk of the industry – the takeover and merger of the fourteen manufacturing companies and nearly 20 subsidiaries – was accomplished in ten months (Ovenden, 140).

3 The BSC: Heritage and Development 1967–73

State ownership and the nature of the watershed

When the fourteen largest steel companies entered public ownership in July 1967 to form the BSC, there was general agreement that a watershed in the history of the industry had occurred. To those at the time (and many since) the principal question was: What kind of watershed? We have seen already that since the 1920s the steel companies had been open to considerable external influence, while protection in the 1930s, followed by a high degree of official control during World War Two, had created a climate where close government involvement in the affairs of the industry had become the norm. In addition, although the Iron and Steel Corporation of Great Britain had remained in existence for only a short period, several of the companies had not returned to the private sector for some considerable time, while in the case of RTB public ownership represented 'business as usual'.

The 1967 Iron and Steel Act set out the powers and duties of both the BSC and the Minister and (perhaps surprisingly given the typically confused nature of the relationship between a Minister and a Nationalised Industry) was relatively explicit on the matter. The BSC was given the duty:

> to promote the efficient and economical supply by the Corporation and the publicly-owned companies of iron and steel products, and to

> secure that such products produced by the Corporation and the publicly-owned companies are available in such quantities, and are of such types, qualities and sizes, and are available at such prices, as may seem to the Corporation best calculated to satisfy the reasonable demands (including those in Northern Ireland) of those who use such products for manufacturing purposes and to further the public interest in all respects.

The emphasis on efficiency and economy stressed the need for the BSC to operate in a proper commercial manner, but at the same time the reference to the reasonable demands of manufacturing industry (what would constitute an unreasonable demand was perhaps inevitably left open), and furthering the national interest, also gives the impression that the BSC was not expected to be the complete master of its own fate. A requirement to consult was put forward explicitly in the section devoted to the Corporation undertaking a review of its affairs. Before reaching conclusions the BSC was required to:

> seek consultation with organisations appearing to them to represent substantial proportions of the persons in the employment of the Corporation or the publicly-owned companies or of any class of such persons.

This was a clear reference to the trade unions, although we shall see that in reality they took little part in BSC policy-making.

Despite the practical importance of the unions in influencing the fortunes of the Corporation, the most crucial relationship for the BSC would involve its dealings with the Minister. Here, the Act allowed the Minister (in a section revived from the 1949 Act) to give directions of a general character:

> as to the exercise and performance by the Corporation of their functions in relation to matters which appear to affect the national interest, and the Corporation shall give effect to any such directions.

Further, the BSC was expected to obtain approval from the Minister on matters concerning reorganisation and major capital investments. Thus it was made clear that, with regard to the BSC's duties, it was the Minister rather than the Corporation who would decide what constituted the national interest, while he was also made the ultimate authority for determining the contents of the capital development programme. In the event, this Ministerial power proved somewhat illusory, for it was the BSC which usually dominated policy, forcing government to be reactive rather than proactive.

The Minister was given further potential power by the reconstitution of the Iron and Steel Consumers' Council. This body (set up by the 1949 Act) had been a victim of the 1953 Iron and Steel Act. It was now to be revived, and given the duty of considering any matter (including prices) affecting the interests of the consumers and making recommendations to the Minister. In his turn, the Minister could act on these recommendations by giving directions to the Corporation. As we will see, there was to be one occasion when the work of the Consumers' Council was greatly to anger the BSC.

The 1967 Act undoubtedly gave the Minister a high degree of potential power over the BSC. In practice, it proved to be a different story. Ovenden makes it clear that the Organising Committee was determined that the BSC should be as free from government 'interference' as it was possible to be. He argues that the government itself did little to challenge the Committee's view (Ovenden, 136–139). Lack of official control over the affairs of the BSC, and its consequences for the fate of the British steel industry, will be one of the most persistent points to emerge from this study. Suffice to say for the present that, with regard to the duties and powers of the BSC and the Minister, the contents of the 1967 Act exerted little influence over later events, and that in this respect the birth of the BSC could not be regarded as the watershed hoped for by the supporters of Nationalisation.

Organisation

If state ownership in itself did not revolutionise the British steel industry, the birth of the BSC did facilitate the final demise of the old company structure. Even here, the change was not immediate, for from 1967–70 the BSC structure was based on geographical groupings which were largely similar to the regional separation between the private companies. The Organising Committee had prepared a Report on Organisation (BSC First Report on Organisation, 1967) and recommended four groups: Midland; Northern and Tubes; Scottish and North-West, and South Wales. The Committee hoped that this structure would encourage competition between the Groups, but as Heal makes clear this competition amounted to little more than a perpetuation of the inter-company rivalries that had afflicted the industry for so many years (Heal, 152–157).

The unsatisfactory nature of the organisational structure soon became apparent to the Chairman Lord Melchett, and he recruited one of the Deputy Chairmen, Monty Finniston (a metallurgist by training and destined to become Melchett's Chief Executive and eventual successor), to chair a committee to inquire into the BSC's structure. The Melchett-Finniston alliance was to prove extremely important for the future of the BSC, for the two men had similar ideas on what they considered to be the radical change necessary if the British steel industry was not to be left hopelessly behind the pace set by its competitors. Ovenden portrays them as representing the 'technocrats' who would succeed the old company 'barons' as the policy leaders in British Steel (Ovenden, 171).

Finniston and his Committee produced two Reports during 1969 (*BSC: Second and Third Reports on Organisation*, 1969). These Reports concluded that the regional groupings impeded the introduction of a rational and integrated policy. Consequently, they recommended a new structure based on product groupings. These new groups were introduced in March 1970 and consisted of General Steels, Special Steels, Strip Mills, Tubes, Constructional Engineering and Chemicals. Each product Division was seen as a profit centre and as giving the Corporation more effective means of:

- rationalising sales where similar products or defined groups of products are involved;
- employing its plants to the maximum benefit of the Corporation as a whole;
- planning its capital investment programme (Third Report on Organisation, 9).

It is reasonable to see the reorganisation as an important step towards a more centralised and integrated structure for the Corporation, and one which in theory at least might be more conducive to the restructuring and modernisation of the industry which had not taken place prior to nationalisation. Thus the BSC itself argued that:

> The authority of the management of each product Division will, however, have to be constrained to the extent necessary to enable the Corporation to fulfill its responsibilities and to act as a single, commercially successful, business... It will accordingly be necessary to reserve to Head Office responsibility for specified key matters, including finance; development planning; raw materials procurement policy; the acquisition and disposal of interests; the overall determination of personnel and social policy and commercial policy; regional policy; and overseas interests (Third Report on Organisation, 19).

This rather centralised structure had implications for the policy process *within* the BSC and as a result had an important impact on the policy decisions themselves. As Pryke suggests, questions of organisational structure can have the effect of diverting management attention from more serious issues directly related to policy (Pryke, 1981, 193). In this sense BSC followed a relatively common pattern within the British Nationalised Industries – a belief once the issue of ownership had been settled, that much attention should be given to organisational structure – if only we can get the organisation right, all else will fall into place. The search for the 'right' organisational structure was greatly influenced by fashion (both internally and externally generated), with a resulting instability. Many publicly owned industries (and indeed other public organisations, such as local government and the health service) have spent an inordinate amount of time on reorganisation, often to little effect.

In the case of steel, Grieve Smith (himself a former Director of Planning at the Corporation) sums up the inherent tensions within an organisation, like the BSC, with partially integrated production systems:

> In such industries the plans of the various operating units impinge on each other and cannot be considered independently as in the true conglomerate. The organisation of such corporations (whether in or out of the public sector) is always a matter of striking a balance between the centralising pressures arising from the need for co-ordination on the one hand, and the need for the maximum decentralisation in the search for efficient and well-motivated operational management, on the other. Whatever the precise form of organisation prevailing at any time, planning has to take place at both divisional or business unit level and corporate level (Grieve Smith, 1981, 26).

As the years progressed, the BSC was to tend towards a more devolved organisational structure, epitomised by the development of 'profit centres'. At the same time, the size of, say, the Strip Mills Division, and its continuing accountability to headquarters, left more potential scope for central control than was suggested by the formal structure.

The new organisational structure represented a considerable victory for the 'technocrats', a triumph which was emphasised by the final dissolution of the old steel companies as a result of the 1969 Iron and Steel Act. Apart from its implications for Corporation policy, the new structure was also to have the political effect of making a second denationalisation a more remote possibility. The Conservative government elected in 1970 did consider at least some degree of denationalisation, but in the event decided against it. We will see that to a large extent they were won over by the arguments of the new 'technocrats', but the government must also have been aware that to reconstruct the BSC in a number of new private sector companies would be a time consuming and politically delicate operation. The new organisational structure of 1970 thus represented a genuine watershed, and was one of the most significant effects of state ownership. Henceforth, the alliance between government and steelmen was to herald a new era.

Capital and finance

The BSC was given a commencing capital debt of £834 million, together with borrowing powers of £300 million (raised to £400 million in 1968). From the outset, the Corporation was concerned about the prospect of repaying this debt, the size of which was generally considered to be on the high side. Bryer, Brignall and Maunders contend that the Labour government was naive in handing over generous compensation to the companies. Their calculation is that the BSC's assets were over-valued by £345 million. In addition, the BSC inherited certain liabilities from the companies, and Bryer *et al.* calculate that these obligations alone took care of the initial borrowing powers of £300 million (Bryer, Brignall and Maunders, 1982, 11–19).

Regardless of the validity of these figures, the most urgent and practical point for the BSC was that the profits could not be earned to repay the debt. We saw in the last chapter that the profits of the steel companies in the early 1960s tumbled to a low level, and the BSC was not able to turn the tide in its early years. In 1968 there was a revenue deficit of £19 million, and in 1969 a revenue deficit of £23 million. The seriousness of the position induced the BSC to propose to the government that £700 million of the £834 million debt be converted into Public Dividend Capital (PDC). The BSC argued that:

> The environment in which the Corporation operates is as commercial and competitive as that of other large industrial undertakings at home and abroad. The industry is one which is susceptible to fluctuations but is nevertheless one capable of earning an adequate return on the capital employed in it (*BSC: Finance for Steel*, 1969, p. 6).

Thus, the provision of PDC would allow the BSC to repay large sums of the debt when business was good, but little or none in the bad years. Bryer *et al.* argue that the Treasury, in imposing a future PDC to debt ratio of 55:45, when the BSC wanted one of 70:30, was not prepared to carry any of the BSC's business risk (Bryer *et al.*, 168). By the mid 1970s, the BSC's mounting debt undoubtedly added heavily to its losses, although in the 1980s government was compelled to write off a large slice of this debt.

The government agreed to go along with this PDC proposal, on the assumption that the BSC would enjoy 'good years' for the repayment to be made over a period of time. As a result, the 1969 Iron and Steel Act (in addition to dissolving the companies) converted £700 million of the capital debt into PDC. In addition, the borrowing limit was lifted to £500 million, with the Minister given the discretion to raise this limit to £650 million.

This new borrowing limit was largely brought about because of the BSC's realisation that the financing of its investment programme could not be generated from its own resources. In 1969 the Corporation calculated that it would require an additional £200 million up to 1976 if it was to meet its investment targets (*BSC: Finance for Steel*, 1969, 9). In reality, this call for investment money was to prove a prelude to a long-running and politically highly sensitive debate on the size and nature of the BSC's investment strategy for the 1970s. By 1970 the BSC had introduced both a new organisational structure which put the management firmly in control of affairs, and had also given itself a more flexible means of repaying the capital debt. It was now ready to launch a strong political offensive, the implications of which were to be felt for the remainder of the BSC's existence.

The ten year development strategy

In 1967–68 the BSC's capital expenditure amounted to £73 million, while for 1968–69 the figure was £74 million. By 1970 the BSC management had become convinced that these sorts of figures were completely inadequate if the Corporation was to be competitive and make a profit. As we saw earlier, the private companies had maintained only a low level of investment during the 1960s. The Benson Committee considered that future investment should be concentrated on large coastal works, employing the BOS steelmaking process, while also importing large amounts of coal and iron ore. After visiting Japan to see the huge coastal steelworks in that country, Melchett and Finniston were persuaded that the BSC would be left hopelessly behind the rest of the world unless the investment strategy was greatly accelerated (and largely financed by government money).

Notwithstanding this new climate, it could not be said that the initial investment strategy in itself was particularly novel. As Heal states, it was in most respects similar to the ideas put forward by the Benson Committee (Heal, 181) and Bryer *et al.* concur with this view (Bryer *et al.*, 23). The BOS process itself was already widely used in Japan, Europe and the US with the result that by 1968 LD steelmaking vessels, and other basic oxygen processes, accounted for 43 per cent of crude steel produced in other OECD countries, compared with 28 per cent in the UK (Pryke, 194).

The BSC investment plan from the Corporation's inception until the publication of the new strategy in 1973 became known as the Heritage

Programme. Heal considers this Programme to have involved three sets of processes: integrating adjacent works; consolidating the output of certain products on selected works; and one single case of a major expansion (the Scunthorpe Anchor project). The plan was therefore incremental in appearance, although by the late 1960s it was clear that a certain pattern of development was emerging. Thus the consolidation involved significant investment at Port Talbot, Llanwern, Ravenscraig and the Lackenby site at Teesside, together with the major expansion at Scunthorpe. Heal makes the significance of these developments clear when he states that the Heritage Programme provided the Corporation with between 26 and 27 million tonnes of modern capacity, which with careful planning could be stretched to 29–30 million tonnes. These figures can be compared with the early 1980s BSC manned capacity of 14.4 million tonnes!

The capacity given to the BSC by the Heritage Programme was of a relatively large size, although it should be noted that it took little account of closures. While the new strategy was being devised scant political attention was given to the obvious implication that development of the major sites would inevitably lead to closures elsewhere. As we will see it was in this area that really radical change was to take place, but in 1970 these events lay well in the future. It could be said that although the 1973 plan was more expansionist than the Heritage Programme, its radical image could be at least partly attributed to the fact that it was only in 1970 that the political, administrative and financial means were available to put it into operation, and also that the climate of the times favoured optimism and expansion. The eventual plan was therefore radical (particularly in its forecasts for steel demand) to the extent that it assumed the BSC would be able to match, and in some cases overtake, its overseas competitors, nearly all of whom were already years ahead in their development of new plant. Despite these disadvantages, by 1970 the BSC had at least removed the old fragmented company structure, and hoped to obtain the finance (through government money) to put its strategy into operation. Politically, it was highly ironical that an alliance between the state owned BSC and a Conservative government was to prove the catalyst for this change!

The Conservative government and the BSC: acceptance and intervention

The arrival of the new Conservative government in 1970 could, of course, have continued the ideological battle over steel. Denationalisation was at least a *possibility* under the Conservative manifesto, although the reference was to 'reducing state involvement in the industry'. In the event the Conservatives did not attempt to denationalise the industry (indeed very little of publicly owned industry was returned to private ownership, and in fact Rolls Royce was nationalised by the Heath government in 1971). In that sense the arrival of the Conservatives in office in 1970 was a watershed in steel policy, as Ovenden suggests (Ovenden, 191), for at least the industry then ceased to be the ideological virility symbol of the two main political Parties. The issue of the ownership of the bulk of the steelmaking industry in the UK was settled (at least for the next decade) some twenty years after the first attempt to nationalise it. This is not to suggest that the Conservative government was unconcerned about ownership, and a number of possibilities were considered – such as the sale of parts of the BSC (e.g. chemical and construction engineering, special steels, etc) back to the private sector. There was also active consideration of the possibility of dividing the bulk steelmaking activities of the Corporation into two competing concerns (Ovenden, 173). A committee was established under the Chairmanship of Sir Ieuan Maddock (Controller, Industrial Technology, in the Ministry of Technology) which came down in favour of retaining the BSC as one business. The committee's decision vindicated the view of BSC management, and the government accepted its judgement (Ovenden, 174). In the event, pragmatism won over ideology (in this sense steel policy began to reflect the characteristic British policy style) and the BSC remained intact.

As time passed, the government became deeply involved with the affairs of the BSC on several fronts, first with regard to prices. In February 1971 the BSC sought permission from the government to increase its prices by an average of 14 per cent. The Corporation was becoming increasingly worried about its inability to make profits, and considered that this price rise would only bring it in line with the prices of its continental competitors. As we saw earlier, the government was compelled by statute to listen to representations from the Iron and Steel Consumers' Council. Perhaps inevitably, the Consumers objected strongly to the BSC's proposals, with the result that the government took the middle course by approving only a 7 per cent average increase. Later in the year, the BSC tried again by proposing selective increases averaging just over 5 per cent (within the voluntary prices control proposed by the Confederation of British Industry). This time the government told the BSC that it must observe a more stringent price regime than the private sector, with the result that the increases were deferred from October 1971 until April 1972, and the size of the increase reduced to below the 5 per cent average (BSC Annual Report and Accounts, 1971–72).

In this case, the government was clearly using a public corporation as a tool of its wider economic strategy to reduce inflation. The problem for the BSC (and the government) was that the hoped-for prospect of a profitable steel industry was becoming a still more distant possibility.

No doubt partly to compensate for its policy on prices, but also to improve the Corporation's profit potential generally, the government introduced the 1972 Iron and Steel Act. Specifically, the Act provided for a reduction in the BSC's debt of up to £350 million (PDC was reduced from £700 million to £500 million, and debts to the National Loans Fund reduced by £150 million). The reserve created was to be used to write off accumulated losses of £240 million, and also for meeting plant closure costs. The Act also allowed for the BSC to increase its borrowing from £650 million to £1,250 million (*BSC Annual Report and Accounts*, 1971–72). This large increase reflected official concern that if the BSC was to increase its capital investment programme to the level considered appropriate for future success, then a large percentage of the money would have to come from external (i.e. government) sources. The stage was now set for the principal debate between BSC and government on the long-term development strategy for the Corporation.

BSC's 1971 Development Plan: professionals v. amateurs

The movement towards a more ambitious investment programme by the BSC, after its Chairman, Lord Melchett, and its Chief Executive, Dr. (later Sir) Monty Finniston, had visited Japan to study their gigantic BOS steelworks developments, was outlined in the Corporation's Development Plan in 1970–71. The report stated that the Corporation must now look beyond the Heritage Programme and produce a long-term programme looking ahead to 1980. The Development Plan envisaged that by 1980 British steel production would be more than 40 million tonnes per annum. A major element would be a massive new 'greenfield' works, producing about 15 million tonnes per annum. The remainder of the steel would be produced at five large coastal works – Teesside, Anchor (Scunthorpe), Ravenscraig, Llanwern and Port Talbot.

The Development Plan promoted closer intervention by the Steel Division of the BSC's sponsoring Department, the Department of Industry, as according to Pryke the plan 'caused alarm and incredulity' within the Department (Pryke, 189). The Department's institutional response was to set up the Joint Steering Group (JSG) to examine the Development Plan. Consideration of the strategy contained in the plan was to prove protracted and the plan, somewhat amended, was not finally agreed until 1972. The Joint Steering Group consisted of civil servants from the Department of Industry and the Treasury, BSC officials, a 'neutral' in the form of the Deputy Chairman of Imperial Chemical Industries Ltd., and was chaired by a civil servant from the Department of Industry. Thus, against a background of worsening financial fortunes for the Corporation, the government instituted what the Minister, the Secretary of State for Trade and Industry, John Davies, described as:

> a deep seated review of the whole of the Corporation, which will, in particular, examine immediately the Corporation's short-term financial position for the coming year, and will subsequently assess the longer-term options for the Corporation in both development and financial terms (Hansard, 18:3:71, Col. 248)

In order to assist them in their work, the members of the JSG appointed McKinsey's, the business consultants, to prepare a report on future world demand for steel (Ovenden, 175). The problem facing the government, in having to assess both the viability of the proposed £3,000 million investment strategy for the coming decade, and the funding implications of such a huge programme, illustrates the classic dilemma in the relationship between government and Nationalised Industries in the UK. Thus the government was, as is usual, put in the position of having to *react* to proposals formulated within the Corporation itself. Moreover, the Department was relatively lacking in expertise and was therefore at a second disadvantage. As C.D. Foster has argued, civil servants in the UK ultimately lack the necessary confidence to say 'no' to a Corporation (Foster, 1971, 63). The battle between the publicly owned industry and its sponsoring ministry (and indirectly, the Treasury), Foster argues, was weighted in favour of the industry.

Though the Minister (in effect the Department) has the statutory power to control investment and borrowing, the political situation puts the Minister at a considerable disadvantage. As Foster puts it:

> when a Board states categorically, as it has been known to do even publicly, that in its opinion some investment is more or less essential if it is to discharge its statutory financial responsibilities to Parliament, this puts the problem of conflicting powers and responsibilities nakedly. A Minister who then denied the Board these funds because, in his judgement, based on the advice he had received from his officials, the investment would not produce a sufficient return, is giving the Board an alibi against a failure in performance (Foster, 58).

In the steel case, not only was the industry still of considerable political salience, it was also of strategic economic importance (indeed this was painfully demonstrated by the mid 1970s when BSC was unable to meet domestic demand in key industrial sectors, such as motor cars) (Pryke, 191). The Department of Industry would therefore be taking a serious political risk if it did deny the funds needed for the development strategy because this might have fundamental consequences for the rest of British industry (the energy industries can and do use the same argument).

In this context, the role of the BSC could be treated by government as being primarily that of a supplier to manufacturing industry. The BSC was inevitably dependent on the commercial success of its customers, and its own assessments for future steel demand were tied in with official growth forecasts in general.

The management of the Corporation was mindful of its duty to serve the needs of manufacturing industry, and sometimes used it to remind the workforce that the industry cannot act in an autonomous fashion, e.g. in a letter to the Iron and Steel Trades Confederation (ISTC), the largest union in the British steel industry, the second Chairman of the BSC, Sir Monty Finniston, wrote in 1976 that: 'The Corporation is not just a Nationalised Industry but a national industry which serves the needs of national manufacturing industry.' This interdependency between an intermediary process industry such as steel, manufacturing industry, and government, can create an enclosed policy environment where optimism is a condition of entry. This appeared to be the case in the early 1970s, when few were prepared to take the risk of forecasting that the future demand for British manufactured goods would not be so high as was being generally predicted.

On the other hand, to go ahead and approve the strategy also represented a policy of high risk because the strategy was a marked change of policy-making style within BSC. Thus the Corporation, in attempting to devise a 'rational' answer to the problems accounting for its poor competitive position, threw overboard the concept of incremental change and replaced it with what Hood has termed 'the great leap forward syndrome' (Hood, 1976, 25). The difficulty with this 'heroic' policy style (for a discussion of the meaning of 'heroic' see Hayward, 1974, 397–410) is that in this case, at least, the uncertainties were very considerable as indeed were the number of clearances and dependency relationships involved in the implementation of the strategy. The paradox of the ambitious development strategy was that the rational 'great leap forward' which it apparently represented was also a leap in the dark. In fact the successful implementation of the BSC's plan was heavily contingent upon at least seven conditions being met:

1. That the enormous capital cost of the investment programme could be financed by the BSC, together with the support and assistance of the government (could the money be found?);

2. That the total concentration of steelmaking at the coastal plants, at the expense of old established inland plants, could be economically justified (did BSC know what it was doing?);

3. That the drastic rationalisation of the older plants would be acceptable to the government and could be achieved with the compliance of those workers threatened with redundancy (was it politically possible?);

4. That the investment at the new works could be brought 'on stream' promptly and achieve maximum efficiency within a short time (was it practically possible?);

5. That the domestic demand would exist for the greatly enlarged steelmaking capacity (what would happen in the rest of the British economy?);

6. That external demand would exist to absorb increased exports of steel (what would the world market for steel look like?);

7. That BSC would be internationally competitive by the time plans were implemented (what would the UK's competitors be up to meanwhile?).

In the event, problems were encountered in fulfilling each of these conditions, with the result that the ambitious strategy of 1971 undoubtedly ensured that the later crisis was all the more serious and difficult to deal with. The Development Plan formulated by the BSC in 1970–71 turned out to be a good example of Wildavsky's dictum that to-day's policies are tomorrow's problems (Wildavsky, 1979, 62–85)! Moreover, it should not be thought that in this respect steel policy is in any way unusual. For example, Hogwood's study of shipbuilding policy in the UK draws similarly pessimistic conclusions about the ability of governments to achieve 'competitiveness' in the industries. That study underlined:

> the extent to which success at achieving competitiveness depends on the appropriate decisions being taken and the levels of performance achieved by a wide range of individuals and institutions (Hogwood, 1979, 273).

In practice, the very concept of the public corporation as 'arm's-length government' intensifies the difficulties which policy-makers have in exercising control over implementation and in steering public programmes (for an excellent discussion of steering problems see Lundquist, 1972). The sponsoring Department (in this case the Department of Industry) is deliberately distanced from the practicalities of implementation (the Treasury even more so), and is thus badly placed to evaluate new policies which inevitably emerge from the experience of implementation. Having decided a policy, it then lacks the necessary steering mechanisms to bring about successful implementation even if external factors (such as world demand, technological developments by competitors, subsidies by foreign governments, etc) do not intervene to disrupt the programme.

The Morrisonian concept, under which the Corporation was charged with developing policies which would secure the maximum efficiency of the industry, while the Minister was given authority to ensure that the general direction of these policies was in line with the responsibilities given to the Corporation under the 1967 Act, may possibly be ill-conceived in the light of what we know about the complexities of the implementation process. If, as in the case with the Nationalised Industries, the government has important financial (and sometimes other) objectives, then it is probably unwise to construct an implementation system which effectively gives a high degree of autonomy to policy implementers. Such a system may be beneficial where central policy goals are deliberately vague (we saw that the work of the organising committee handed the initiative to the industry in devising its relationship with government) and where a considerable latitude in goal interpretation (and ultimately goal substitution) is considered to have few serious consequences. For example, Rawson's study of the Tennessee Valley Authority suggests that agencies can be very successful in implementing a nebulous statement of goals (Rawson, 1980, 1108–18). Indeed he suggests that an unfortunate dilemma for central policy-makers is that they may have to accept a degree of drift in both means and goals or face implementation failure (Rawson, 1116). It is difficult to argue that such a key industrial policy as steel should be operated in this way because of the financial consequences to the government, the political salience of steel as an issue, and the impact that steel policy can have on other policy sectors.

The consequences of allowing steel policy to be formulated in a relatively closed and segmented policy community are potentially serious for other policy sectors, and hence there should be, ideally, effective analysis of cross-sectoral implications. Even if a 'rational' steel policy was to emerge from the steel sector, this 'sector rationality' could be in conflict with what elsewhere we have termed 'co-ordinated rationality' (for a detailed discussion of sector and co-ordinated rationality see Gustafsson and Richardson, 415–36). For these reasons alone, it might be argued that a more appropriate model for operating steel policy would be that suggested by Montjoy and O'Toole, in which the benefits present in applying specific conditions may outweigh the disadvantages (Montjoy and O'Toole, 1979, 465–76). Their analysis of implementation includes a typology of intra-organisational problems that includes a categorisation of legislative 'mandates' to agencies as either general or specific, as well as whether new resources accompany these mandates. Four types of implementation situation are thus created: (a) resources and general mandate; (b) resources and specific mandate; (c) no resources and general mandate; (d) no resources and specific mandate.

Though not fitting exactly, the relationship between the government and BSC more closely resembled situation (a), whereas one might argue that it should have been closer to situation (b). In practice, the relationship between sponsoring Departments and their public corporations does vary across the whole spectrum outlined by Montjoy and O'Toole, and Departments have to strike a difficult balance, in assessing investment proposals, between being far too general in their oversight, and being too specific and appearing to do the corporations' jobs for them.

In the steel case, the JSG had to tread this difficult and delicate road. The internal politics of the JSG have been fully analysed elsewhere (Ovenden, 175–84), but in an attempt to overcome the inherent disadvantage which we suggest the Department was under, it commissioned private consultants, McKinsey's, to make a detailed analysis of the future trends in the world market for steel. The dilemma for a ministry which employs this tactic is that while it may 'buy' technical knowledge, and hence increase its own authority, it can also alienate the senior management of a public corporation who could consider that their own position is seriously undermined as a consequence of the government's lack of faith in the collective expertise to be found within the corporation. The length of time during which the JSG was in existence, from March 1971 until March 1972, indicated both the general lack of rapport between the DTI and the BSC, and the singular reluctance of the Corporation's representatives to reduce their massive capacity target.

The very existence of the JSG touched a raw nerve in the BSC. Thus in its evidence to the Select Committee on Nationalised Industries in the Autumn of 1972, it openly admitted that it was hostile to the JSG (House of Commons Select Committee on Nationalised Industries 1972–73, para. 748). The reason for the hostility was easy to understand – the Department, through the JSG, was quite obviously second-guessing the Corporation to an unusual degree and was undoubtedly invading what the Corporation felt was its territory (for a discussion of organisational territory and policy space see Downs, 1967, 211–22). In particular, the Corporation argued that the technical, commercial and financial assessment of the investment programme were matters for the Board: 'that is what we are appointed to do'. The BSC believed that there was no:

> function for the Board of this Corporation if it did not have responsibility for setting the plans for the Corporation itself in commercial, technical and financial terms (Ovenden, 177).

The JSG was therefore an example of jockeying for policy space (see Downs, 215), the outcome of which determined the shape of the industry in the decade ahead.

A further significant feature of the JSG was the exclusion of the Trades Union Congress Steel Industries Consultative Committee representatives from its membership. The TUC Steel Committee was set up at the time of nationalisation in order to be representative of all the nationally recognised unions in the BSC (seventeen unions in all). Two factors in particular, however, severely limited the ability of this Committee to exert any significant influence on the BSC's development strategy. Firstly, individual unions were reluctant to sacrifice their own autonomy. The largest union within the steel industry, the Iron and Steel Trades Confederation (ISTC), was continually suspected by other unions of attempting to extend its power by enlarging its membership. This rivalry among the steel unions was also exemplified by the fact that the first national joint union conference was not held until March 1973, a year after the JSG had completed its work. Although the conference discussed BSC policy at length, the TUC Steel Committee did not have the authority to interfere with the autonomy of any individual union, which resulted in the lack of a co-ordinated union response to the Corporation's development and rationalisation plans. Secondly, the steel unions were faced with a dilemma in their attitude towards rationalisation of the industry. On the one hand modernisation would bring greater prosperity to their members at the large coastal works, but conversely would also entail redundancy for many of those working at the plants scheduled for closure.

The unions and their place within the steel policy process

Before returning to the main theme of the evolution of BSC policy, we should perhaps examine briefly the origins and nature of attitudes adopted by unions within the steel industry, and how these have affected their relationships with the BSC. The relative moderation (say compared with unions in the motor industry) of the steel unions can be attributed to traditional social and economic factors, together with the political complexion of the ISTC. Vaizey points out that wages in the steel industry have been relatively high, partly because of productivity bonus payments, but also because (particularly when compared to the coal industry) labour represents a small (about 30 per cent) proportion of total costs. In addition:

> The firms employed men in small groups, many of them recruited in teams, and paid directly by the group leader. Consequently there was a strong pressure to conform to the norms of the small group (Vaizey, 28).

The ISTC itself traditionally has been under the tight control of a moderate, politically right-wing leadership. Vaizey admits that it is not easy to trace the source of these union attitudes, although obviously the wage scales might have been an important factor. At the same time, it does appear that union leaders were prepared to identify themselves with the objectives of management, even in the case of policies which appeared to be detrimental to the interests of their members. During the 1920s a sliding scale was adopted which adjusted wages according to the financial state of the industry. As the decade progressed this sliding scale led to a relative decline in wages, but unlike the mining union (whose members were generally employed by the same companies as the steel workers) the ISTC did not adopt a militant position. Vaizey gives an example of the ISTC leadership's dislike of strikes which:

> occurred at Consett in 1925, when E.J. George became general manager, and ordered the melters to work at weekends to keep up ingots output while a furnace was being rebuilt. The men struck; the unions sent them back to work; and ultimately virtually on George's terms (Vaizey, 28).

An even more revealing example of the ISTC's ability to exert central control is given by Brannen *et al.*, who instance a case in 1947 when the union concluded an agreement with the employers which entailed weekend working in order to bring British practice in line with foreign experience. One thousand workers refused to accept the new arrangements, with the result that the union threatened them with expulsion (tantamount to dismissal) if the agreement was not implemented. A union official interviewed by Brannen *et al.* took pride in the method employed to enforce compliance, and claimed: 'I know of no other union who would have dealt with it in that way' (Brannen *et al.* 1976, 73).

Like Vaizey, Brannen *et al.* see the relatively high wages and the small working group as prime factors in shaping union attitudes. They also stress the importance of the widespread use within the industry of conciliation boards, which in many cases were well established by the turn of the century. These boards have the effect of forestalling strikes, while also enhancing the status of the unions and possibly allowing the unions influence over the actual working of plants.

Given the general passivity of the steel unions at the national level, it is perhaps not surprising that the government and the BSC considered that their representatives could be excluded from the JSG with impunity. At the same time, even allowing for the moderation of the ISTC in particular, it appears strange that the unions were not invited to take part in the JSG, for their presence would have enhanced the legitimacy of the strategy which emerged. The treatment of the unions by the BSC and the government thus in one sense reflected traditional attitudes, but at the same time indicated the

Corporation's tendency to distance itself somewhat from the national union leaders, and to eschew the more collaborative approach adopted by the steel companies.

The 1970s were also to see a change in the relationship between union leaders and their members. When steelworks were threatened with closure, a number of local action committees were set up which, although moderate in their approach, were not prepared to be so compliant as in the 1947 example described above (one of these groups will be examined in Chapter Four). It is therefore possible to see a contrast between rather compliant unions at the national level (lacking influence over policy) and a tradition of significant union power at the local level. The case of the JSG thus presaged a period when the leadership of the steel unions found themselves generally excluded from the principal forums of policy-making in the steel sector, while also encountering increased difficulty in exerting control over their members. In this sense it could be said that the leadership of the ISTC in particular has yet to find a coherent role within the modern steel industry.

In 1977 the first steps were taken towards the formation of a 'Steel Contract' between the BSC and unions. This would have allowed for the creation of a 'Steel Council', where BSC management would discuss future policy with union leaders, together with increased consultative machinery at divisional and plant levels. In the event, divisions within BSC management itself, and the deterioration in relations between the management and unions during the late 1970s, as a result of drastic rationalisation within the Corporation, led to the collapse of the 'Steel Contract' discussions. In January 1980 the first national steel strike began, and was not settled for three months (see Chapter Six). This strike could be attributed partly to the lack of communication between the unions, the BSC management and government. This breakdown within the steel community indicated a lack of consensus for change, although subsequently the ISTC returned to its basic passivity.

Given the moderation of the steel unions, it is perhaps not surprising that the BSC was one of the first organisations in Britain to appoint worker directors. Section 31 of the 1967 Iron and Steel Act, which set up the BSC, required the Corporation to make available to members of consultative committees within the industry:

> at a reasonable time before the discussion is to take place, such information in their possession relating to the subject as, after consultation with those persons, appears to the Corporation ... to be necessary to enable those persons to participate effectively in the discussion.

Brannen *et al*. make it clear that the first Chairman of the BSC, Lord Melchett, was particularly keen to see this Section of the Act effectively implemented (Brannen *et al*. 89–90). The Labour Government of the time also saw the newly created BSC as offering a means of experimenting with new forms of industrial relations. The first worker directors were selected by the Chairman of the BSC from lists of names submitted by the steel unions. Originally, the worker directors were required to resign any lay trade-union offices on their appointment, but after 1972 they were allowed to retain their union links. The first worker directors were also restricted to membership of Divisional Boards (although the Main Board did have one shopfloor representative from the beginning), but from 1978 they were represented also on the main BSC Board. All main Board directors of the BSC were appointed by the Secretary of State, but only after consultations with the Chairman.

Despite the high hopes expressed for the worker director scheme on its inception in 1968, it has not proved to be a success. In their evaluation of the early years of the scheme, Brannen *et al.* describe how many of the worker directors felt themselves to be mistrusted by their own union officials and colleagues at work (the appointments were part-time), while at the same time they lacked the authority and power to influence BSC decision-making. The Board had been portrayed to them as being the main locus of power within the Corporation, but many discovered that they were there merely to 'rubber stamp' decisions already made by senior management (in this respect the worker directors were usually placed in the same position as Ministers and Departmental civil servants!) (Brannen *et al*. 99–116). In discussions on some issues (e.g. conditions of work), the worker directors discovered that they could exert some influence, but with regard to key elements of policy, such as capital investment, plant configuration, and productivity, their power was minimal. The existence of the worker directors could also enable management to claim that employees were being consulted, even when the unions were excluded from a body such as the JSG. The existence of worker directors could therefore give a superficial appearance that a steel policy community was in operation, when in reality such a structure may not have existed.

In a more recent re-appraisal of the scheme, Brannen concludes that the worker directors inevitably became integrated into the boardroom structure:

> Worker directors of necessity enter worlds already established in terms both of formal roles and processes of custom and practice, of values and language. The social dynamics of those worlds strongly favour the encapsulation of worker directors within the pre-existing boardroom ethos and organisation, and within, though in a limited way, the pre-existing organisational categories of information and analysis (Brannen, 1983, 136).

In one sense it could be said that by acquiring the management ethos the worker directors could be in a better position to influence events (i.e. by speaking the language of management their views could be given greater weight). The chief problems appeared to be that the BSC Board in general was not a major policy-making body, and also that the unions themselves did not attach great significance to the views of worker directors.

Given these circumstances it is perhaps hardly surprising that worker directors eventually disappeared from the main Board of the BSC (from a representation of six in 1981), although they were still to be found on the Divisional Boards until the last of them disappeared in October 1989. Ian MacGregor, Chairman of the BSC from 1980–83, was anxious to reduce the size of the Board, and Conservative ministers were prepared to go along with this policy (active civil servants were also removed from the Board). This policy was achieved by not replacing the worker directors when their term on the Board had expired. Initially, the ISTC reacted in a hostile manner to this policy, but in its usual moderate way the Union quickly appeared to accept the situation. It might also be said that, because of their ambivalent relationship with union officials, the worker directors were something of an embarrassment to their own colleagues. It is perhaps a typical comment on British industrial relations policy (and also the status of state enterprise Boards), that an experiment in partnership should be viewed almost with disdain by both sides of the industry (for an analysis of board appointments in the Nationalised Industries see Centre for Policy Studies, 1984).

The authority of the unions themselves has been further weakened by a considerable amount of inter-union rivalry. It is ironic that union officials, who are often the epitome of moderation in their relationships with management, can become extremely militant and aggressive in their dealings with fellow trade unionists. A traditional tension exists between the ISTC (representing mainly production workers) and craft and white-collar unions. The ISTC has often been suspected by the other unions of attempting to increase its membership at their expense, while the ISTC itself is fearful that, as the steel workforce contracts, its identity and influence within the union movement as a whole will decrease. Nearly all the steel trade unions are represented on the Trades Union Congress Steel Committee, but have frequently found difficulty in acting in concert.

The ten year strategy

Though the JSG was itself controversial, it is an example of a very restricted and exclusive policy community in operation, notwithstanding the historical political salience of steel as an issue. By 1971 nationalisation of steel was a dead issue and steel policy could be formulated according to the 'standard operating procedures' which have become the characteristic feature of the British policy style. Thus the issue of a new strategy for steel was internalised and privatised and processed within a committee structure on which those interests who were thought to 'matter' were represented. It is perhaps a significant comment on the then relatively closed nature of the British policy process that the unions were not considered to be an 'insider group' in this case (for a discussion of the concept of an 'insider group' see Grant, 1978. Also Richardson, 1977). Even more significant is the fact that the unions seemingly did not press for insider status on an issue which so directly affected their members. Evidence in other policy areas suggested that relatively little 'pressure' is needed to gain entry to policy communities (see Jordan and Richardson, 1982, 80–110), and it seems reasonable to argue that the reason for the exclusion of such a key actor as the trade unions is as much a reflection on the unions themselves as it is on the steel policy process as a whole.

After much delay (see Ovenden, 180–185), the government announced the results of the review (and then only in the barest outline) in May 1972. On 8th May the Minister for Industry, Tom Boardman, announced that iron and steelmaking investment would be concentrated on the five main coastal works, and stated that the JSG had recommended a 1980 BSC capacity in the range of 28 to 36 million tonnes (Ovenden, 180). The detailed BSC programme for modernisation and expansion was not to be announced until after consultations with the government on strategic decisions. The wide range of the 28 to 36 million tonnes capacity target indicated the lack of a consensus between the DTI and the BSC on the future demand for steel, while the government registered its concern about the capital cost of the programme and the likely future demand for steel by imposing an upper limit which was 4 million tonnes less than the BSC's original target (McKinsey's, in their Report, had advocated a target of 23 million tonnes per annum) (Ovenden, 179).

During the months immediately following Mr. Boardman's announcement, the government and the BSC appeared to make little progress in evolving a future strategy for the Corporation. The deadlock was finally broken in October 1972 when Peter Walker replaced John Davies as Secretary of State for Trade and Industry. Mr. Walker was generally regarded as possessing a more adventurous spirit than his predecessor, an attitude which inclined him to accept an ambitious investment strategy for the BSC (Walker's arrival and impact illustrates yet again the unpredictability of the policy process and the key role that some individuals can play within it. Walker's 'heroic' policy style – his penchant for the bold radical solution – was earlier exhibited when he became Secretary of State for the Environment and sanctioned a radical reorganisation of Britain's water industry) (see Jordan *et al.* 1977, 317–34.

Also Headey, 1974). Walker's optimism for the future of the Corporation was reflected in his announcement to the House of Commons in December 1972 (Hansard, 21:12:72, Cols. 1576–78), which outlined the BSC strategy and also in the White Paper, *Steel. British Steel Corporation: Ten Year Development Strategy*, published in February 1973 (HM Government, 1973), which elaborated on this statement.

It had been decided that the BSC would require a programme of investment amounting to £3,000 million. This plan involved a major development of the five main Heritage sites and of special steel plants in the Sheffield–Rotherham area, in addition to the development of a major new steel complex. It was also expected that two 'mini' electric arc furnace plants would be constructed in Scotland (possibly a political decision to gain the support of the Secretary of State for Scotland). The decision to accept the BSC's strategy represented a great act of faith by the government, which had committed itself to a BSC capacity target at the upper limit of the JSG's recommendation, thereby supporting the BSC's contention that the future demand for steel would justify such large-scale expansion.

In approving the £3,000 million investment plan, in a form only slightly modified from the plan originally envisaged by the Corporation, the government admitted that 'the strategy was not free of risk' (HM Government, 1973, 5). The White Paper also noted that BSC production in 1972 was 22.9 million ingot tonnes. Yet it went on to accept certain key assumptions about world demand and the rate of economic growth in the UK. For example, it envisaged that an improved rate of economic growth in the UK would be such that steel demand in the UK in the 1970s was likely to grow 50 per cent faster. This was despite the depressing evidence of the past. Between 1955 and 1970 the trend growth in UK steel demand averaged 1.7 per cent per annum (HM Government, 1973, 6). Moreover, the Nationalised Industries had all been told, in 1967, to base their investment plans on an assumed 3 per cent growth in GNP and that too had not materialised. It seems reasonable to argue that both the BSC and the government failed to take adequate note of the past performance of the UK economy and, as has often been the case, plans were justified on the basis of hoped-for improvement. The fact that Britain's major competitors were also planning considerable expansion of their steel industries was also cited in the White Paper as justification for approval of BSC's strategy.

Yet the fact that all of the major steel producers were planning big investment programmes could equally have been put forward as a reason for caution and a more incremental approach. Thus the government and the BSC were aware of extra-UK influences which could have an adverse effect on BSC's fortunes, as these forces were likely to increase the degree of international competition which BSC faced (again a parallel with shipbuilding suggests itself). As Hogwood noted, in considering extra-UK influences on policy outcomes:

> it is almost certain that at least some competitors will improve their performance. Many competitors have behaviour patterns more conducive to securing maximum output from new equipment than yards in the UK. Because of this, the problem cannot be regarded as the UK trying to pass a fixed level of performance, because the threshold of profitable performance is constantly advancing (Hogwood, 272).

Experience of steel should have suggested to the government that there were powerful internal factors – such as trade union structure, attitudes towards changes in work practice, quality of management – which if not tackled would ensure that the UK steel industry would indeed find difficulty in 'securing maximum output from new equipment' (as we shall see, this is in fact exactly what happened). Prophetically, the White Paper noted that if excess capacity developed then this could endanger employment. Because of this: '*the Government had paid particular attention to the robustness of the proposed strategy if the market for steel turned out to be either better or worse than assumed*' (HM Government, 1973, 10. Our emphasis). Because the government was concerned about the element of risk to long-term employment resulting from a bold, high investment course, it:

> welcomed the emphasis placed on flexibility in BSC's proposals. Within the agreed strategy decisions will be taken in the light of developments over the decade, including market trends and the changing efficiency of individual works (HM Government, 1973, 11).

This flexibility was to be operationalised, within the Department of Industry, through an annual review of the five year rolling programme of investment based on the strategy. The government would review the progress of the strategy before giving firm approval to capital expenditure for the second year of the programme and provisional approval for the third year. A detailed approach would be made to each project in accordance with the principles which had been laid down for all of the Nationalised Industries in an earlier (1967) White Paper, *Nationalised Industries: A Review of Economic and Financial Objectives* (HM Government, 1967), (by 1976 it was clear to everyone that these guidelines were not working). (For a good analysis of the developing approach to financial targets, see Likierman, 1979, 16–32). The government appeared to believe that it could both adopt a very bold, radical, even heroic, policy, with admitted high risks, yet could also pursue an incremental policy through the annual review of the programme and through the detailed approval of individual projects.

There was thus a failure to recognise the inherent difficulties which Departments have in making their appraisals and the political difficulty of halting such a huge programme once it was under way. Why then did the government take the risk? The short answer is that, ultimately, the government was not responsible. In largely upholding what proved to be a miscalculated case, the government could gain comfort from the knowledge that should things go wrong then the Chairman of the BSC and his senior executives were always expendable. Indeed, the White Paper was perfectly explicit on where the responsibility lay when it pointed out that:

> the *Corporation's considered judgement* is that the medium term outlook for the world steel industry in general, and for the BSC in particular, is promising and justifies substantial investment in further steelmaking facilities in Britain (HM Government, 1973, 7).

Not only were there advantages to the government in shifting ultimate responsibility to the BSC for the 'great leap forward', it was also anxious to avoid responsibility for the more immediate consequences of the development strategy. Prior to the emergence of the strategy it was estimated that BSC would have to shed some 20,000 jobs. As a result of the strategy, which would involve some plant closures, an additional 30,000 jobs would be lost, mainly in the second half of the decade. The loss of some steelworks as a result of the rationalisation would, the government calculated, mean that by 1980 180,000 jobs would remain in the BSC compared with 230,000 in 1972 (in fact as at March 1981 the number of BSC employees had dropped to 120,900). The government gave relatively sparse details of the threatened works, preferring to entrust the BSC with the responsibility for authorising plant closures. Clearly there was no political advantage to the government in becoming too closely involved in discussions which could safely be left to the BSC. This was at least one aspect of the relationship between a public corporation and the government where the Morrisonian doctrine could be followed.

Clearly the ten year strategy, agreed for BSC in 1972, was a tremendously important landmark in the development of the UK steel industry (probably more so than nationalisation/denationalisation). More importantly for our study, it proved to be, as suggested earlier, of central importance in determining the scope and nature of the steel 'crisis' in the late 1970s and early 1980s. Pryke has described the programme which the BSC presented to the government in February 1971 as 'totally unrealistic and wildly expensive', and has accused the BSC, in identifying the optimum output within the 28–36 million tonne range suggested by the JSG, as resorting to 'statistical legerdemain.' He concludes that: 'It is difficult to avoid the conclusion that the scales were deliberately weighted against the less ambitious strategies' (Pryke, 189–90).

However, as we have seen, the ambitious programme, with the Minister's active approval (and notwithstanding his own Department's doubts), went ahead. The White Paper, of February 1973, in fact brought to a climax the Heath government's involvement in BSC policy-making. In contrast, for the men at the works threatened with closure, the main battle was about to commence.

4 The Closures Programme: Implementation Problems

Introduction

The years of the early and mid-1970s were a highly distinctive period for the BSC. On the one hand, there was optimism that the enormous investment programme would bear fruit by the end of the decade. On the other hand, there was a realisation that implementation of the strategy would involve extensive closures at plants outside those areas marked for major development. Before the alleged benefits of the new strategy could be gained, the costs had to be distributed. The Conservative government was unwilling to become involved in the closures programme, and left the matter in the hands of the BSC (as far as the unpleasant side effects were concerned, the arm's-length relationship could work). The BSC itself knew full well that it would need to face the closures issue sooner or later, yet when we look at activity during the period 1971–74 we see a general reluctance on the part of the Corporation to bring matters to a head. The problem for the BSC was that until it pushed through at least a certain amount of its development programme, it needed the steel from the remaining works in order to maintain production levels. Indeed, in 1974 there was a steel shortage, with consumers complaining that supplies were not coming through quickly enough.

Implementation of the closures programme was also delayed as a result of the return of a Labour government in March 1974. A major review of all proposed BSC closures was immediately undertaken by a government Minister, and was not concluded until 1975. Ironically, the completion of this review coincided with a general slump in UK and world demand for steel, and heralded a crisis for the BSC which continued for more than a decade.

In this chapter we will be concerned chiefly with examining the activity of one of the local action committees set up to resist the closures programme, as this provides a practical illustration of the nature of the implementation process, and also of the broader constraints facing the Corporation. (The material in this chapter is taken chiefly from two sources: Dudley, 1976 and Dudley,1979). The Shelton Action Committee (SAC) was founded in 1971, some time before the BSC had announced its ten year strategy. In this respect it was ahead of other threatened works in mobilising political support for its cause. Thus although the Shelton works in Stoke-on-Trent was one of the smaller outposts in the BSC's empire, it assumed a political importance disproportionate to its size, not only by the intensity of its activity, but also through the pioneering nature of its lobbying techniques. In conducting its campaign, it was undoubtedly assisted by the pluralistic nature of the process associated with proposed plant closures in the period up to 1977.

Although there was never any doubt about the BSC's intentions, there was a long series of consultative meetings in which the SAC put forward its case for retention of steelmaking at the works. In the course of these meetings, the SAC was perpetually moderate and constructive in its approach. The Committee was anxious to stress that its case was based on a rational economic argument for development, and not just on social grounds. At the culmination of the government review it appeared that the SAC had won its case, but subsequent events proved this victory to be illusory.

Nevertheless, in reviewing the work of the SAC up to 1975 we can gain a flavour of the general political mood of the times. On all sides, there remained an optimism amounting to complacency that the future of the BSC offered a rosy prospect, and that the closures programme could be implemented both gradually and painlessly. In this sense the sparring between the SAC and the BSC from 1971–75 resembled a gentle game of cricket in a high Edwardian summer. After 1975 the more appropriate analogy is with a kind of warfare – not so much the trench warfare of 1914–18, but rather a type of 'ultimate weapon' which not only obliterated a large part of BSC capacity, but also reduced the workforce by nearly three quarters.

The commencement of the Beswick Review

The Labour government which took office in March 1974 lacked an overall majority in the House of Commons, but despite these difficulties the new Secretary of State for Industry, Tony Benn, wasted little time in fulfilling a manifesto commitment to hold a review of all proposed steelworks closures. Mr. Benn had previously had contact with the SAC, and only a week after taking office sent a telegram to the Committee. This telegram confirmed that the review would commence shortly, and concluded:

> I will want to satisfy myself that the proposals for Shelton are soundly based and in the best interests of the country as a whole. I would not accept that the Corporation should reach any final decision unless and until I am so satisfied.

As we have seen, there was no doubt that the 1967 Iron and Steel Act gave Mr. Benn the power to undertake this review. From a purely political perspective, there was also no doubt that nearly all the proposed works closures lay in areas of traditional Labour support, and in such a delicate electoral climate the government could not afford to alienate these areas by apparently agreeing to immediate closures.

From the BSC's point of view, the review had ominous implications. The Corporation had already been frustrated by the protracted work of the Conservative government's JSG, but now it saw that its implementation plans could be threatened by a new government, with new political priorities of its own. Nevertheless, the BSC's initial objections to the review (and its eventual findings) should not be exaggerated. In 1973–74 the BSC made an operating profit of £56 million, and as we have noted the Corporation was not in any hurry itself to close down the threatened works. It was perhaps the very fact of government intervention and its implications for future viability, rather than its immediate effects, which most concerned Monty Finniston, now the BSC Chairman.

There was of course no suggestion on Tony Benn's part that the BSC should modify its investment strategy. Indeed, if condemned works were given a reprieve, then it was likely that the BSC would produce more steel than even it intended to do! Thus the Labour government was not likely to be any less optimistic about the future of the BSC than its Conservative predecessor. The review could also be placed in the general political context of Labour's broad industrial policy with its commitment to conclude planning agreements with industry (a policy strongly supported by Tony Benn). As the BSC would clearly require a high proportion of its massive investment programme to be financed by government, it was not in a strong position to resist the principle of official intervention. Consequently, the BSC could be seen as something of a trial run for the planning agreements which the government hoped to make with British industry as a whole.

Tony Benn delegated responsibility for the review itself to Lord Beswick, one of the Ministers of State at the Department of Industry (DoI). Lord Beswick made it clear that he considered it part of his responsibility, when reviewing the proposed closures, to take into account the practical validity of the BSC's strategy. He was nevetheless well aware of the political implications of his work, and rejected the idea of an independent review on the grounds that these political pressures might not be taken into account.

Lord Beswick's reference to the validity of the BSC's strategy was linked to a developing concern that the major steelmaking sites might not be worked to the level of efficiency hoped for them. In 1973 BSC output reached 23.9 million tonnes when 25 million tonnes had been hoped for (potential capacity was over 27 million tonnes). In 1974, as a result of the three day week caused by the miners' strike, and strikes at two of the major coastal works (Port Talbot and Llanwern), production fell to 19.2 million tonnes. The shortages which resulted inevitably led to a rising import bill for steel (unfortunately for the BSC, the UK's entry into the EEC in 1973 made it easier for consumers to obtain steel from the rest of Europe).

The steel shortages obviously had an influence over the work of the Beswick Review (see Pryke, 191). Even in 1974, however, there was already public criticism of the manner in which the BSC had devised its strategy. A Young Fabian pamphlet judged that the ten year strategy had been: 'a classic example of corporate planning as done by central head office by non-steelmakers and divorced from steelmaking' (Young Fabian Steel Group, 1974).

Thus despite the purely political aspects of the Beswick review, it did have practical problems to confront. The shortage of steel, the efficiency of the large plants and the capital required to develop them were all major problems to be considered. In turning to the work of the Shelton Action Committee, it is important to bear in mind all these considerations.

The development of Shelton and the work of the SAC

Prior to nationalisation, the Shelton works was owned by the John Summers Company, whose main operations took place at Shotton in North Wales. In the view of Heal, Summers purchased Shelton (where blast furnaces had been operating continuously since 1841) in order to obtain a supply of cold pig-iron for Shotton (Heal, 64). By the late 1950s a crisis had developed at the works, for the Open Hearth furnace and the mills were out of date. Closure could have taken place at that time, but Summers decided to invest in a new plant, a decision which Heal suggests was partly activated by sentimental considerations, as well as by objective criteria (Heal, 116). Thus, a sinter plant was built in 1958, followed by the installation of two 55-ton Kaldo converters which were chosen in preference to the LD process (a decision which was to prove a mistake). Finally, the primary mill was replaced by four continuous casting machines (which greatly improved efficiency), and a new Universal or 'Demag' mill. The total cost of this work, completed in 1964, was £20 million, and, as a result, it increased the capacity of the works to 0.35 million tonnes of finished steel per annum. This represented about 1.5 per cent of total British steel output at that time. Shelton thus became a modern, integrated 'mini' steelworks. Its steel was mainly produced for the construction industry, and the close proximity of the majority of its customers in the West Midlands and North West of England, together with the flexibility of the new mill, meant that new orders could be met promptly.

Although productivity and profitability at Shelton compared favourably with the other BSC works, by the late 1960s the local management there was concerned that the strategy for development of steelmaking only at the large coastal works would leave their own plant out in the cold. In order to match these new developments, the Shelton management recommended to the BSC in 1969 that electric arc furnaces be installed at the works. These would replace the existing iron and steelmaking processes (sinter plant, blast furnace and Kaldos) and increase production.

The recommendation for arc furnaces was highly significant. Electric arc furnaces are charged entirely with scrap metal. The steel is made molten by an arc of electric current passed between the metal and carbon electrodes placed immediately above it. The use of scrap rather than iron ore, and the smaller size of the furnaces compared with the BOS process, gives arc steelmaking a distinctive character. As Heal states, arc and BOS steelmaking represent not so much alternatives to steelmaking which can be rationally evaluated, but rather differing philosophies (Heal, 197). As the years have gone by, several critics of BSC strategy have advocated that the Corporation should have placed more emphasis on the development of 'mini' works (e.g. Bryer *et al.*, base some of their chief criticisms of the BSC on this point). In the early 1970s, however, the 'big is beautiful' movement was still in the ascendant, and it was generally thought that works such as Shelton had little or no future.

When the BSC announced the closure of the 18 inch mill at Shelton in 1971, this prompted the formation of an action committee to fight any further closure proposals by the BSC. The membership of the SAC consisted of all 53 trade union delegates at the works, together with four local Labour MPs (there were no Conservative MPs in Stoke-on-Trent). Like this book, most political analyses of UK steel policy are chiefly concerned with the implications of events at a national and European level. Important as these are, we should always bear in mind that it is the activity at each individual works which goes to make up the whole. Despite technological advance, a given steel plant is not a monolithic structure. It contains several sections, each of which has a distinctive character of its own. The internal organisational characteristics of the SAC offer a good opportunity to gain an insight into the effects of the BSC's strategy at the 'grassroots' level. Initially, the BSC was content to allow the SAC and other action committees the opportunity to fight closures. As the years passed, and the BSC crisis deepened, the Corporation resorted to by-passing the local Action Committees and exploiting their latent internal differences in order to kill off resistance and speedily achieve its closure aims.

The organisational characteristics of the SAC

In addition to the full SAC, a policy sub-committee of nine was also elected. This sub-committee consisted of two representatives of middle management, who were able to provide technical expertise which would not otherwise have been available to the Committee. The position of the senior management at Shelton was extremely delicate, in that their official duties required them to implement the decisions of BSC central management. Notwithstanding this commitment, to expect Shelton management to stay loyal to BSC headquarters was rather like asking a condemned man to pay the hangman. Their own objectives were much closer to those of other workers at the plant, and they therefore gave a great deal of assistance to the SAC. At one point the Chairman of the BSC, perhaps doubting the loyalty of Shelton management, impressed on them that even if Shelton closed they need not fear for their future employment within the BSC, but his words had no effect. Apart from their strong local loyalties, they had, as professionals, a different view of how the future of the steel industry might develop. Complementary to this internal unity was a determination to remain independent of other interests. Basically, the SAC did not believe that it could rely on wholehearted support from either national trade union officials or other steelworkers. As we will see later in this chapter, this internal unity and independence on the part of the SAC created tension both locally and nationally.

The character of the leadership of the SAC was also highly significant in shaping the nature of the campaign. The Chairman, Ted Smith, brought an emotional and evangelical spirit to his task which, in addition to maintaining momentum, was also highly effective in bringing eye-catching publicity from all sections of the media. One statement in particular made by Ted Smith (taken from a theatre documentary on the work of Shelton) illustrates particularly well his whole attitude to the campaign. It came when he spoke of his feelings immediately after Lord Melchett's visit to the works in 1973, when it had been announced that steelmaking at Shelton was likely to cease in two years:

> I'm coming to work on the Friday morning and another friend had died recently ... a little old man, he was store-keeper in there, and I was walking down there and [thinking of] these men that had died ... I got off the bus and I was thinking to myself, what'd they do? and the message came to me, they'd fight, and that was the only message that came across and I could hear these voices, mate, and loud and clear they were, you know, men, Tommy Cooper, you know that had been killed. George Rushton, his head crushed. What would that man do? And this all ... came through my mind ... and I came down there and energy surged through me ... and I was on my way from that point on mate, and I thought, the bastards aren't having my job. Those bloody men aren't giving their lives up for me and me let 'em down ... The tears did roll down my bloody cheeks mate ... tears for bloody joy, because I was prepared to put my bloody life down same as they had then (*Fight for Shelton Bar*, 1977).

It is no doubt easy for the dispassionate observer to dismiss such words as being melodramatic and over-emotional. They do nevertheless represent a genuine sentiment, and remind us that, for steelworkers generally, the closure of a works does represent the death of a long-standing community which for many of them proves to be irreplaceable.

Despite the emotional nature of Ted Smith's commitment, the SAC's campaign generally was conducted with great moderation, and reflected the traditional trade union attitudes referred to in the previous chapter. This approach was summed up by the Secretary of the SAC, Bill Foster, when he was asked about the attitude of the Shelton workforce towards BSC management:

> Well, it's certainly moderate. I mean we don't go marching around ... because we think we should treat people with respect when we meet them – that if we invite somebody to come and meet us at our works, that they should have every courtesy extended to them. I think this is only right. I think it's the way we would expect to be received in a similar situation (Fight for Shelton Bar, 1977).

The manifestation of these words was shown when Lord Melchett, and later his successor as Chairman, Monty Finniston, paid visits to Shelton.

Basically, the object of both occasions was to tell the workers that steelmaking at Shelton had no future, yet each Chairman was presented with an expensive piece of pottery to commemorate his visit! The amicable SAC style, together with its rational approach to lobbying, was no doubt an unnerving experience for BSC management.

The SAC and selective incentives

Ted Smith provided a strong focal point for the SAC's activity, and at times forceful leadership was required in order to quell latent tensions between different sections of the works. These tensions came about because the SAC leadership was serving the interests of one particular section of the works, while also seeking to mould the interests of workers in other departments. This situation can best be appreciated by analysing the effects of the campaign on each section of the works.

1. The sinter plant and blast furnaces were scheduled for closure and would not be replaced. As arc furnaces are fuelled entirely with scrap, the iron produced by the blast furnaces would not be required, with the result that the men in these sections had no obvious reason for supporting the SAC's campaign.

2. The Kaldo steelmaking plant was also due for closure. The BSC wished to bring cold blooms from Scunthorpe for re-rolling, which would have the effect of making redundant the men working in the steelmaking departments. The installation of arc furnaces was intended as a direct replacement for the Kaldos, and the men working on these sections of the works had most to gain from the success of the SAC's campaign.

3. The rolling mill was given an assured future by the BSC, and would also be required if arc furnaces were installed. The men employed on this section of the works, therefore, appeared to have no particularly pressing incentive to support the SAC, for their future was secure.

In the case of the SAC, Olson's concept of 'selective incentives' is particularly appropriate. Olson has claimed that, where some members of a group have a greater interest in the collective good than others, the members of the greatest interest would bear a disproportionate share of the burden of providing the collective good. Only if some special institutional arrangement were made, did he consider that a different situation would be possible.

These arrangements were termed by Olson as 'selective incentives', which would operate: 'not indiscriminately, like the collective good, upon the group as a whole, but rather selectively towards the individuals in the group' (Olson, 1965, 34). These 'selective incentives' could be either positive or negative:

> in that they can either coerce by punishing those who fail to bear an allocated share of the costs of the group action, or they can be positive inducements offered to those who work in the group interest (Olson, 34).

The position within the various sections of the Shelton works indicated that the SAC would have to apply 'selective incentives' if it wished to obtain support from the whole workforce. The leadership offered a 'selective incentive' to the sinter and blast furnacemen, in the form of an assurance that if Shelton received arc furnaces, every effort would be made to avoid compulsory redundancies at the works. As the SAC was in no position to give guarantees on job security, this was a relatively weak incentive. Although it was sufficient to assure support for the SAC from the blast furnacemen in the earlier years of the campaign, we will see in the next chapter that the BSC was able to exploit this latent division within the SAC as a means of initiating closures at the works.

In order to obtain support from the men working in the rolling mill, the SAC leadership applied a coercive type of 'selective incentive'. This technique was illustrated in 1973, when the BSC produced a brochure setting out the future of Shelton. The BSC was no doubt aware that by placating the fears of the mill men it could weaken the unity of the SAC, and so the brochure emphasised that the future of the mill was guaranteed throughout the decade. In order to maintain support from the mill men, Ted Smith turned this paragraph on its head by claiming that beyond 1980 the mill would be under threat of closure! The SAC Chairman also declared that the men at the mill should never allow the mill to be operated in isolation. This caused great disquiet at the mill, and Ted Smith was compelled to deliver a number of emotional rallying calls to these men. The Chairman's efforts were eventually rewarded with support from the mill workers, but this incident illustrated that the leadership could never completely rely on wholehearted support for its campaign from the works as a whole. As long as the BSC refrained from imposing coercive pressure on the works this inherent tension was of little material consequence, but (as we will see in the next chapter) once the BSC turned the screw it was of decisive importance.

The SAC and the Conservative government 1971–74

The SAC's contacts with the Conservative government had been generally brief and of less significance than those with the BSC or the Labour government which took office in 1974. The explanation for this relatively distant relationship lay in the government's determination to make the BSC entirely responsible for a decision on the future of the threatened works. Thus the one direct contact between the government and the SAC during this period concerned the possible purchase of Shelton by a Norwegian steel company, Christiana Spigerverk (CS). The pragmatic SAC realised that a privatised Shelton could appeal to the political instincts of a Conservative government, and for a brief period in 1972 it did appear that the government was sympathetic to the idea.

The general significance of the CS decision to invest in the UK was that it illustrated how, in the early 1970s, an alternative to the BSC's strategy was being implemented by the private sector. In addition to CS, a Canadian backed mini-mill was constructed at Sheerness, while a similar arc furnace mini-works was opened at Dudley in Worcestershire. All these ventures required government approval, which was readily given. No doubt a Conservative government was pleased to encourage private sector steel developments, but the BSC itself appeared to have no objections to these new works. This acquiescence on the part of the BSC suggests that they were content to concentrate all their efforts on the large BOS works, and regarded the mini-mills as being of only peripheral value. Yet, Bryer *et al.* point out that, from having 80 per cent of the domestic market for ingots, billets, blooms and slabs in 1967, by 1977 the BSC's share of the same market was only one half (Bryer *et al.*, 143).

The private sector obviously had more faith in the arc furnace mini-mill concept than the BSC, and Shelton saw that perhaps its only hope was to escape from the control of the BSC. In later chapters we will see that in the 1980s several of these mini-mills themselves came under financial pressure but, as Bryer *et al.* argue, the case for mini-mills was always stronger than the BSC believed (Bryer *et al.*, 149–152). In the specific case of CS, the SAC lobbied both the company and the Department of Trade and Industry, but in the event Ministers were anxious that the plant should go to a Development Area, and its eventual location was Manchester. For the SAC, the CS episode was no more than a brief diversion from its main task of lobbying the BSC. Shelton could not escape to the private sector, and this example illustrates the extent to which all the BSC works were now tools of the strategy outlined in the 1973 White Paper.

The SAC and the BSC 1971–74

The SAC's strategic advantages

The relationship between the BSC and the SAC during the years leading up to the Beswick review consisted mainly of a long series of consultative meetings, eventually developing into an almost continuous dialogue. The discussions centred around the question of Shelton's future as either a 'mini' works making its own steel by means of arc furnaces (as advocated by the SAC), or as a plant containing only a rolling mill, and importing all its steel from Scunthorpe (the plan proposed by the BSC). In the event, discussions on such complex topics as the rival costs of steelmaking at Shelton and the Anchor works at Scunthorpe, in addition to the question of the future supply and price of scrap, failed to reach any definite conclusion.

Although the economic case for arc furnaces at Shelton could not be proved conclusively, the SAC was successful in producing financial data which could not be rejected by the BSC. The SAC undoubtedly regarded the BSC as being the main target for its pressure, for it saw that the fate of the works rested in the hands of the Corporation rather than the government. In this respect the BSC is multi-functional, in that at various times it acts as both a target of group pressure and as a pressure group itself. Thus, when the Corporation lobbies the government to provide resources in order to finance an investment programme, it is acting as a pressure group. Conversely, in assessing Shelton's case for new investment it takes on the role of a target – it is, itself, subjected to the kind of pressure which it applies to government departments. In a discussion of what constitutes a public target, Wootton concludes that 'public is as public does', on the model of 'handsome is as handsome does' (Wootton, 1970, 26). When examined in relation to Shelton, Wootton's conclusion signifies that the BSC was a public target in the same category as the government, for the Corporation undoubtedly did have a large public-policy role in deciding the configuration of its works.

In presenting its case to the BSC, the SAC had important strategic advantages. As Hood points out: 'the individual case or small opponent may have much more time in which to prepare his moves than his large adversary' (Hood, 89). As the Shelton works represented only a small part of the BSC's empire, the SAC was in a position to give greater attention to its own case than the senior management of the BSC could afford. Olson makes a similar point to Hood when he contends that 'only in very rare situations will a decision command the attention of all the potential participants' (Olson, 1972, 48). In the case of the SAC the intensity of their campaign enabled them to develop their tactics at a faster rate than the BSC. Within the process of what Hood terms, 'reciprocal learning' (i.e. learning against an environment which is also learning) (Hood, 79), the SAC frequently moved one step ahead of the BSC.

In the presentation of statistical documents which countered the BSC's arguments, the SAC could call upon a large amount of specialised expertise. Firstly, senior and middle management possessed a great deal of technical data concerning the efficiency of the works. Secondly, lecturers at the North Staffordshire Polytechnic also provided expert help. Finally, at a later stage in the campaign Hambros Bank acted as consultants to the SAC, and had an important effect on the technical quality of the SAC's case. The assistance of Hambros was particularly significant, in that to associate the name of a well-known merchant bank with that of the SAC added to the political prestige of the campaign. In its lobbying of the BSC the SAC was, therefore, not in a disadvantageous position with regard to the possession of technical and financial knowledge. The depth and range of expertise which it had at its command, was at least equal in quality to any which could be put forward by the BSC.

Consultative meetings 1971–73

Prior to the publication of the White Paper in February 1973, the SAC, in a series of consultative meetings, attempted to persuade the BSC that an arc furnace plant at Shelton would be both profitable and efficient. It was argued that Shelton lay in the midst of an area where scrap was plentiful, and that the works was also in a central position in respect of communications, with approximately 60 per cent of home trade deliveries within a fifty mile radius. In 1971 it was estimated by the SAC that the capital expenditure required for this scheme would be approximately £4.4 million, a relatively tiny amount in relation to a projected BSC investment expenditure of £3,000 million. Further factors in Shelton's favour were good labour relations, with no major stoppage since 1926, and the profitability of the works, claimed by the SAC to be more than £2 million in 1972–73, (although it should be noted that by re-allocating orders, the BSC could drastically affect the profitability of any works).

The main theme of the consultative meetings revolved around the prediction of cost comparisons between steel made at the Anchor works, and transported to Shelton for rolling, and steel obtained from arc furnaces at Shelton. The principal point of contention concerned the allocation of capital costs to steel manufactured at Anchor. The BSC contended that as the capital cost of the Anchor development had already been committed, it need not be considered as part of the cost of producing steel from that works.

Conversely, the SAC considered that cost comparisons could only be fair if the capital costs of both Anchor and arcs for Shelton were included. Throughout the campaign, the BSC refused to concede this argument to the SAC, for to have done so would have been tantamount to an admission that steel produced from arc furnaces at Shelton would be cheaper than that produced from the Anchor works.

Difficulties in cost comparisons were compounded by errors in the guess-work about the future price of scrap (needed to fuel the arc furnaces). The price of scrap fluctuates widely and projections of its future cost to Shelton could only be highly speculative.

Despite intensive lobbying by the SAC, the BSC could not publicly commit itself on the future of Shelton until after the publication of the White Paper. As noted previously, the Conservative government did not wish to become embroiled in the implementation of the BSC's rationalisation plans, with the result that the White Paper was non-committal about Shelton's future.

The visits of Lord Melchett and Dr Finniston to Shelton

More decisive news for the plant arrived when Lord Melchett visited Shelton in March 1973. He announced that steelmaking at Shelton would probably cease in mid-1975, although a definite date could not be given until the BSC had officially announced the guaranteed two years notice of closure previously agreed with the trade unions. The chairman of the BSC regretted that the Corporation could not accept Shelton's case for arc furnaces, although he considered it was possible that the BSC would require an electric arc plant in England towards the end of the decade. This decision was entirely consistent with the BSC's strategy to restrict steelmaking to coastal sites, although the lack of any detailed economic case in support of its closure plan suggested that the BSC was sensitive about challenging the SAC's arguments. A further point in Shelton's favour was a sharp increase in demand for steel during 1973, which explained why Lord Melchett was reluctant to issue an official closure notice for a works from which the BSC might require future supplies of steel, at least for a limited period.

The second major event in the Shelton campaign during 1973 was the sudden death of Lord Melchett in May of that year. His successor, Dr. Monty Finniston, accepted an invitation to visit Shelton on 27 November.

A fortnight prior to Dr. Finniston's meeting with the SAC, the BSC issued a brochure outlining the case for closure. This document was to form the basis of discussions between the Chairman and the SAC. It was comparatively short, but for the first time it set out a more detailed explanation of the BSC's argument. An interesting addition to Lord Melchett's statement was contained in the first paragraph. It was now revealed that the construction of a second small arc furnace plant, in addition to the one planned for Scotland, 'had been largely pre-empted by the decision of GKN to build one at Cardiff, as announced on 7 August 1973'. As GKN's plans could not have been known at the time of Lord Melchett's visit, it suggested that the Corporation had only recently thought out a more detailed argument against Shelton. Once again the BSC omitted capital costs for Anchor in its cost comparisons and, not surpringly, concluded that Anchor steel would be cheaper than arc furnace steel from Shelton by £7.36 per tonne. In reply the SAC produced a cost advantage for Shelton steel of £7.93 per tonne.

It was at this point that Hambros Bank was brought into the campaign. At the invitation of the SAC Hambros produced an appraisal of the rival documents. The Bank concluded that only if ideal conditions concerning Anchor's development could be proved by the BSC, was exclusion of capital costs and closure of steelmaking at Shelton justified (it will be described later how the BSC experienced considerable difficulties in implementing its investment projects at the coastal works). Hambros' report did not give unconditional support to the SAC, but threw sufficient doubt on the BSC's total strategy to suggest that the Corporation's plans for Shelton were not built on particularly substantial foundations.

Like his predecessor, Dr. Finniston was reluctant to issue a closure notice for Shelton, with the result that what he termed his 'consultative' meeting with the SAC was inconclusive. At the conclusion of the meeting Dr. Finniston stated that, as the discussions could not be completed immediately, he proposed to return for a further meeting in June 1974. At this time the BSC was in a dilemma with regard to Shelton. It wished to retain the works in the short term, while at the same time insisted that steelmaking at the works had no long-term future. This mood of indecision resulted in the BSC being reluctant to employ severe pressure in an attempt to close the works, and also enabled the SAC to buy time. It will be seen later that when demand for steel slumped, the BSC was free to exercise absolute authority over Shelton.

The commencement of the Beswick review meant that Monty Finniston's second visit to Shelton was postponed, and for the next twelve months government became the principal target of SAC pressure. Before examining the Beswick review, however, we can briefly look at the SAC's relationship with other groups. These interactions (or often lack of them) demonstrate the real tension which existed within and between trade unions and steelmaking communities.

The SAC, trade unions and other interests

National trade union officials

The TUC Steel Committee was expected to achieve a co-ordinated policy on capital development and proposed closures, but failed to fulfil this objective. The steel unions generally are reluctant to sacrifice their own autonomy (e.g. the TUC Steel Committee plays no part in wage bargaining), while the BSC management was usually able to exclude the Committee from its policy-making discussions with impunity. In the early 1970s the Committee faced a dilemma, in that modernisation and rationalisation of the BSC would bring greater prosperity to its union members at the major coastal works, but would also entail redundancy for many of those working at plants scheduled for closure. Inhibited by the problem of serving these conflicting interests, the TUC Steel Committee was unable to offer direct help in support of the SAC or any other action committee.

In contrast to the coal mining industry (at least until the aftermath of the miners' strike during 1984–85), the steel industry possesses a fragmented union structure which gives great scope for mutual suspicion and divisiveness. Ironically, the moderation of the unions towards the BSC was not always reflected in their behaviour towards one another! Each one of the union leaders zealously guards the status of the group which he represents, and is watchful for any attempt which the ISTC (the largest union in the industry) might make in order to extend its power by enlarging its membership. Apart from the ISTC, in the period of the Shelton campaign, the only other union represented on the TUC Steel Committee with members working exclusively in the industry was the National Union of Blast Furnacemen (NUB), (the NUB eventually merged with the ISTC in 1985).

An example of the rivalry between the ISTC and the NUB occurred in September 1975, when the NUB threatened a national strike if the BSC did not meet a pay claim by the men who were to operate a new blast furnace at Llanwern. Fulfilment of this claim would have made the Llanwern blast furnacemen the highest paid workers in the steel industry, placing them above ISTC members working on the basic oxygen converters. In order to retain the existing differentials, the ISTC was unwilling to support the NUB's claim. The ISTC also held an advantage in that its membership included the blast furnacemen at Ravenscraig. In making the claim at Llanwern, the NUB possibly feared a take-over by the ISTC, and was therefore anxious to prove to its own members that greater benefits could be obtained by remaining independent of the larger union. Agreement was eventually achieved on rates comparable to those paid to other unions, but the £27 million blast furnace was a year late in being commissioned. This incident illustrates something of a contradiction in the role of the unions. They were very moderate in terms of national steel policy, but their actions at plant level often had damaging consequences.

Other unions represented on the TUC Steel Committee are the GMB (formerly the General and Municicpal Workers Union, and from 1982–89 the General Municipal Boilermakers and Allied Trades Union) and the Transport and General Workers Union (TGWU). The remaining members of the Committee represent the National Craftsmen's Co-ordinating Committee (NCCC), which brings together a number of craft and white-collar unions. As we saw earlier, the TUC Steel Committee greatly resented its exclusion from the JSG in 1971–72. Notwithstanding the Committee's antagonism to this snub by the government and the BSC, it was not prepared to encourage militant action against the Corporation's closures strategy. The dissatisfaction of union members, at the lack of action taken by the leadership of the steel unions with regard to proposed closures, was discussed at a TUC Steel Committee meeting in July 1972. The General Secretary of the ISTC and Chairman of the TUC Steel Committee, Sir Dai Davies, claimed that it might be necessary for the Committee to consider recommending to their Executives some kind of combined action in order to prevent further closures. As examples of this, he suggested a systematic ban on overtime, a boycott on removing machinery from any plant which was due to close, or on using it at any plant designated to receive it.

Sir Dai Davies then followed up these militant remarks by performing an immediate about-face! He claimed that industrial action would inevitably affect the BSC's costs, and might be used by the BSC and the government as an excuse for closing marginal plants even more rapidly. He also recalled that the General Council of the TUC had agreed that each union would retain the right to decide for itself what action to take in opposing closures. With this statement, the Chairman of the TUC Steel Committee had apparently ruled out the possibility of militant action, or the formation of a co-ordinated strategy by the unions with regard to proposed closures.

Given that the TUC Steel Committee felt itself unable to resist the BSC strategy, it was necessary to evolve a commitment which would at least suggest some sort of support for the threatened works, without entailing any action. The details of this policy were spelt out by Sir Dai Davies at a special conference called by the steel unions to discuss plant closures. Sir Dai stated that:

> Before a plant is closed, adequate alternative employment opportunities must be provided. They could be provided by the government, the BSC or some other agency. We don't mind ... It is not a policy of opposition to change, but of insisting that change is phased so that as jobs are lost in steel, other jobs are created in new industries. (Minutes of National Representative Conference, 7:3:73)

This strategy suggested that Sir Dai was prepared to accept a considerable reduction in the BSC's workforce. The only policy question on closures appeared to concern their exact phasing. It also posed the problem of whether the decision on the provision of 'adequate employment opportunities' at the threatened works should be left to the unions at local level, or be made by the TUC Steel Committee. An individual action committee, such as the SAC, was unlikely to agree to any suggestion at the national level that adequate alternative employment existed to the extent that closure of the works could be justified. In this situation there would be the likelihood of a direct conflict between the local trade unionists and their national leadership. A crisis of this nature eventually arose in March 1975 when, at the conclusion of the Beswick review, it was announced that no reprieve would be granted to the steelmaking plant at Ebbw Vale. The workers at this plant considered that they had been betrayed by the Labour government, particularly in view of the fact that the MP for the area, Michael Foot, was also Secretary of State for Employment. As a result, an extremely militant mood was created at the works. Lord Beswick therefore expressed his gratitude to Sir Dai Davies, whose roots were in Ebbw Vale, for speaking to the workers and attempting to convince them that sufficient employment opportunities existed for redundant steelworkers. In this case, the TUC Steel Committee had become incorporated as an arm of government, helping to 'deliver' unpleasant decisions.

The basic acquiescence of the TUC Steel Committee in the BSC's strategy, illustrated by its mainly passive role in the Beswick review, was indicative of the manner in which a national body could be made impotent in its ability to influence the policy process by the impossibility of effectively representing the conflicting interests of its membership.

The power of the TUC Steel Committee was further reduced by the independence of each individual union within the steel industry. A prime example of this disunity was provided at Shelton itself, when Sir Dai Davies visited the works in 1973. The date of his meeting with the SAC coincided with a similar visit by two representatives of the blast furnacemen. It had been assumed by the SAC that a joint meeting would be held between the union officials and the action committee. It therefore came as a shock when Sir Dai informed Ted Smith that he would only meet delegates from his own union, the ISTC. At the meeting, Sir Dai explained that his reason for this separation was his loyalty to his own members. He added that the ISTC's interests lay solely in the steel industry, whereas other unions in the TUC Steel Committee had outside connections. This disregarded the fact that the members of the NUB also worked exclusively in the steel industry, but in a wider context it was strange behaviour from a man who, as Chairman of the TUC Steel Committee, was expected to act as a national spokesman for the workers in the steel industry. The splitting up of the SAC membership in order to hold separate meetings simultaneously was far removed from the usual procedure at Shelton, and exposed the manner in which a national union leader could become as equally out of touch with affairs at the local level as any member of BSC management.

Another aspect of union disunity was illustrated during the course of the Beswick review in 1974, when the SAC submitted to the Secretary of the TUC Steel Committee the list of representatives it proposed to send to a consultative meeting at the Department of Industry. The Secretary immediately objected to the inclusion of a delegate from the Steel Industry Management Association (SIMA), a middle management union which was not represented on the TUC Steel Committee (from 1981 SIMA became a section of the Electrical Electronic Telecommunications and Plumbing Union). The SAC stood its ground, but the SIMA representative was only accepted a few minutes before the meeting commenced, and then only after the personal intervention of Lord Beswick.

The SAC and other interests

The SAC's separation from the national trade union leadership was also reflected in the manner it deliberately distanced itself from other interests. Thus when a National Action Committee (NAC) was formed, in the hope that a more co-ordinated approach could be adopted towards the fight against closures, the SAC took no further part in its affairs after attending only two meetings. It was generally considered by the SAC that the NAC was 'a bit of a shambles', although there was also concern that the NAC could usurp some of the authority of the SAC leadership.

Even at the local level the independent nature of the SAC created considerable tension. Local full-time trade union leaders, in particular from the TGWU and AUEW, attempted to become more closely integrated with the work of the SAC, but were frequently rebuffed. Inevitably there was an element of personality clashes in these disputes (Ted Smith was not a leader who took kindly to others telling him how to run his campaign), but there was also a deliberate calculation on the part of the SAC that the group's isolation gave the campaign an originality and flexibility which it was considered would be lost if other interests were included. Thus in a similar manner to the local union officials, the local authorities were only brought into the campaign when it suited the interests of the SAC (in particular Stoke-on-Trent City Council paid the Hambros Bank fee).

The SAC could therefore be described initially as an 'outsider' group (see Grant, 1978), which alienated many sections of what might have been considered its likeliest supporters. Ironically, this policy ultimately proved a great asset, for there is no doubt that the work of the SAC made a considerable impression on Lord Beswick, and played a significant part in his decision on the future of the works. By this time, the SAC had become a part of the official policy community (see Richardson and Jordan, 1979, 99–100), but had achieved this position largely through by-passing the normal channels of political communications.

The Beswick Review

In May 1974 Lord Beswick announced that BSC proposed closures would be reviewed case by case. The review of each closure was to begin at a tripartite meeting consisting of representatives designated by the BSC, the TUC Steel Committee (it was the Committee's authority in this respect which caused the tension with the SAC over the SIMA delegate) and the government. After this initial meeting, and while each side was considering the points at issue, Lord Beswick was to consult local MPs, local authorities and other appropriate bodies who had made representations to the government. A further tripartite meeting would, if necessary, then take place, following which the Secretary of State would hope to reach a decision on each closure proposal. Priority in the order of review was to be given to those works whose closure dates were relatively imminent, and to those which were implicit in the BSC's investment strategy. The order was, therefore: Hartlepool iron and steelmaking; Cleveland ironmaking; Workington; Ebbw Vale; East Moors; Shelton; Bilston; Shotton; the Scottish closures and the proposed Redcar Plate Mill Development and associated closures, together with other closures in England.

The statement concluded by revealing that, in the course of the review, Mr. Benn would look into the possibility of other changes in the Corporation's general strategy providing for improvements to, or alternative Corporation activities at, plants now proposed for closure. Mr. Benn also intended to continue work on the provision of alternative job opportunities (House of Lords, 23.5.74 Cols. 1676–8). These vague references were the only examples of the possibility of a wider 'options search' being considered. The latter part of the statement was vague on the question of whether the BSC production target would be achieved by new investment or retention of the older works. It was clearly impracticable, however, that all of the works should be spared. Therefore reference was made to alternative job opportunities for works which could not be saved.

In addition to the government's concern that redundancies in the steel industry should be kept to a minimum, and doubts that the BSC's organisation and strategy were justified, there were external economic pressures on the government in 1974 which encouraged a reprieve for some of the threatened works. The rise in the annual rate of inflation to 20 per cent in 1974 led many to believe that the level of unemployment would be bound to rise. The Irlam steelworks had closed in June 1974, leaving severe local problems of redeployment. Further closures in the near future could only have created new political difficulties for the government. It should also be noted that cost savings to the BSC, and thus indirectly to the taxpayers, as a result of closures, could be largely offset by redundancy payments, social security benefits and government aid to encourage replacement industry in areas where closures took place. An additional factor was the likelihood of a severe cutback in the programme of public expenditure, with possible repercussions on the BSC's investment programme.

It could also be argued that, in determining the outcome of the review, the government's prestige was at stake. If it merely rubber-stamped the BSC's strategy, it would be accused by its followers of capitulating totally to the Corporation. Conversely, the government's image would also have suffered if it alienated the BSC by rejection of the Corporation's strategy. Lord Beswick had to chart a narrow course of compromise, in which the authority of the government must be asserted, while at the same time the credibility of the BSC strategy could be retained, i.e. despite some doubts the government retained a basic faith in what the BSC wanted to do in terms of expansion.

Although the BSC was clearly unhappy about the Beswick review, there was some degree of ambivalence in its response. As described earlier, this period was a relatively prosperous time for the BSC, with a 1973–74 operating profit of £56 million. The profit represented a return on capital of 6.8 per cent (still below the 8 per cent target), and for the first time a dividend was to be paid on public dividend capital. There was even a steel shortage, and so the BSC could afford to be comparatively relaxed on the prospect of closures being delayed for a while. At the same time, Monty Finniston gave an intimation of his views on the Beswick review by stressing that, unless the government completed its investigation by 1975, the BSC's target of 35 million tonnes capacity by 1980 could only be met by going 'hell for leather'. It cannot be stressed too firmly that at this time, in contrast to later events, the BSC saw itself as being one of the first British industries to come to terms with the modern industrial age. This attitude was well summed up by Monty Finniston's remark that someone (meaning the BSC) would have to make the money to subsidise the government's other activities (*The Sunday Times*, 4:8:74).

The strained relationship between the government and the BSC as a result of the Beswick review came into the open in August when *The Financial Times* revealed that Dr. Finniston had sent Mr. Benn a letter in which he claimed that the government could not force the BSC to change its strategy, and that the BSC had no intention of making any compromises over the proposed closures. Such an unequivocal statement was typical of the BSC's Chairman, who did not believe in concealing his views behind a camouflage of diplomatic nicety. After a distinguished scientific career as a metallurgist, Finniston became a Deputy Chairman of the BSC at its inception in 1967 and Chief Executive in 1971. After Lord Melchett's death in 1973, he was the most obvious candidate as his successor. Not being a politician by inclination or experience, he tended to be openly contemptuous of government intervention into the Corporation's affairs. At the time of his letter to Mr. Benn, it was reported that Finniston was 'not afraid of being sacked' (*Financial Times*, 5:8:74).

The DoI under Tony Benn, however, was unlikely to capitulate by granting Finniston his demands. Operating from his base on the left wing of the Labour Party, Benn had also won for himself a reputation as an independent thinker, who was prepared to stand firm in the face of opposition. Politically, the Benn-Finniston relationship had all the ingredients of the irresistible force meeting the immovable object (in the next chapter we will see how matters eventually came to a head), but in the case of the Beswick review the ultimate collision was avoided. In 1974 Finniston could afford to be flexible, and indeed he was always anxious to point out (correctly) that the BSC was more thorough and conscientious in the task of redeploying redundant workers than any other Nationalised Industry. For his part Tony Benn, like Finniston, was undoubtedly an expansionist, who was firmly in agreement with the Corporation's high capacity target for the 1980s.

By the Autumn of 1974 Lord Beswick had completed his review, but it was to be some time before an official announcement was made on the future of the threatened works. The delay was caused partly by the General Election of October 1974, but could also be attributed to the difficulties in achieving a Benn-Finniston compromise. For Shelton, news came in the form of an obviously official 'leak' in *The Economist* during January 1975. It stated that the government would make an interim statement, within a couple of weeks, on the review of English and Welsh closures. The results of the Scottish review would be announced at a later date. The report went on to say that although closure dates at several works would be put back, only Shelton would be saved altogether. The SAC had convinced the government that the plant could be profitable on a long-term basis, with the installation of arc furnaces. The nature of the 'leak', together with its contents, demonstrated that Shelton had become the 'special case' of the Beswick review. Approval of the arc furnace scheme (calculated to cost £6 million at 1974 prices), offered the government an opportunity to assert its authority over the BSC without causing any major disruption to the Corporation's strategy.

The results of the Beswick review, as it concerned England and Wales, were announced on 4 February 1975. On Shotton, Lord Beswick deferred a decision on future development until further studies had been made. Nevertheless, closures of steelmaking at the works were deferred until at least 1980–81. A battle had developed between Shotton and Port Talbot as to whether Shotton should be developed at the expense of a portion of the South Wales works. The government was clearly unready to commit itself on this controversy. Two other works were reprieved for a few years but no new investment was recommended. The East Moors closures date was deferred from 1976 until 1980, and Hartlepool from 1976 until 1978. In the case of Cleveland and Ebbw Vale, however, the closure dates of 1975–78 and 1975–77 respectively were accepted by Lord Beswick. To counteract the Ebbw Vale closure, the development of the works as a major manufacturer of tinplate was to continue and in addition, intensified efforts were to be made to provide new job opportunities. It was considered that BSC involvement in job creation would introduce a new dimension into redundancy problems in the public sector. In addition, there was the decision on Shelton. The government estimated that, in total, the review would save 13,500 jobs at the threatened works for from two to four years or more.

The Beswick review decision on Scottish closures did not arrive until August 1975. By this time the BSC's financial position had already begun to deteriorate, while the more cautious Eric Varley had replaced Tony Benn as Industry Secretary. There was no hint of doubt about the future, however, in Varley's claim that by the early 1980s the BSC would be producing about 37 million tonnes of steel. This continued optimism was reflected generally in the decisons on Scotland. Open-hearth steelmaking at Clydebridge, Dalzell, Lanarkshire, Ravenscraig and Glengarnock were to go once alternative supplies became available, but to counteract this loss six new projects were confirmed. These included electric arc plants at Hunterston and Ravenscraig, with increased capacity at Hallside. In addition, two direct reduction plants (an alternative to scrap for arc furnaces) would be built at Hunterston. Elsewhere, the decision on Shotton was still delayed.

Thus the Beswick review was concluded, and with it went the end of a kind of political Arcadia.

5 The BSC and the Labour Government 1975–79

New realities

From April 1975, until the return of a Conservative government in May 1979, the chief concern of both the BSC and the Labour government was to deal with the increasing sense of crisis within the steel industry. At first, the official stance was that the existing problems were mainly transitory, and that by the end of the decade the BSC would still require an output of around 37 million tonnes. Only towards the end of 1977 was it officially acknowledged that there had been a great sea change in the whole manner of the BSC's affairs, and that the large-scale strategy would have to be modified. In the course of this chapter we will be examining these momentous changes in the political climate surrounding the steel industry. Initially, there were differences between the BSC and government as to the pace of change. As we will see shortly, a brief, but heated, debate on this subject took place between Tony Benn and Monty Finniston. This highly publicised interlude proved to be only of passing importance, for as the crisis deepened both the government and the BSC became increasingly dubious about the whole value of the 1973 strategy. Eventually, a White Paper in 1978 greatly reduced the scale of investment proposals, and also largely reversed most of the Beswick review decisions on delayed closures.

This closing of ranks by the government and BSC was to some extent countered by the activity of the Select Committee on Nationalised Industries

(SCNI). The Committee produced several reports which were highly critical of how both the Corporation and the government were dealing with the crisis. In the course of its work the SCNI also found itself in a highly publicised dispute with the BSC over papers which the Corporation refused to release to the Committee. The SCNI thus had the effect of at least bringing to general public attention the fact that the BSC was no longer in a position to 'earn the money to pay the subsidies'. Instead, the reverse was now the case.

The commercial failure of the BSC

We described in the last chapter Monty Finniston's misgivings about the whole concept of the Beswick review, together with his eventual acceptance of compromise. The truce was only temporary, for an even bigger row was about to break out in public.

In the first half of 1974–75 the BSC made a profit of £82 million, but from this point on the position began to deteriorate rapidly. By the beginning of 1975 production cutbacks in several product areas had taken place (particularly the strip mills) and prices were being raised in the hope that this would make up for the loss in revenue. During 1975 the steel crisis became a worldwide problem, mainly due to the rapid increase in oil prices which had restricted industrial growth generally. In addition during this period, developing countries such as Brazil, Mexico, India and Korea were quickly expanding their own steel industries. and so did not require supplies from traditional steelmakers such as Britain. These external factors were undoubtedly of great importance, but were only part of the story. The general performance of the BSC itself was disappointing, particularly on matters of imports, exports and product delivery. On imports, the situation had become particularly worrying. Pryke illustrates how, between 1970 and 1975, BSC's share of the UK home market declined from 70.4 per cent to 54.1 per cent. During the same period, the private sector's share increased slightly from 24.1 to 25.2 per cent, but imports shot up from 5.5 per cent to 20.7 per cent. From his figures, Pryke comes to the conclusion that approximately two-thirds of the decline in the Corporation's performance between 1968 and 1978 was due to the fact that it lost ground and became less competitive (Pryke, 195). Thus in 1974 Britain became a net importer of steel for the first time. On exports, Pryke calculates that between 1968 and 1978 UK net exports of steel slumped from 2,125,000 product tonnes to under 550,000 product tonnes, a decline which he attributes chiefly to the rise in imports from the remainder of the EEC (Pryke, 195).

Related to the export/import balance is the whole question of BSC's ability to meet the requirements of its customers. Pryke places this factor at the head of the Corporation's problems and gives a damning resumé of the organisation's performance (Pryke, 195–200). In particular he sees the steel shortage of 1974 as being the critical factor in many customers resolving that in future they would 'second source' their supplies from steelworks outside the UK. The whole question of BSC's relationships with suppliers and customers is obviously of great importance, yet it receives relatively little political attention. In Chapters Seven and Ten we will be looking at some of these major inter-relationships, particularly with regard to how the structural nature and consequent development of this network may have affected BSC's fortunes.

As one example of the BSC's fragile relationship with its customers in the mid-1970s, the Ford Motor Company complained early in 1975 that it would look abroad for supplies unless the quality of steel produced at the Llanwern works improved. Ford had been disturbed about the quality of Llanwern steel for two years, with the result that 50 per cent of its steel requirements were now imported. BSC attributed its failure to maintain quality at Llanwern to the fact that the workforce had yet to become familiar with the new machinery (*Financial Times,* 29:1:75).

The obvious problems at Llanwern must have been of great concern to the government, and perhaps contributed towards delayed closure decisions in the Beswick review, yet the official line was still that everything would be all right once the new plant was built and in operation. In retrospect, it appears strange that at the time so little political attention was directed towards BSC's disastrous performance in its own home market, even where the new works such as Llanwern were concerned. There was apparently an act of faith on the part of both government and BSC that once the development strategy had been carried out, then all the 'nuts and bolts' problems, so evident at Llanwern, would magically disappear.

It was also hoped that the recession would be relatively short-lived, and so attention was given to stockpiling. Thus in March 1975 the BSC set up its own stockholding group, British Steel Service Centres (BSSC), which incorporated existing stockholding interests together with the newly acquired Lye Group. In addition to countering cyclical change in demand, the BSC also hoped no doubt that BSSC could become a strong force in the steel stockholders' business. As an intermediary between steelmaker and customer the stockholder is in a potentially powerful position in the market (e.g. a customer buying steel from a stockholder will usually care little where the product comes from, so long as it supplies his needs), and the BSC needed to make sure that its own products were well represented in this sector (we will be examining the power of the stockholders in more detail in Chapter Ten).

The Benn-Finniston dispute

By April 1975 the extent of the steel crisis was beginning to become clear. Some BSC plants were working at 50 per cent or less capacity, with the result that overtime and shifts worked were severely reduced. In 1974–75 BSC made a profit of £73 million (compare this with the six months profit of £82 million), but within the BSC there was already little hope that a heavy loss could be avoided in 1975–76. Despite his public acceptance of the Beswick review, Sir Monty Finniston was still seeking radical change, and saw the new crisis as a means of achieving his objectives.

Politically, he made the shrewd decision of announcing his proposals to a meeting of industrial journalists, thus ensuring that his remarks would gain maximum publicity! Sir Monty was particularly concerned to tackle the problem of overmanning, and so he declared that the BSC planned to reduce its 220,000 workforce by 20,000 within a matter of months. More dramatically, he hoped that in the long term the BSC would employ 50,000 people to produce 37 million tonnes of steel (as a point of interest, in 1986–87, 52,000 BSC employees produced 11.7 million tonnes of steel). With regard to productivity he drew highly unfavourable comparisons with Japan, where production rates of 750 tonnes per man were achieved, compared with 150 tonnes per man in the BSC. Regardless of the arguments over capacity and product quality, there is little doubt that Sir Monty had identified a key weakness in British industry as a whole. Much of his efforts during this period were directed to driving the simple truth home to workers, politicians and the public, though it was to be some eight years later (and two Chairmen later!) before the BSC finally began to approach the productivity levels of its competitors within the EEC, let alone Japan.

Sir Monty scrupulously refused to state in public his immediate proposals for a reduction in the workforce, but it was widely reported (clearly the information had come from the BSC itself) that the management was seeking an immediate reduction of 10 per cent in the workforce of more than 220,000. Given the nature of later events, the proposed 22,000 reduction was not a particularly radical proposal. Nevertheless it aroused a great political furore, and in particular caused a confrontation between Finniston and the Secretary of State for Industry, Tony Benn. Despite Benn's response, it was unlikely that Sir Monty's prime aim in making his statement was to provoke the Secretary of State. Pay negotiations between the unions and BSC were due in May, and the Chairman was no doubt setting out markers in advance. For Benn, however, Finniston's remarks were seen as a direct challenge to his authority, and he promptly announced that he would arrange for the BSC Chairman to come and meet him in order to discuss the matter. He was upset that he had received no advance notice of the speech, which he saw as breaching the conclusions of the Beswick review.

The meeting took place on 28th April, amidst much speculation about Finniston's future with the BSC. In the event, Benn drew back from directly challenging Finniston's comments, for the meeting lasted only twenty minutes, and was described by Finniston as being remarkably amiable (*Financial Times* 29:4:75). The device chosen by Benn was to switch the debate on to a more general and philosophical plane than the immediate prospects of the BSC, and instead try to discover Finniston's opinion on the roots of the government/BSC relationship. Consequently, Benn asked four questions of the BSC Chairman:

1. How does the Corporation now see the role of the government in the review of the steel plant closures upon which we have been engaged?

2. Apart from the steel plant closures, how many redundancies, permanent and temporary, does the Corporation now believe to be necessary at individual steel plants in England and Wales and Scotland, and what is the Board's best estimate of the build up of these redundancies, showing the timetable for each?

3. What is the Board's view of the correct role of government in respect of the BSC, and of the Corporation in relation to government?

4. How does the Board believe that the policy upon which the present government was elected last year might best be realised in respect of the BSC?

Benn expressed this policy thus:

> We intend to socialise existing nationalised industries. In consultation with the unions, we shall take steps to make the management of existing nationalised industries more responsible to the workers in the industry and more responsive to their consumers' needs. (*Financial Times*, 29:4:75)

Sir Monty took these questions away with him in order to consider his reply. They clearly offered much food for thought, and represented one of the very few occasions on which a confrontation between a Minister and a Nationalised Industry Chairman has been brought out into the open. In so doing, Benn called into question the whole concept of the generally accepted Morrisonian concept of an arm's-length relationship between government and a Nationalised Industry. Question 2 was obviously seeking to push the BSC into making explicit its intention on rationalising the industry, but the spirit behind the other three queries appeared to be one of attempting to force the BSC into a corner, where it would either have to accept that (as in the Beswick review) the government would have the final say, or instead assert its own authority to determine future strategy.

When Sir Monty's reply arrived a few days later (published in *The Times* 6:5:75), it refrained from openly challenging the authority of the government, and was perhaps rather disappointing for students of government/ Nationalised Industry relations who hoped to gain a definitive description of the respective powers and duties of the parties involved. In the main, Sir Monty chose to give rather vague answers to Tony Benn's questions, and his tone was considerably more moderate and conciliatory than in his speech a few days earlier.

In answer to the first question, of the government's responsibilities with regard to the Beswick review, Sir Monty readily admitted that he had been content to agree to the review's conclusions. He also stressed that the Corporation was doing everything it could to help redundant steelworkers, but stressed that the government should support the BSC in the work of producing a modern steel industry (i.e. limited government intervention was just about tolerable, but Ministers must know their place!)

On the second question of the scale and timing of redundancies, Sir Monty firstly referred to the reduction in the workforce of 40,000 which had been outlined originally in the 1973 White Paper. Sir Monty then used the ploy of separating long-term and short-term employment prospects. He considered that the White Paper proposals represented the long-term plan. In contrast, the current crisis required prompt short-term measures (presumably this referred to the 22,000 reduction 'leak'), and so negotiations with the unions were about to take place. As we will see, the long-term and the short-term would eventually become indistinguishable on the matter of redundancies, but for the moment the BSC was content to treat its new proposals as being a response to a temporary crisis.

The third question, on the relationship between the BSC and the government, brought an even more non-committal response from Sir Monty. He made the point that the BSC was concerned with long-term objectives, while the government was obliged to deal with the pressures of today (in fact the BSC was entering a crisis where the 'pressures of today' would require nearly all their attention!) At the same time, Sir Monty expressed the hope that a clarification could be made of the relevant statutes. It could be said that in his letter the Chairman had an ideal opportunity to attempt this clarification himself, but his comments went no further. 'Clarification of relevant statutes' remains perhaps the most common plea from Chairmen of public corporations. Yet the day of enlightenment never seems to arrive. It is perhaps asking too much of both sides that it should, for both Nationalised Industries and government usually find it convenient to adapt the statutes to their needs at the time. Governments had no qualms about initiating such interventions as the JSG and the Beswick review, while the BSC was equally happy to push through a radical investment and then a rationalisation plan which ultimately both Conservative and Labour governments were powerless to prevent. We will see in Chapter Eight that in the 1981 Iron and Steel Act the Conservative government sought to increase its authority over the BSC, but the evidence suggests that greater political control requires more than just a new statute to bring it about.

The fourth question, on Benn's concept of Labour policy, was perhaps the most difficult for Sir Monty to answer, but he was at least able to point to the existence of worker directors, and that the Board also contained two representatives of steel consumers. The Chairman also referred to a recently framed proposal for a National Joint Council, where management and unions could get together on a formal basis (we will shortly be looking further at this proposal, which was part of what became known as the Steel Contract).

In sum, the conclusion with regard to Sir Monty's letter could be that, for a self-confessed non-politician, it represented a good politician's answer! As a reply to Benn's questions it was both conciliatory and vague, and contained nothing of which the Secretary of State could take offence. Benn's specific response to the letter was to propose tripartite meetings between government, BSC and unions in order to discuss matters further. Sir Monty replied that he would discuss the matter with his Board, but also proposed an end to public correspondence! By mutual consent the 'great debate' was now over. If it brought little enlightenment on the vexed question of Nationalised Industry/ government relationships, it did at least bring briefly into the open matters which are usually considered too sensitive for public ears, and for even this moment of openness both Benn and Finniston deserve credit.

As events transpired, the tripartite discussions were never to take place. Benn's tenancy of the Industry Department had only a short term to run, and his successor, Eric Varley, was not disposed to pursue the matter. In any case, the unions themselves were not enthusiastic about bringing the government into their discussions with the BSC in case it should jeopardise their own decision-making autonomy. Consequently, it was union/BSC talks which produced a response of sorts to the deepening crisis.

Response to crisis

Between May 1975 and January 1976 the BSC and unions hammered out three agreements designed to improve productivity and reduce losses. Nevertheless, in this period BSC losses rose from £2.5 million per week to £8 million per week, with no sign of an early improvement. Despite the

BSC's initial hopes, the agreements did not entail specific manpower cuts, although targets were set with regard to hoped-for savings. The truth was that the BSC's order book had deteriorated to the point where nearly all its plants were no longer viable, and that relatively small-scale productivity improvements could only nibble away at the problem.

The first agreement came in May – a few weeks after the Benn-Finniston debate. Finniston himself, perhaps calculating that the nature of his earlier remarks might have offered a provocation to the unions, was absent from the talks. Instead, he left matters in the hands of Bob Scholey, his Chief Executive. A manager at United Steel prior to nationalisation, Scholey acquired the nickname 'Black Bob', chiefly because of his bluff and sometimes forbidding Yorkshire manner. Over the years Scholey became a highly influential figure within the BSC. He still occupied the same post in 1985 (although by now his power had increased because the Chairman held only a part-time appointment). In 1986 he assumed the office of Chairman, and his own steel background could be contrasted with that of all except one former Chairman (Ian MacGregor).

After a meeting of ten hours between BSC management and the TUC Steel Committee, an agreement was achieved on several points, although the Corporation had dropped its hints of a 22,000 cut in the workforce (it appears likely that at this time it was something of a negotiating tactic). The agreement included the elimination of all unnecessary overtime; a commitment by the unions to try and clamp down on absenteeism; strict controls on recruitment; the maintenance of the guaranteed working week (the BSC had been pushing for its abandonment); an acceptance of voluntary redundancy, provided it was agreed with the unions that overmanning existed; and a commitment by the BSC that plants threatened with closure would still receive their fair share of orders. The BSC hoped that these measures would save £100 million over twelve months, although such phenomena as unnecessary overtime and absenteeism would obviously require a great deal of heart-searching at the local level. The savings which could be made thus had the appearance of being largely notional, and based more on hope than expectation.

In July, Finniston announced that he was disappointed with the manner in which the package agreement was being implemented at the local level, and also revealed that the Corporation was now losing £5 million a week. New negotiations were entered into with the TUC Steel Committee in an effort to bring savings back up to the £100 million mark. Shortly before this meeting Sir Dai Davies had retired, and had been replaced as General Secretary of the ISTC and Chairman of the TUC Steel Committee by Bill Sirs. At the time of his appointment, it was thought that Sirs might be somewhat to the left of his predecessor, and thus not be so pliable in his dealings with BSC and government. In the event, although Sirs proved to be less autocratic than

Sir Dai, he maintained the strong steel union tradition of moderation and a general willingness to co-operate with the strategic plans of management.

The results of the July meeting reflected this continued moderation, for the new agreement took the earlier six-point plan somewhat further: workers with less than twelve months service would face compulsory redundancy on a selective basis; workers over sixty would be considered for redundancy on a selective basis; local talks would take place on jobs considered by management to be surplus to requirements; the guaranteed week agreement (which gave workers eighty per cent of normal earnings even when not enough work was available to keep them occupied) would be waived subject to local agreement; flexibility in plant loadings would be subject to local discussions. The emphasis on compulsory and voluntary redundancy, together with the potential undermining of the guaranteed working week, indicated that the BSC was becoming desperate to cut costs. The unions were prepared to go along with this plan, despite obvious reluctance on certain points. Within the ISTC there was a move to strike out the agreement on compulsory redundancy after twelve months, and also the waiving of the guaranteed working week, but these proposed amendments were both defeated by ten votes to three.

Although the new agreement was much stricter than the first, it still obviously depended on acceptance at the local level it if was to be implemented successfully. In addition, the number of job losses required was not specified, and so the proposals still had something of a vague appearance; hoping that events at the local level would somehow evolve in a way which could improve the fortunes of the BSC as a whole. There was also no suggestion of plant closures, for any reversal of the Beswick decisions would have been politically unthinkable at that time.

By November the BSC was losing £5 million a week, with losses amounting to £125 million in the first part of 1975–76. In these circumstances, the BSC considered that it must cut costs by a further £200 million in each of the next two years, with most of the economies (£170 million) to come on the labour side. After abortive discussions with the unions in December, the BSC decided early in January 1976 to unilaterally suspend weekend working, with the threat that the guaranteed working week would also be removed in the near future. The end of weekend working led to a complete shutdown at Port Talbot, together with other Welsh steelworks. Unofficially, Bob Scholey was quoted as wanting to cut 44,000 jobs 'for starters' (*Sunday Times*, 4:1:76), but as with the BSC throughout the 1975–76 period, this sort of figure appeared to be plucked out of the air for public consumption, and played no real part in the union negotiations proper.

The attitude of the unions during this period was not helped by the government ruling that productivity deals with manual workers in the steel industry, which had been signed after July 11th 1975, breached pay policy, and so would not be valid. As Secretary of State for Industry, Eric Varley appeared to be less amenable to pressure from the unions than had Tony Benn, and was not a man likely to conduct a public debate with either the unions or Sir Monty Finniston, the government's attitude at this point appeared to be to sit tight on the results of the Beswick review, and (at least publicly) stay out of negotiations between the BSC and unions.

A provisional agreement between the BSC and unions was achieved (in talks lasting twenty-one hours) during January, which involved the possibility of compulsory redundancies and the formal end of weekend working. The union representatives refused to sign the agreement until they had consulted their respective executives, and at these meetings the new agreement was on nearly every occasion decisively rejected. Once again, the union leadership appeared to be out of touch with their members, who were much less inclined to go along with the BSC's plans.

For a brief period it appeared that the BSC might risk a serious confrontation with the unions and pursue unilateral action. The Corporation was now reported to be losing £8 million a week, with the Strip Mills Division contributing two-thirds of these losses. After another long negotiating meeting, however, agreement on another package was achieved on 23rd January. The BSC now promised to provide detailed figures for the manpower reductions required at each plant. The unions agreed to voluntary redundancies, but also accepted that the BSC could seek compulsory redundancies if the required manpower cuts had not been achieved in twelve weeks. The unions also conceded in principle that week-end working should only apply where it was necessary, but retained the right to be consulted on this matter at local level. The guaranteed working week was to remain in operation, and there was still no suggestion of premature closures at any of the Beswick plants.

Despite this new national agreement, it was clear that, like the other two, this third settlement depended entirely on co-operation at the local level. In this sense these meetings at national level, despite attracting a great deal of political attention, were only a prelude to the main task of actually implementing manpower cuts and changes in working practices. In essence, the BSC had cleared the way for its plans by obtaining agreement from national union leaders that compulsory redundancies may be necessary. Given the inspired 'leaks' emanating from the BSC on the job cuts it required, it is interesting to note the actual figures. At the end of 1974–75, when Sir Monty Finniston made his statement about wanting a rapid decrease of 10 per cent in the workforce, the BSC employed 228,300. At the end of 1975–76 the figure was 210,200. This loss of just over 18,000 was 4,000 short of the 10 per cent, but still represented a significant reduction. In contrast, despite the agreement of January 1976, the reduction for 1976–77 was only from 210,000 to 207,900. Thus the BSC did not find policy implementation an easy job during this period.

As Bryer *et al.* point out, the BSC's problems were not helped by the Treasury changing the PDC to debt ratio from 55:45 to 45:55. In other words, the Treasury's growing impatience with the BSC's problems caused it to impose a still heavier debt burden on the Corporation. Bryer *et al.* calculate that this change caused the BSC's debt burden to rise by 61 per cent and 56 per cent in 1975–76 and 1976–77 respectively (Bryer *et al.*, 169). In the event, the government was only heaping trouble on itself, for higher debt required higher cash limits. From 1978 the BSC was mainly financed by what became known as 'Section 18' capital, which to all intents and purposes was a government subsidy to the Corporation.

We will shortly be looking at events during 1976–77, a period when the BSC's fortunes deteriorated still more rapidly. A whole political era of anticipated expansion culminated in 1978 with official acknowledgement that both the 1973 strategy, and the outcome of the Beswick review, were no longer viable propositions.

A return to the regions

In August 1975 the BSC produced a document which proposed that a new organisational structure should be put in place by April 1976 (BSC Organisation Review, 1975). Ironically, the 'new' concept involved a return to the regional concept which had been rejected at the end of the 1960s. Two reasons were given for this change of heart. Firstly, it was considered that multi-product plants made it impracticable to adhere to a system of grouping plants into Product Divisions. Secondly, it was acknowledged that the Product Divisions had required more central direction than had been envisaged back in 1970.

The return to Regional Divisions would at least allow each Managing Director to concentrate on the problems at a particular works. As we described in the last chapter, there were great problems for the BSC in commissioning the new blast furnace at Llanwern, and indeed (as we will see later in this chapter) production problems at Llanwern continued to multiply during 1976–77. The BSC was coming to realise that it was not just a question of installing new capacity and sitting back to reap the benefits. New investment required new expertise and techniques, and plant performance was crucial if the BSC was to work more efficiently (Divisional management could now give authority for individual investment projects costing up to £2 million).

Thus the chief responsibility for the Regional Divisions (Scottish, Scunthorpe, Sheffield, Teeside and Welsh) was to ensure continuity and consistency of production at the lowest cost and to the highest quality standards. Sales, order handling and plant loading functions were to be different matters altogether. Here, the Divisions would still be organised on a product basis (Billet, Bar and Rod Products, Plates, Sections, Strip Mill and Narrow Strip Products), and the Divisions reported direct to the Commercial Managing Director at Head Office. Thus the real autonomy at the local level was to be strictly controlled, for there was to be no question of Regional Divisions competing for orders against one another. In addition, Head Office could also ensure that the Beswick plants were not given the orders to make them profitable!

In addition to the 'main line' steel products, several separate Profit Centres were set up for other BSC products and services. These included Tubes, Stockholding, Stainless, Tinplate, Light Products, Associated Products and Forges, Foundries and Engineering, Redpath Dorman Long Ltd, and BSC (Chemicals) Ltd. Unlike the larger Divisions, these Profit Centres were to be given responsibility for both manufacturing and selling. The nature of these Profit Centres gave them a more unitary and autonomous appearance than the fragmented Regional Division – Product Unit structure, with the result that the Profit Centres represented a truer form of decentralisation. In later years the BSC was to extend the Profit Centre concept to all its products (and once more abandon a regional structure), as a prelude to what a Conservative government hoped would be privatisation of the Corporation.

An important addition to the BSC structure during this period (particularly in political terms) was the setting up of BSC (Industry) Ltd. This company is a wholly owned subsidiary designed to plan and co-ordinate all BSC action aimed at attracting new and expanding industry to steel closure areas. Although the Company would clearly need to work closely with central government, local authorities, trade unions and other job creating agencies in these closure areas, the existence of an agency of this type, operated by the BSC itself, was an undoubted boost for the prestige of the Corporation. Politically, it suggested that the BSC had a 'heart', and was not just a soulless organisation set only on casting aside thoughtlessly those workers it no longer required itself.

A study of the work of BSC (Industry) lies beyond the scope of this volume (for a detailed analysis of post closure implementation see Young 1987, 369–415), but the organisation has been active in all the steel closure areas since 1976. With the BSC Chairman (Sir Charles Villiers) himself heading the company, by 1983 BSC (Industry) had been active in fifteen areas, and claimed that since 1978 some 1,250 companies had been helped to start-up, expand or relocate, and that these companies had created more than 19,000 new jobs (it should be noted that betwen 1978 and 1983 the BSC workforce fell from 197,100 to 81,100). In itself, BSC (Industry) could never hope to create the number of jobs required to make up for the BSC's own manpower cuts, but in political terms it was an asset of great value.

The culmination of an era 1976–77

During 1976–77 the mismatch between hope and reality continued for the BSC. Superficially, there was an improvement, for the loss (after tax) of £255 million in 1975–76 was reduced to £95 million in 1976–77. In addition, investment continued to steam ahead, with new peaks in capital expenditure of £530 million in 1975–76 and £579 million in 1976–77. On the debit side, long-term borrowings for these years were £315 million and £375 million respectively, while the figures for Public Dividend Capital were £345 million and £490 million respectively. The Corporation was thus being required more and more to rely on borrowing in order to finance its activities, and the interest on the borrowing would inevitably be an increasingly important component in determining the BSC's future financial fortunes. In order to accommodate this new borrowing, the Iron and Steel (Amendment) Act increased the BSC's statutory borrowing limit from £2,000 million to £3,000 million, with provision for a further increase up to £4,000 million. By the end of 1976–77 it was already becoming clear that these previously undreamed of limits would undoubtedly be required if matters did not improve soon, for total borrowing already amounted to £2,285 million (BSC Annual Report and Accounts, 1976–77). In 1976 the BSC also became subject, together with the Nationalised Industries as a whole, to the new cash limit regime. These External Financing Limits (EFLs) were intended to bring public expenditure under control. In 1976–77 the BSC kept within its EFL of £950 million (at £931 million), but in future years the Corporation's mounting losses chased the limit in an upward direction, and these sums tended only to highlight the deepening crisis.

With regard to production, 19.7 million tonnes of steel were produced in 1976–7, compared to only 17.2 million tonnes in 1975–76. This improvement sadly turned out to be a false dawn, but the temporary upturn probably contributed towards the general passivity which was evident in this period. During 1976 there was a feeling of drift on the part of government, BSC management and unions, with the impression being that no-one was able to take a firm grip on events.

On investment, the government continued to delay a decision at Port Talbot and the future of steelmaking at Shotton. In May 1976, two Department of Industry civil servants told the Select Committee on Nationalised Industries (SCNI) that the delay was due to social and employment considerations

(House of Commons SCNI, HC 26, 1977–78, para 181). In reply to accusations that they lacked the expertise to evaluate BSC's plans properly, they responded that sufficient expertise was available, although on one occasion they had called in a consultant on a temporary basis (House of Commons, SCNI 1977–79, HC 26, para. 173). We will be examining the Departmental role in more detail in Chapter Eleven, but it is fair to say that over the years great doubt has been expressed by a succession of Select Committee Reports on the ability of Departmental officials to properly appraise and pass judgement on BSC strategy. For its part, BSC management took little trouble to disguise a dismissive attitude, sometimes bordering on the contemptuous, towards the judgement of both Ministers and officials. Sir Monty Finniston himself epitomised this attitude in his 'debate' with Tony Benn, but successive Chairmen have taken a similar view.

Despite the problems of Departmental expertise, it must be said that the position over Port Talbot in 1976 gave justified grounds for scepticism on the part of the government. Notwithstanding the assertion by officials that the delay in a decision on Port Talbot was due to social and employment considerations, there was doubt on the part of the Department of Industry about the continued practicality and potential viability of the 1973 strategy. Sir Monty himself told the SCNI in July 1976 that the £3 billion programme would now cost £10 billion, but was understandably vague as to the extent the BSC could finance this investment itself (House of Commons SCNI, 1977–78 ,HC 26, paras. 349–54). Thus a large percentage of the now enormous sum required to finance the strategy would have to come from government, but the ever optimistic Sir Monty was still talking in terms of the next boom and the crucial importance of the BSC being able to meet the required demand.

By the time he gave evidence to the SCNI, Finniston was in his last couple of months as Chairman of the BSC. In March 1976 the government announced that his contract would not be renewed when it ran out in September. After a succession of confrontations with the government this news could hardly have come as a surprise to Sir Monty, but as a principal architect (together with Lord Melchett) of the 1973 strategy his departure did signify the end of an era. After 1976, expansionist talk at BSC headquarters quickly went out of fashion.

Finniston's replacement was named as Sir Charles Villiers, a merchant banker (like Lord Melchett) who in the 1960s had been Managing Director of the Labour government's Industrial Reorganisation Corporation, and also Chairman of the Northern Ireland Finance Corporation. His record of service to a Labour government was no doubt considered an asset by Ministers. It was reported that before approaching Sir Charles, the government had been seeking a new Chairman amongst the ranks of management in the private sector of the steel industry, but had found no takers (*Financial Times*, 19:3:76). By his nature, Villiers was to prove a less outspoken Chairman than his predecessor. On the other hand, we will see that the course of events placed Sir Charles as the central figure in a number of politically inflammable situations, with the result that he had no option but to make himself a public figure.

Before his departure Finniston was to have one more galling experience, for in July 1976 Eric Varley, the Secretary of State for Industry, announced that the government had still made no definite decisions on the future of Port Talbot and Shotton. Instead, the BSC was to be asked to make a further review of its proposals (Hansard, 19:7:76, cols 1283–4). Although the delay in a closure decision on Shotton came as no surprise, it had been expected that the government would sanction the BSC's plans to increase Port Talbot capacity from 3 million to 6 million tonnes. It was estimated, however, that the cost of the Port Talbot development would now be £580 million, and there was obvious (valid) doubt on the part of the government as to whether demand would justify the outlay. In order to ameliorate the BSC's disappointment, the government did agree to spending a substantial amount (£350 million) on improving existing facilities at Port Talbot. The fact that at the time an expenditure of £350 million was almost ignored on the grounds that it was of incidental importance illustrates:

- the extent to which large-scale BSC expansion was still regarded as virtuous in itself;
- the success of the BSC in winning support for its view that the only 'solution' to the steel problems lay in the development of massive coastal steelworks;
- the general confusion in separating the necessity for investment which would quickly improve BSC's poor record in meeting its customers' needs, from investment which was rooted more in grandiose visions of the future than the hard reality of solving existing problems.

Thus press comment on Varley's statement almost entirely supported Finniston's view that the government's decision on Port Talbot represented another 'disastrous delay': *The Sunday Telegraph* argued that the Port Talbot development should have gone ahead in 1973 when it would have cost £350 million rather than £580 million (25:7:76). Similarly, *The Financial Times* considered that the BSC's obsolete plant was in itself a valid argument for pressing ahead with the full Port Talbot development (20:7:76). The strength of this consensus (shared by government, BSC and unions), even as late as the end of 1976, represented an accepted wisdom which few were prepared to question.

In an attempt to boost its revenue, the BSC introduced three separate rounds of price rises for a wide range of its products during 1976. As we have seen, BSC's steel production for 1976–77 was an increase on the previous year, and for a brief period during 1976 it did appear that the traditional cycle of demand was at work, with an upturn about to begin. This was not to be the case, for throughout Europe demand was generally depressed, and in any case, the BSC's problems were more deep-seated than those to be found in other EEC countries. As demand sagged the BSC was unable to enforce the price increases. Penetration of the EEC steel market by third countries, and in particular Japan, was placing great strain on EEC producers, who in turn looked to other markets, such as the UK, to sell off their surplus. This over-supply led to widespread discounting, and inevitably to accusations and counter-accusations of dumping. Politically, this chaotic situation was the signal for the EEC Commission to step in and attempt to create a more regulated market for steel.

Firstly, they negotiated voluntary trade agreements with third countries such as Japan, which reduced imports into the EEC. Secondly, the Commission introduced increasingly more widespread and enforceable controls on steel production and prices within member countries. Thirdly, they attempted the even more Herculean task of removing subsidies to steel producers. The breadth and intensity of the Commission's work deserves a study of its own, and in Chapter Eight we will be examining this activity with particular reference to its effects on the BSC.

In spite of (some might argue because of) the importance of the Commission's work, the fact of central importance is that by 1976 BSC's share of its home market had declined from 70 per cent in 1970 to 53 per cent, and until the late 1980s failed to improve on this depressing position. In addition, although exports increased to 2.7 million tonnes in 1976–77 from 2.2 million tonnes in 1975–76, it was still one million tonnes below the Corporation's objective (BSC Annual Report and Accounts 1976–77). Without an improvement in these home market and export positions, it was inevitable that the BSC would suffer serious problems in maintaining a stable financial position.

As with all governments, Labour in 1976 had a vested interest in maintaining (at least publicly) an optimistic view about the BSC's prospects. Thus the official view was that the cycle of demand would soon bring about an upturn, while investment in new plant would improve the BSC's efficiency and product quality, bringing a revival in the Corporation's commercial position. Despite obvious doubts about the need for the Port Talbot development, the government turned its hopes for the future into a more tangible form by financing BSC stockpiling schemes in both 1975–76 and 1976–77. In 1975–76 it had been intended that 800,000 tonnes, costing £70 million, would be stockpiled. In the event, because of difficulties in getting the scheme under way, only 282,000 tonnes costing £39 million was put away in readiness for the expected upturn in demand. In 1976–77 £50 million was provided by government for stockpiling.

In addition to the government's own view on this subject, they were supported by the sector working parties of the National Economic Development Office, where it was reported that as many as nineteen of these committees had expressed concern about steel supplies during 'the recovery' (*The Financial Times*, 5:8:76). No doubt this concern expressed by consumers was activated by their experiences in 1974, when steel shortages were widespread. During this period, 'second sourcing' had become common. In this sense the consumers had answered their own question, for in the future, 'recovery' or not, they would bring in steel from abroad rather than rely solely on the BSC.

In 1976, it was surely this politically delicate question to which the government should have been addressing itself, rather than stockpiling in the vague hope that matters would soon improve. It was not just a problem of the efficiency of the older works and overmanning in general, for even in the depressed year of 1976 there were steel shortages in some BSC product areas! Thus the Society of Motor Manufacturers and Traders complained that imports of cold reduced sheet steel had increased from 15 per cent to 45 per cent because BSC supplies were insufficient and of poor quality. Llanwern, the BSC's chief producer of cold sheet steel, however, was operating well below its 3.5 million tonnes per year capacity, despite a large decrease in the previously high level of strikes (we described in the previous chapter the twelve-month delay in commissioning a new blast furnace). It appeared that, in addition to industrial relations, there were real problems in actually getting the plant to work as it should. The ultimate embarrassment for the BSC occurred in September 1976, when the new Llanwern blast furnace overheated and had to be closed down for nineteen days while repairs were undertaken. The furnace was twice as large as any previously built in the UK, and clearly posed technical problems which the BSC had not anticipated.

The BSC's watershed 1977–78

Superficially, it appeared that in 1976–77 the BSC's fortunes were on the upgrade. After all charges the loss amounted to £95 million (compared with £255 million the previous year), with liquid steel production up 2.5 million tonnes to 19.7 million tonnes. On the other hand, capital investment had risen to £579 million, while funding from all sources had risen by the huge sum of £931 million. Despite the general support for the BSC's investment programme, there was obviously no chance that the Corporation could finance this investment from its own resources, while the issue of Public Dividend Capital (amounting to £490 million) was now no more than a government subsidy to the BSC.

Despite the continuing crisis, the ‘old world’ of economic and political optimism still had one brief moment to reassert itself. In March 1977 the government announced that it had approved BSC’s plans to double the steelmaking capacity of Port Talbot (from three million to six million tonnes) at a cost of £835 million. In return for this approval, the BSC agreed to retain steelmaking at the Shotton works. Given the parlous state of the BSC, it stretched credulity to the limit that the Corporation would in the future require 1.85 million tonnes per annum of steel from Shotton in addition to the six million tonnes from Port Talbot. It was a political deal between a Labour government and the BSC on the grand scale. A few years earlier the façade might have been maintained, but by 1977 the real world would not allow such indulgences.

The fragility of this new edifice was illustrated by the mundane way in which it was destroyed. Only two weeks after the official announcement, 520 Port Talbot electricians went on unofficial strike for pay rises which, if paid, would have breached the Social Contract agreement between the government and the trade unions. As we have seen, it is arguments over pay differentials and inter-union rivalry which have caused more problems for the BSC than basic union opposition to its strategy. A few days after the electricians began their strike the BSC was forced to close the whole of the Port Talbot works. Despite repeated appeals from the TUC the electricians refused to return to work (the strike finally ended in June). After the strike had continued for nine weeks the BSC announced that the £835 million Port Talbot investment had been deferred. The Corporation estimated that the strike had cost £60 million, but it was likely that general doubts about the project had already become deep-rooted.

The new BSC approach to investment was indicated on 25th May 1977, when Sir Charles Villiers gave evidence to the Commons Select Committee on Nationalised Industries. He told the Committee that he could not see the BSC requiring a capacity of more than 30 million tonnes by 1980, instead of the 35 million tonnes set out in the White Paper. It was calculated that this modified programme would cost £5 billion over the next five years. (HC SCNI 1977–78, HC 26, para.942). Even with this reduced plan, it was still not explained what effect a one billion pounds a year investment programme would have on the BSC, when it was already borrowing on a massive scale. During the summer of 1977 the BSC’s financial position deteriorated at a highly alarming rate. The speed of this decline, and the apparent reluctance of the BSC and government to make it public, was (as we will see later in the chapter) to become the subject of a heated argument between these two parties on the one side, and the Select Committee on Nationalised Industries on the other. In July, Villiers announced that the BSC was in danger of losing £225 million in 1977–78 (in the event the figure proved to £443 million). Papers eventually obtained by the SCNI from the BSC disclosed that on 14th July (five days before Villiers' statement) the Chairman had been informed that losses of about £450 million were now indicated for 1977–78. The management of the Corporation nevertheless decided at a meeting on 17th July that the public announcement would only refer to a loss equivalent to that made in 1975–76 (£255 million) (HC SCNI 1977–78, HC 238, para 8). Villiers later told the Select Committee that the reasons for concealing the £450 million figure were that: (a) this figure was unestablished; (b) if policies changed the position would improve; (c) if a huge loss was forecast and thus regarded as inevitable, then management would do little to fight against it becoming a reality. On this third point, it could of course also be argued that only by knowing the true seriousness of the situation was management likely to be able to take the necessary measures! (HC SCNI 1977–78, HC 238, para. 9).

If the BSC was less than frank in its public statements during this period, then the behaviour of the government was perhaps even more ambivalent. The SCNI revealed that on 18th July, Villiers met Secretary of State Varley and Minister of State Gerald Kaufman. Varley asked Villiers if the prospective loss for 1977–78 would be £250 million or £350 million (the figure given to the government in May 1977). Villiers replied that, by taking action, he hoped the losses could be contained at the lower figure, although he could not yet see a significant upturn in demand or prices.

Apparently both Ministers took this statement at face value (it seems strange that Villiers did not give them at least some hint of the new loss forecast at £450 million) and had no wish to probe further. On July 22nd, Kaufman introduced an order into the Commons to increase the limit of the BSC's borrowing powers from £3 billion to £4 billion (the limit allowed under the 1976 Iron and Steel (Amendment) Act). Perhaps understandably, given Villiers' statement to him a few days earlier, Kaufman told the Commons that a 1977–78 BSC loss of £255 millions might be expected. More dubiously, he also claimed that a higher level of demand was likely later in the year, when Villiers had told him that no such upturn was in sight (Hansard, 22:7:77, Cols. 2135–9).

Remarkably, on the day Kaufman was addressing the Commons, BSC Board Members were given the £450 million prospective loss figures by the Chief Executive. By this time, the figure must have become semi-public, yet at a meeting between the Corporation and Department of Industry officials on 27th July, it was still not mentioned, and it was only on 15th August that the Department of Industry was fully informed (HC SCNI 1977–78, HC 238, para. 17).

The Select Committee was understandably scathing in its comments on this saga, in particular on the apparent lack of communication between the BSC and the Department of Industry; the apparent lack of sensitivity at the depth of the crisis; and the reluctance of Ministers to press the BSC on its figures immediately prior to raising the Corporation's borrowing limit by one billion pounds. Clearly, the BSC had been disinclined to tell the government the whole truth. After nearly a decade of carefully constructing an ambitious and optimistic development programme, management was obviously reluctant to admit that the whole strategy had fallen about their ears, and was now unworkable. For its part, it might be said that the government was not eager to be told the truth. Like its predecessors, the government's credibility depended on presenting a picture where optimism for the future was a *sine qua non*. If an organisation such as the BSC was on the point of collapse, then what were the implications for the whole of British manufacturing industry? In the short term, and more specifically, the government was politically committed to maintaining steelmaking at the 'Beswick' plants for several years to come, but the BSC's position made that stance highly vulnerable.

The period July–September 1977 undoubtedly encompassed the major turning-point in the BSC's history. Although the official White Paper was not to be published for another six months, it was during this period that the BSC decided (with government approval) to make a new assault on the 'Beswick' plants.

At the time of the Beswick review in 1974–75 the BSC had required the steel from its older plants to meet the demand at that time. By 1977 the position had completely changed, and virtually all these plants were now considered surplus to requirements. As we saw in the Shelton case, the BSC was not able necessarily to prove that all these plants should be closed, but it was already so far down the road of developing the five major coastal sites that to abandon any of these was politically unthinkable.

From the point of view of the Labour government, it was clearly alarmed at the BSC's financial position. At the same time, to openly involve itself in the closure of the 'Beswick' plants was a highly embarrassing prospect, for nearly all of them lay in areas of strong Labour support. The solution arrived at was to adopt a 'softly, softly' approach, where each of the works would be picked off one at a time by means of attractive redundancy payments offered to its workers. In this way, the government could leave the messy business of implementing the closures in the hands of the BSC, while also claiming that the voluntary acceptance of redundancy terms by the workers concerned demonstrated the lack of coercion. The government's reluctance to admit publicly the nature of the new strategy is well illustrated by Eric Varley's assertion to the SCNI in December 1977 that there was no cause for panic: 'Just to turn around as though you were a chicken with its head chopped off, deciding that you would close plants and do it extremely quickly would cause calamitous problems for the steel industry and, I think, the nation' (HC SCNI 1977–78, HC 127, para. 344). It might be said that for the chicken, headless or not, its problems had come home to roost.

By the time the Secretary of State had made his remarks, the first closure under the new procedure had already been implemented. The Clyde Iron Works shut down in September 1977, with the loss of 430 jobs. It was followed in December by Hartlepool iron and steelmaking, with 1,300 made redundant. In May 1978 it was the turn of the East Moors works in Cardiff to be closed, and here 2,820 were made redundant. In June 1978, the axe finally fell on Shelton (in the event the arc furnace development never got beyond the drawing board stage) with 1,330 lost jobs. In August 1978 steelmaking at Ebbw Vale ceased, with 1,520 redundancies. This was followed by closure of the Scottish Glengarnock works in December 1978, with 680 made redundant, and the Bilston works in May 1979, with 1,300 job losses. All of these closures occurred under a Labour government which, under the Beswick review, had hoped to prolong the life of these threatened works.

That all these closures were secured with relatively muted opposition, and the minimum of political embarrassment to the goverment, can be attributed partly to the incremental approach (where works were closed almost before the workers had realised what was happening), and partly to the lure of generous redundancy payments which (at least initially) softened the blow of redundancy.

Initially, as so often in the past, union leaders at the national level were apparently content to go along with BSC closure plans (as we will see in the next chapter, the backlash was to come in the steel strike of 1980). In February 1978 the ISTC agreed to the early closure of loss-making plants in return for a ten per cent wage increase. Thus at the 'Beswick' plants union officials were prepared to act as agents of the BSC in persuading their members to accept the redundancy payments. The role of union officials inevitably created tension at times between them and local action committee members.

This animosity was well illustrated at Shelton, when in January 1978 the blast furnacemen were visited by their General Secretary, Hector Smith. The NUB members voted to examine ways of obtaining the best possible redundancy payments. We described in the previous chapter how the blast furnacemen did not have the same incentive to support the Action Committee as other workers at the plant, as they would not be required if an arc furnace was installed. When he arrived to speak to his members, Smith was jostled by a 600-strong crowd. Fights broke out and Smith had to be given police protection. This type of response was a far cry from the courteous manner in which BSC Chairmen had been received at the works a few years previously, but by this time it was clear that the era of gentlemanly consultation was over. By alienating the NUB from the other Shelton unions, the BSC was able to create a feeling within the works that the old unity had gone, and that closure had become inevitable. Four Shelton NUB officials did resign from their posts in protest at the decision by their members, but the die was cast. The new NUB Chairman of the Shelton lodge indicated the new approach:

> Promises have been made that there will be jobs for us in future. But the Action Committee are in no position to guarantee any jobs. They do not speak for the majority of production workers. In fact, the NUB men do not recognise the Action Committee.

Later in 1978, the SAC did receive support at the national level (chiefly from TGWU officials), but in the event the BSC announced a June closure of steelmaking. The government, now itself committed to closure of the 'Beswick' plants, refused to intervene, and the closure went ahead without any final agreement on redundancy terms. After this *fait accompli*, agreement on these redundancy payments came in September, at a cost of about £7 million (partly financed by government and EEC grants).

The BSC (with the tacit agreement of government) was prepared to offer relatively high redundancy payments as the means chosen to close the 'Beswick' plants, and later to reduce its workforce. This inevitably placed further strains on its finances, but the Labour (and later a Conservative) government was apparently quite prepared to accept this burden, and in effect to eventually write it off. The existence of BSC (Industry) was also of considerable political help to government, BSC and the unions, for it could be claimed that active measures were being taken to find new work for redundant steelworkers.

The road to viability

The government abandoned what was still, at least ostensibly, a non-interventionist approach during 1977 when it became clear that the BSC's financial position was deteriorating rapidly. Partly as a result of pressure from the Corporation itself, the government undertook a study of the medium and longer term position of the BSC, and after consultations produced a White Paper entitled *BSC The Road to Viability* (HM Government, cmnd. 7149), in March 1978. It was now for the first time accepted that the decline in demand for steel was of a long-term nature and that some markets might be lost forever. Some development projects were 'deferred' in favour of a 'step by step' approach which would retain the flexibility to adapt to unexpected changes in the situation (HM Government, 1978, para 12).

This 'rephasing' of capital projects brought with it substantial reductions in capital expenditure (HM Government, 1978, para 6). The Port Talbot rolling mill development, the proposed Teesside plate mill, and further tinplate developments were all deferred, although it was considered that the construction of continuous casting facilities at Port Talbot should go ahead.

The new White Paper marked an important change in the approach to making steel policy, i.e. in 1978 there was a change in what Dror terms 'metapolicymaking' – a policy on how policy should be made (Dror, 1973, p.8 and pp 172–3). The BSC and the government had become convinced by events, that:

> it would be unwise at this low point in the present recession to establish a 'blueprint' for future capacity, over a period of years, of the kind set out in the Ten Year Development Strategy of February 1973 (HM Government, 1978, para 12)

A specific target was, therefore, abandoned. It was also considered that the 'Beswick' plants had no future and plans were quickly put in hand to secure the closure of steelmaking at all these works. With the agreement of the government, the BSC adopted a more aggressive approach to the older plants, and the offer of large redundancy payments in all cases ensured the compliance of the workforce. In fact the White Paper revealed that:

> while the study was taking place the government had agreed that the Corporation should seek to negotiate terms for the early closure dates of high cost plants with the TUC Steel Committee and the local workforces concerned. (HM Government, 1978, para.9)

One other important element of the White Paper concerned finance. The interest bearing Debt/PDC combination was now abandoned in favour of subscription of capital under Section 18(1) of the 1975 Iron and Steel Act. The White Paper stated:

> The immediate prospects for profits are poor and no dividend on the new capital is expected before the reconstruction. All the new capital subscribed under Section 18 will, however, require to be remunerated by dividends after a reconstruction has taken place (HM Government, 1978, para 29).

In the event, when the reconstruction finally arrived via a Conservative government in 1981, £3 billion of this 'Section 18' capital was written off, followed by another billion in 1982. In 1988 the BSC's 'Section 18' capital stood at £3.98 billion, and as we will see in Chapter Nine, was used imaginatively to assist in the privatisation of British Steel. It could be said therefore that 'Section 18' capital, totalling almost £8 billion, represented the price government was prepared to pay in order to keep the BSC in business.

Thus in two important respects the policy outlined in the White Paper was indicative of cyclical change. Firstly, the cutback in the development programme suggested a return to the incremental development favoured in the 1960s. Secondly, the government's reluctance to become involved in discussions on closures was reminiscent of the policy outlined by the Conservative government in the White Paper of 1973. Needless to say, there had to be some justification for the shift away from radical and heroic policy-making, and the White Paper pointed out the tremendous consequences of the slump in world demand. It also noted that:

> a huge amount of modern highly productive new capacity was either completed or under way in Europe, Japan and the developing world before the Middle East War of 1973 and the steep rise in energy prices that followed (HM Government, 1978, para.2)

The irony, of course, was that this very expansion by Britain's competitors was seen as one of the justifications for approving this bold Development Strategy in 1973! The White Paper also laid some emphasis on the question of BSC's very bad productivity record. It pointed out that in the preceding three years BSC had produced 100 tonnes of liquid steel per man year as compared with 1976 figures of 150 tonnes in Germany and 120 tonnes in France. Moreover, it stressed that:

> the differences cannot be explained away only by differences in the pattern of plants or products. They are due more to management and manpower practices. Very substantial improvements are needed during the next five years if the Corporation is to reach present levels of manpower productivity in Germany (HM Government, 1978, para.31).

In highlighting the degree to which BSC's serious situation was self-inflicted, the government was in effect recognising the power of the trade unions. As we saw earlier, in general the steel unions can be characterised as moderate and having little impact on steel policy. There have been damaging local strikes in the industry and, in 1980, a very damaging national strike (see Chapter Six), but the strike record has been better in steel than in many other UK industries. The unions, at least at the local level, also had an impact on delaying closures, though once the market collapsed BSC was able to be much more assertive (with at least government compliance) in forcing closures through. In a particular sense it could be argued that the unions have exercised considerable power over the way in which the industry actually operates. The power of inertia should be recognised as an important determinant of the outcome of public policies. As Wildavsky suggests, the most difficult policies are those which attempt to change the behaviour of individuals. Quoting Etzioni, he notes that human beings are not very easy to change at all (Wildavsky, 1979, 49). If policies are based on the assumption that there will be radical shifts in behaviour, and that deeply entrenched working practices will change as a result, then they are probably doomed to failure. Policy-makers would do well to remember the rural expression 'There's nowt so strange as folks'!. All the government funding of BSC, and all the effort to produce a 'rational' policy could achieve little without fundamental changes in attitude by managers and workers within BSC.

In addition to the White Paper, a statement of what proved to be of vital importance was included by the BSC in its own supplementary document, *Prospects for Steel*. The paper included a hostage to fortune in the statement that:

> The Board of the Corporation has set management a financial objective of operating on a break-even basis by the end of the financial year 1979–80 (BSC, 1978, 17).

This ill-considered directive was to come back and haunt the BSC, for in 1979 (as we will see in Chapter Six) it allowed the new Conservative government to impose an unrealistic objective on the Corporation, and so played a significant part in causing the national steel strike of 1980.

The role of the Select Committee on Nationalised Industries

That steel policy was seriously adrift was recognised by the Select Committee on Nationalised Industries, in several reports published in the Session 1977–78 (House of Commons SCNI 1977–78, HC 26, HC127 and HC238). The Select Committee on Nationalised Industries (SCNI) had been monitoring BSC's developing financial crisis since early 1976, and it saw fit to express strong criticism of both government and BSC. For example, in January 1978 it suggested that BSC should concentrate on raising productivity at existing works rather than incurring higher costs by introducing new facilities; BSC should improve its monitoring and control procedure for capital projects; both BSC and the Department of Industry should improve financial forecasting; BSC's captial spending should be approved only for specific projects promising adequate returns; negotiation on manning levels should begin as soon as possible.

Though hard hitting and very much to the point, the impact of the Parliamentary Committee was relatively small. The SCNI is more to be noted for its dramatic clash with the Chairman of BSC, Sir Charles Villiers, than for its actual impact on policy. Thus in November 1977 Sir Charles was involved in what is now seen as a classic dispute over the rights of Parliamentary Committees to have access to information. In fact both Mr Varley, the Minister, and Sir Charles crossed swords with the SCNI on the same point of principle. Mr Varley refused to provide the Committee with all papers that had passed between him and the Chairman of BSC during the previous year. Sir Charles refused 'short of being sent to the Tower' to provide details of plans being devised to reshape the BSC and to steer it through its existing crisis. It is important to note that the issues involved could hardly be described as relatively minor ones best left to confidential negotiations between interested parties. For example, BSC was then losing £1.5 million per day, while its spending on fixed assets between 1977 and 1981 was estimated to be £4,000 million at out-turn prices. The Committee appeared to recognise its own relative impotence in its pessimistic comment that it feared that:

> the remedies proposed after due consultation between the Government, the BSC and the Trade Unions, will be an unsatisfactory compromise involving each of the three courses of action (manpower reductions, reduced capital spending, and capital reconstruction) and that as a result of the elements of financial reconstruction in the compromise, the true nature of the crisis in the Corporation will be concealed only to reappear in yet more virulent form at a later date (House of Commons SCNI 1977–78, HC 127, XIV)

In fact the SCNI had identified a common weakness in the relations between governments and Nationalised Industries in the UK – namely that a 'capital reconstruction' is used to paper over the serious financial and strategic cracks in the industries concerned. There seems to be an unwritten rule – if the debt gets too high, write some of it off at the taxpayer's expense!

In practice, the SCNI mobilised the rest of Parliament in the attempt to get access to the necessary documents, much to the consternation of ministers and civil servants alike! It was claimed that if all the correspondence was published: 'it would be explosive. All the shoddy compromises between the Government and BSC would come tumbling out' (*The Observer* 8:1:78).

This statement was hotly disputed by Departmental and BSC officials. In the event Sir Charles did in fact comply with the SCNI's request, but only after a representative of the Serjeant at Arms (an official of the House of Commons) was sent to BSC headquarters, amid much publicity. The papers gave the quarterly revisions of BSC forecasts submitted to the Department of

Industry between 1st January, 1976 and 30th September, 1977. Sir Charles claimed that the Committee had not asked for these papers previously, but members of the Committee countered that the papers had been requested on numerous occasions. The efforts by the SCNI merely served to illustrate the fact that it managed to learn of the extent of the steel crisis some time after it had been recognised by BSC and the government. It seems that both the government and BSC knew of the extent of the crisis in July 1977, some six months before it was revealed to the Select Committee of the House of Commons, though this view is contested by Sir Charles Villiers who commented on TV that it was 'scandalous' for MPs to claim that BSC was misleading them (*The Guardian* 12:11:78). If we are to look for an impact on policy from the SCNI it might be that, in reinforcing the shift in policy that had already been decided between government and BSC (namely a reduction in manning, closure of old plants, and a revision of the investment programme) they contributed to a sense of realism in the industry. Thus the Parliamentary influence was at best to add to the growing climate of opinion, rather than to change that climate, or indeed specific policies.

A new course : 1978–79

In the final year of the 1974–79 Labour government, the BSC began to implement the provisions of the 1978 White Paper. As we have seen, the course of the policy had been set several months before the White Paper's publication, but it was now openly acknowledged that the old development era had gone. Both the BSC and government had to put a brave face on the new situation, and hope that at least the Corporation's losses and trading position would remain stable. In the event, the losses were contained after a fashion, but the trading position became worse.

In 1977–78 BSC losses after tax amounted to £443 million, but by the following year they had fallen to £309 million. The most striking change during 1978–79 was that capital expenditure fell from £476 million the previous year to £318 million. This sharp decrease reflected the fact that the Port Talbot and Redcar developments had finally been shelved. Work already begun, such as at Redcar and Ravenscraig, had to be finished, as these developments had commenced when the mood was more optimistic. BSC capital development was never again to reach the heights it had attained in the mid-1970's, and some have been critical of the decisions made (e.g. Bryer *et al.* argue that the Port Talbot development would have improved BSC product quality). There could well be some validity in this argument, but the problem for both the BSC and the government was that if Port Talbot was developed further then, with BSC markets crumbling, other major steelworks, such as Llanwern and Ravenscraig, would probably have had to close. By restricting Port Talbot capacity, it was more likely that all the big plants would remain open.

The BSC's financial position was also improved by virtually all its loan capital being made available under Section 18(1) of the 1975 Iron and Steel Act. As, not surprisingly, it was envisaged that no dividend was likely to be paid on this capital, then it basically amounted to an interest free loan to the BSC, to be paid back at some indefinite date in the future (in the event, the Conservative government was later to write off much of this capital).

The BSC also improved its position during 1978–79 by reducing the size of the workforce from 196,900 to 186,000. This loss of just over 10,000 in the year does not appear particularly dramatic in relation to reductions in later years, but it began a radical change in BSC employment levels. Earlier in this chapter, we listed the plants which were closed during 1978–79. Politically, these closures were somewhat embarrassing for a Labour government, but it could offset complaints about its policy by pointing to the fact that at all the works, redundancy terms had been accepted voluntarily. On the debit side to the savings made, these redundancy payments were to be a heavy drain on BSC finances for several years to come (e.g. at Shelton redundancy payments to 1,250 workers amounted to £7 million, while at some works individual payments went as high as £20,000). Although obviously a 'once for all' cost, the number of closures to be made meant that they were spread over a number of years, and so could not be dispensed with quickly. Nevertheless, the readiness with which the redundancy payments were offered suggested that both the government and BSC were convinced that the benefits to be gained by the closures outweighed the costs.

As we have seen, the national union leaders were generally prepared to accept the closures, so long as the local workforce was in agreement. At times, this acquiescence inevitably placed a strain on the union leaders' relationship with the BSC. The principal example of this occurred in July 1978, when it was announced that the Bilston plant in the West Midlands (a plant not originally on the 'Beswick' list) was due for closure in August. The leadership of the ISTC immediately reacted and threatened a national strike. At this point the BSC withdrew the closure notice, and the strike threat was lifted. It was purely a tactical withdrawal on the BSC's part, however, for closure came a few months later in May 1979.

The national union officials were also restricted in their actions by the negotiations on the Steel Contract. In 1978 the BSC–TUC Steel Committee joint working party produced its proposals. These included a National Steel Trades Union Joint Advisory Conference. This NSJC would be a non-voting body, but in its turn would advise a National Joint Executive Committee chaired alternately by management and unions. The NJEC would comprise the TUC Steel Committee and an equal number of BSC executives. The decisions of the NJEC would be binding, with similar machinery at Divisional and works levels (although BSC management was anxious to stress that they would still be in charge of day to day decisions). With regard to the BSC Board itself, concrete action was taken during 1978 with the appointment of six worker directors and two civil servants.

The Steel Contract was a scheme particularly close to the heart of Sir Charles Villiers (other members of BSC management, particularly Bob Scholey, were less happy about the prospect of greater consultation with the unions), but it was never to become a reality. Ironically, it was the type of scheme which might have won wider acceptance in the early 1970s, when all parties could have agreed on the expansionist policy. By the late 1970s, however, it was never likely that the unions would wish to be openly associated with the closures programme and the general reduction in the BSC workforce (there was also some inter-union rivalry in the form of the ISTC wishing to have a greater share of the seats on the various joint committees).

Politically the atmosphere was also about to change. The Labour government favoured greater employee participiption and encouraged negotiations on the Steel Contract. In contrast, the Conservative government elected in 1979 was generally inclined to give greater authority and power to management. Villiers' successor as BSC Chairman, Ian MacGregor, was the living embodiment of the new values, with the result that the worker directors and the civil servants gradually disappeared from the Board, and all talk of a Steel Contract was swiftly buried.

Union opposition to the reduction in the size of the steel industry was to culminate in the national strike of 1980, but in 1978–79 the closures continued with little opposition. The government and the BSC could take heart from these developments, but the actual performance of the Corporation left no room for encouragement. Liquid steel produced by the BSC in 1978–79 amounted to 17.3 million tonnes, compared with 17.4 million tonnes in 1977–78. BSC's share of its home market in 1978–79 was still only 54 per cent, and showed no sign of returning to the figure of 70 per cent achieved in the early 1970s. Export levels were also stagnant at 2.9 million tonnes. This disappointing performance led to net borrowings of £715 million, which included £850 million of 'Section 18' capital. By the end of 1978–79 the BSC's outstanding borrowings added up to £3,710 million. In view of this continuing deterioration in the Corporation's financial position, the Iron and Steel (Amendment) Act 1978 increased the statutory borrowing limit from £4,000 million to £4,750 million, with provision for a further increase up to £5,500 million.

Given the huge size of the BSC's debt by 1979, it must be doubted if either the Corporation's management or the government had any realistic hope that the money was ever likely to be paid back. Thus when the Labour government left office in May 1979 there was still great uncertainty concerning the BSC's ability to even survive, let alone prosper. Consequently, the likely cost of change, both political and economic, could only be the subject of inspired guesswork. In its turn, the Conservative government which inherited these problems was soon to discover that returning the BSC to viability was to be a far longer job than initially it had supposed.

6 The 1980 Steel Strike

Commercial pressures and further decline

Two cornerstones of the new Conservative government's philosophy were that the Nationalised Industries should be privatised to the maximum possible extent; and that those state industries which remained in public ownership should rapidly cease to be a drain on the public purse. We saw in Chapter Three how the Conservative government of 1970–74 had come to accept a state owned BSC as a *fait accompli*, and had gone on to give its approval for the development of the massive capital development programme.

By 1979 much had changed, both for the BSC and the Conservative Party. The BSC had by this time basically abandoned its plans for major capital development projects, and indeed was closing down steelworks at an unprecedented rate. For the new government, under the leadership of Margaret Thatcher, commercial viability, rather than industrial regeneration and expansion for its own sake, had become the chief policy aim. Thus the Conservatives (like their Labour predecessors) had no expectation that the BSC would reverse its decline in terms of steel tonnage produced. Instead, the emphasis was to be on reducing costs and improving productivity in order to make a profit.

The government hoped that, at least in the long term, a profitable BSC would be a candidate for privatisation. In the event, these hopes were not to be fulfilled in the early 1980s. By 1986 the BSC had made its first profit since the far-off days of 1974–75 (the *Road to Viability* had indeed been a long and hard one), but by the mid-1980s only comparatively small sections of the Corporation had been returned to the private sector. Indeed, in the early 1980s the government was propping up the BSC on a scale unprecedented in British industry, and this after a damaging national steel strike lasting three months.

In this chapter we will look at the events of the strike in detail, not only because it was a major event in BSC (and UK industrial) history, but also for the reason that the strike as a political event in itself demonstrates how the chief protagonists i.e. government, BSC management and unions, can behave amongst themselves and toward each other in moments of crisis. The strike itself was nominally about pay, but in reality it could be said to have operated on two different levels.

Firstly, there were the elements of symbolism and myth for both the unions and the government. For the unions, the symbolism was represented in the strike seen as a reaction against government economic and industrial policy, and also in the fight to save steel industry jobs. For the government, the steel strike represented an example to other public sector unions that strikes would not force Ministers into swift concessions, and that the Nationalised Industries in particular must be made to face commercial reality. Although this symbolism is of considerable political significance (strengthened as it was by the coal strike of 1984–85), the reality is somewhat less dramatic, but possibly more revealing about UK industrial policy.

The strike was caused chiefly by the BSC being panicked into attempting to impose on the unions a virtual pay freeze at the national level (rises were to depend on the implementation of locally negotiated pay agreements), and also publicly announcing their intention to seek 52,000 job losses in eight months. This panic was caused partly by the deteriorating finances of the Corporation, but chiefly by the government's refusal to fund BSC operating losses after March 1980. When the strike was over, the government was forced into a complete U-turn on this policy, and pumped several billion pounds into the BSC in order that it should survive. As for the strike itself, after a long series of tortuous negotiations, and heated debates on the legality of a private sector steel strike and the practice of secondary picketing, it was finally settled by the comparatively simple expedient of appointing an independent arbitrator. After the heat had evaporated, the BSC strategy on job losses and a reduction in output remained intact, but the Corporation had lost a large slice of its market, and its finances were in a still more parlous state.

The politically confused and messy reality of the strike in itself makes a definitive evaluation of its events a difficult matter, but an examination is worthwhile not only because of the wide spectrum of British political interests it touched, but also for what it reveals about the difference between political values and reality.

Prelude to the strike

The BSC was quickly to experience the impact of the new Conservative administration when the Secretary of State for Industry, Sir Keith Joseph, announced that the BSC's operating losses would not be financed by the government after March 1980. As we have seen, since publication of the 1978 White Paper, the BSC's objective had been to break even by March 1980. In this sense the new directive from Sir Keith did not cause a fundamental change in the Corporation's long-term strategy. Nevertheless, the initial political zeal of the Conservative government to make the state owned industries profitable was an obvious incentive for the Corporation to push through its rationalisation plan as quickly as possible. In reality, there was never the slightest chance that the BSC would meet its break even target, yet in June 1979 Sir Charles Villiers maintained the mood of public optimism which consistently characterised his Chairmanship of the Corporation, but was unfortunately in direct contrast to the actual course of events. After a meeting with Sir Keith Joseph, the Chairman revealed that he had told the Secretary of State that the BSC could meet the target date for breaking even and added: 'We are going to be operating at a rate of break-even from the time I mentioned (March 1980). I see no reason at the moment why we should not do it' (*Financial Times*, 22:6:79).

In strange contrast to this optimistic outlook, on the same day Sir Charles told the annual conference of the ISTC that the BSC was still making 'huge losses'. He also warned the steelworkers that the rising level of inflation, coupled with an upward movement in the value of the pound, were looming shadows over the BSC. If these factors continued, then the Corporation faced a flat market for its products both at home and abroad (*Financial Times*, 22:6:79). This was a dire warning from the Chairman of things to come, and was surely more representative of his true feelings than the brave words offered to Sir Keith Joseph.

During this period Sir Charles also suffered a personal defeat within the BSC itself. He had proposed an organisational restructuring which would have entailed the creation of three large regional divisions, together with fifteen managing directors within these regions, who would each have been made responsible for the fortunes of his own profit centre (although responsibility for purchasing and sales would have remained with head office). Some members of the Board (notably Bob Scholey) opposed this plan, apparently on the grounds that the profit centres would not be truly decentralised units (in the event by 1982 the BSC had divided its structure into profit centres, each with responsibility for sales). It was also likely that the government would not have looked favourably on Sir Charles' scheme. Ministers were keen that, at least in the long term, the BSC could be privatised. In order to achieve this aim, it was important that individual units could operate autonomously, so that they might be disposed of in separate lots. In contrast, Sir Charles' scheme retained a substantial element of 'old style' central control. Thus the relatively independent Profit Centres which eventually emerged were undoubtedly more to the government's liking, although the privatisation plans were to be consistently frustrated.

After this setback for his plans, Villiers placed organisational change in the background, and the energies of BSC management were once more directed to a new series of closures. In July 1979 the government re-emphasised that it would not finance BSC operating losses after March 1980, although it was prepared to continue discussions on a capital reconstruction for the Corporation. In effect this meant writing off a slice of the BSC's debt. The significance of this debt is illustrated by the figures for 1978–79, for in this year interest payable accounted for £207.7 million out of a final loss, before extraordinary items, of £309.4 million. In contrast, the trading loss after depreciation amounted to £129.3 million. By the end of 1978–79 the BSC owed £767.5 million in long-term indebtedness to the Secretary of State (although it should be noted that foreign loans were at the slightly higher level of £779 million).

From the government's point of view, the size of the BSC's debt was a huge obstacle obstructing its privatisation plans for the Corporation, but more immediately the size of the interest payable on the debt created a vicious circle, where the Secretary of State was being forced to grant the BSC more money in order that the sums due to him could be paid back! Thus the interest on this long-term indebtedness was largely financed as part of the non-interest bearing 1975 Iron and Steel Act Section 18 (1) capital advanced (in effect given) to the BSC. In 1978–79 this capital totalled £850 million. In 1979–80, under the new, market-orientated 'make the Nationalised Industries stand on their own feet' Conservative government, this capital amounted to £905 million, and in 1980–81 totalled £1,233 million, surely a classic example of political rhetoric not matching reality! As we will see in Chapter Eight the capital reconstruction itself did not arrive until 1981, by which time the BSC had almost reached its statutory borrowing limit of £6,000 million (from April 1979 the BSC's assets were written down by £1,141 million in order to match the lower capacity from 1980 onwards).

In the summer of 1979 the BSC was still losing about £1 million each working day, and despite Sir Charles Villiers' superficial optimism there was no concrete evidence to suggest that an improvement in trading conditions was in the offing. Consequently, when he met the TUC Steel Committee on 12th July, Villiers told them that the BSC intended to cease iron and steelmaking at Shotton and Corby, with a possible total loss of more than 12,000 jobs. The BSC estimated that these closures would save £80 million a year. Inevitably, the action committees at both Shotton and Corby strongly contested the BSC's figures (for a detailed outline of Corby's case see Bryer *et al.*, 199–237). As in the earlier cases of Shelton and the other 'Beswick' plants, however, the crucial point was not the actual or potential viability of Shotton and Corby as individual plants, but the BSC's long-term commitment to the five coastal works. The John Summers company had implemented a considerable capital development programme at Shotton during the 1950s, but the BSC intended that in future the plant would receive most of its steel from Ravenscraig (the plant was also to receive some work envisaged for Llanwern). Much of the development work envisaged for Ravenscraig as part of the 1973 strategy was already well under way before the 1978 policy change, and so by 1980 £300 million had been spent in giving the major Scottish plant a planned capacity of three million tonnes per annum. There was, of course, no chance that this capacity would be required in the market conditions which prevailed in 1979–80, but by taking over Shotton's iron and steel requirements it could at least be hoped that Ravenscraig would achieve a higher rate of utilisation (in 1979–80 Ravenscraig produced 1.2 million tonnes of liquid steel, against the modest target of 1.75 million tonnes a year, while the Scottish Division as a whole made a loss of £96 million after interest). Chapter Five described how in 1977 Shotton's iron and steel making had been given an assured future at least until 1983, but two years later this largely political decision was seen to count for nothing.

The decision to close iron and steelmaking at the Corby tube works was something of a landmark, for unlike Shotton and the 'Beswick' plants its demise had not been envisaged in the 1973 White Paper. There was also considerable poignancy in the decision as a result of Corby's position in the British steel industry. It had been the one major new plant of the inter-war years, and based as it was on the use of home ores, together with the creation of a new town (in 1979 11,000 out of Corby's population of 57,000 were employed by the steel plant), it had become something of a showpiece. Thus the name of Corby had become synonymous with the steel industry, and its loss as a steelmaking plant was an emotive political issue. Like Shotton, the people of Corby were overwhelmingly dependent on the BSC as their main employer, and redundancies on the scale envisaged by the Corporation were seen as social disasters in both areas.

The problem for Corby was that, once the BSC had made the decision to rely on imported ore, its inland situation made it vulnerable. The BSC planned to bring in Corby's steel from Teesside, where the phase IIB development project had just been completed at a cost of over £260 million, but in 1979–80 the Teesside Division made a loss of £137 million after interest. The loss figures for both the Scottish and Teesside plants (although these figures include the period of the national strike) illustrate that by 1979 the BSC was making heavy losses in nearly all its areas, and Shotton and Corby represented just the start of a new rationalisation plan.

During the Autumn of 1979 the ISTC in particular became more concerned about the nature of the BSC's plans. By this time BSC management was hinting darkly that closure of a major works could be on the horizon. It was able to emphasise its case as the result of yet another inter-union dispute, which prevented a new capital development coming into operation. The £100 million Hunterston ore terminal on the Scottish west coast (designed to serve Ravenscraig) was due to be brought into use in May 1979, but a manning dispute between the TGWU and ISTC prevented this happy event taking place. As the argument dragged on, the BSC threatened that its continuance could results in the permanent closure of Ravenscraig.

After six months, the dispute was finally settled when the two unions agreed on an allocation of work and the government decided to make Hunterston a dockworkers 'scheme' port i.e. its employees were given statutory job security (this is another good example of a government apparently committed to facilitating a freer labour market being prepared to change course in a case where political circumstances dictated otherwise). Despite the opening of Hunterston in November 1979, and the lifting of any immediate threat to Ravenscraig, BSC's financial position continued to worsen. During November, it was announced that the Cleveland works on Teesside would close, with a loss of 1,200 jobs. By this time it must have been completely obvious to both the BSC and government that the March 1980 break-even target was hopelessly out of reach.

For several months Industry Department and Treasury officials had been keeping an almost daily check on BSC accounts in order to ensure that money earmarked for capital development was not being syphoned off to subsidise losses. This understandable concern on the part of the government was nevertheless a King Canute-like exercise, for Sir Keith Joseph was now in the politically embarrassing position of either sanctioning a complete policy reversal, or else presiding over the collapse of the British steel industry.

At the end of November the BSC announced a half-year loss of £145.6 million, and at the same time Sir Charles Villiers confessed that there was no chance of the Corporation achieving its March 1980 break-even target (we will shortly see that the government appeared not to have received this message). Two days after revealing this loss the BSC's Chief Executive, Bob Scholey, met representatives of the ISTC and told them that the Corporation wished to cut its workforce by at least 52,000 over the next eight months, while reducing manned capacity from 21.6 million tonnes per annum to 15 million tonnes per annum. The figure of 52,000 included a reduction of 20,000 already announced, i.e. the closures at Shotton, Corby, Cleveland and other smaller closures in Lanarkshire and on Teesside.

The political significance of Scholey's action cannot be over-estimated, for it was the catalyst which led to the strike a month later. For the previous two years the BSC had been highly successful in implementing closures by adopting a piecemeal approach which was deliberately vague on future manning and production targets. Now the policy had changed, for Scholey's attitude was reminiscent of Sir Monty Finniston's dramatic attempt at accelerating the closures programme in 1975, which led to the confrontation with Tony Benn. On that occasion, the union leaders themselves had appeared less inclined than the Secretary of State to provoke a conflict, but this time the situation was different. The Conservative government was quite content to support the new round of closures, but for the ISTC leadership it acted as a rallying call to action. It appears that Scholey's radical approach reflected the desperation of a BSC management which knew that the hoped for viability, even in the medium-term, had disappeared completely, and yet faced a government which apparently refused to let it off the hook. Tactically, it was a serious mistake for the BSC to turn away from its 'softly softly' methods, but the pressure of events had driven it to adopt a confrontational approach.

Four days after Scholey's meeting with the ISTC, the union received another severe shock when the BSC offered only a 2 per cent national pay increase for the following year (in reality this 2 per cent was itself only a consolidation of previous agreements). The Corporation proposed that further increases should depend on locally negotiated productivity deals, up to a maximum of 10 per cent. On behalf of its 90,000 members the ISTC negotiators completely rejected this offer, and recommended that the union's National Executive committee should call a total strike.

The BSC offer represented another radical policy change, for it now intended to virtually by-pass the national union leadership and negotiate directly at plant level. In this way, it was hoped that considerable improvements in productivity could be achieved at all the plants, including the 'big five'. The national ISTC leaders greatly resented (and no doubt feared) their exclusion from pay bargaining, and this was one of the explanations for the strike call.

At the same time, the strike threat could have fallen flat if, at the local level, the traditionally moderate steelworkers had not wished to answer their leaders' call. On this occasion, however, there were two reasons in particular which made it likely that the strike decision would fall on receptive ears. Firstly, even with local productivity bonuses, steelworkers' pay would have fallen well below the annual rate of inflation, then running at over 17 per cent. The steelmen resented the fact that other Nationalised Industry workers, in particular the miners (who had accepted a 20 per cent pay rise), were not being asked to take a cut in real wages. In addition, the government had made an Election commitment to implement public sector pay awards recommended by the Clegg Committee, and the majority of these awards were above the then annual inflation rate. Thus, despite the BSC's parlous financial state, the ISTC felt that it was being picked on as a pathfinder in government attempts to restrict public sector pay.

The second major factor encouraging strike action was that the BSC's new demands for manpower and steel capacity reductions were not just applicable to the older works scheduled for closures. It was now clear that the 'big five' would have to accept their share of the misery. Indeed, there were widespread unofficial suggestions that a major plant (Llanwern was the usual name mentioned) might be due for closure. We have seen that in the past there was no unity at plant level on action to be taken against the closures programme, but now it was evident that almost all BSC jobs were potentially at risk. In this atmosphere, it was not difficult for the ISTC to secure a united front (in contrast to the National Union of Mineworkers in their strike of 1984–85).

In addition to the BSC and ISTC, the other crucial actor in the period leading up to the strike was, of course, the government. The problem in examining the government's behaviour during this period was its own claim that the new turn of events within the BSC was none of its business! In a debate on the steel industry in the House of Commons on 10th December (by which time the ISTC had officially decided to call a strike for 2nd January), Sir Keith Joseph made the standard government statement on these occasions that closures and pay offers were both matters for BSC management to decide, and were not the responsibility of government. More significantly, he refused to give any hints that the government would raise its 1980–81 EFL for the BSC above the £450 million already offered (in the event the EFL for this year was eventually set at £1,121 million!) and apparently still expected the BSC to eliminate trading losses during the following year (despite Villiers' public statement that there was now no chance of the Corporation achieving this aim) (Hansard 10:12:79, Col. 899). What conclusions can be drawn from the Secretary of State's apparently quixotic behaviour? Evidence which came to light during the strike suggests that the government's role during this period was not entirely the passive one it claimed for itself. Indeed, it would be highly surprising if it did not have some part to play in the affairs of a publicly owned industry at a time of extreme crisis.

Documents leaked to the ITV 'World in Action' programme during February 1980 suggested that the BSC had planned to offer pay rises of between 10.5 and 13 per cent until the tight borrowing limits had been announced (*The Guardian*, 5:2:80). In this case, the government did not have to issue an official instruction to the BSC on job and output reductions, for the borrowing limit itself passed the message over nicely! With such limited room for manoeuvre, the BSC had little choice but to restrict its pay offer. Throughout the strike there were many accusations from the Labour Party and the steel unions that the government had deliberately engineered a confrontation with the steel unions. Certainly the Conservative MP, Nicholas Ridley, had produced an internal party document in 1978 which suggested that the steelmen were one group of workers who would find it difficult to withstand a tough government policy, and during the early 1980s the government's attitude with regard to the miners was more circumspect than with the steelmen.

In addition, towards the end of the strike, Sir Charles Villiers told a private meeting that he had warned the government as early as October 1979 that a steel strike was likely, and that steel stocks had been built up accordingly. Even discounting these facts, given the government's philosophy on making the Nationalised Industries pay their way, it is understandable that it should have wished to push BSC management into taking radical action. The alacrity with which most steelworkers were prepared to accept redundancy pay, both before and after the strike, suggested that there would be no insurmountable obstacles in implementing job losses. Finally, although the three month strike had some effects on industry, imports and the level of steel stocks ensured that the damage was kept to a minimum.

Regardless of any Machiavellian strategy it might have had (and it should be noted that the new BSC policy course was largely set under a Labour government in 1977), it was entirely consistent with Mrs. Thatcher's political philosophy that the BSC should be made to 'mend its ways'.

The chief problem for any government wishing to precipitate a strike in the steel industry was that the policy bore great political and financial costs. Throughout the strike there was a general consensus of opinion amongst many Conservative MPs (and several Ministers) that the BSC's handling of the dispute had been unsatisfactory. The two per cent offer was seen as a particularly unfortunate tactic, while towards the end of the dispute the policy capital to be made out of its continuance became decidedly thin. In the end, the government itself appeared relieved to find a means of achieving a rapid conclusion to the strike. Once the strike was over the government was left to pick up the bill for its cost.

Probably the most damaging criticism that can be made of the government's handling of affairs during the strike was that, just before his departure from the BSC, Sir Charles Villiers revealed that he had written to Sir Keith Joseph both before and during the strike. In these letters he warned the Secretary of State that the BSC's EFL of £450 million for 1980–81 would be totally inadequate, and that the Corporation's financial problems were almost beyond control. In October 1979 he had warned the government that it would be advisable to plan for further adversities, and then in February 1980 (when the strike was at its height) he told Sir Keith that the over-run on the cash limits for 1980–81 was likely to be between £300 million and £500 million. Yet the government continued to maintain in public that its £450 million EFL for 1980–81 would be sufficient. Only in June 1980 did Sir Keith Joseph finally reveal the true position to a shocked House of Commons, and concede that the government had no choice but to offer the BSC a blank cheque to cover its losses. Thus the steel strike in itself could not be judged as a vindication of the government's approach to dealing with the BSC.

The breakdown of negotiations

Two days after Sir Keith's statement, the BSC elaborated on its planned cutbacks. Two definite plans for works closures were announced: Consett in Durham with the loss of 4,000 jobs, and the Hallside plant in Scotland with the loss of 600 jobs. Elsewhere, there were a number of options for the two major Welsh plants at Llanwern and Port Talbot, ranging from the complete closure of one works to a joint reduced capacity of 2.75 million tonnes per annum, compared to the existing figure of five million tonnes (job losses 11,000). Scunthorpe capacity would be reduced from 4 million to 3 million tonnes per annum (job losses 2,800). In addition, Redcar steel production would be set at 3.5 million tonnes per annum, Ravenscraig at 2 million tonnes per annum, and Sheffield at 3.5 million tonnes per annum. It was proposed that these surviving plants should lose 12,000 jobs, with another 2,500 jobs disappearing at a variety of rolling mills (it should be noted that the steel strike totally failed to prevent these plans being carried through).

Perhaps the most vivid illustration of the BSC's demoralised state at this time was its assumption that the new manned capacity of 15 million tonnes per annum (3 million tonnes capacity per annum was to remain unmanned in the hope of future recovery) would include a static, or probably falling, level of exports. Chiefly because of the rising value of the pound, it was considered that the majority of export orders offered no hope of profit. In other words, despite the BSC losing nearly 20 per cent of its home market share to imports during the 1970s, and despite the huge amounts of money ploughed into capital investments, the Corporation had given up hope of increasing its penetration of overseas markets in any way! For a brief period, there was even a suggestion that the BSC might withdraw from as much as two-thirds of its export markets, but even the Corporation had its limits for pessimism at this time, and the plan never came to fruition.

During the remainder of December several attempts were made to avert the strike, but to no avail. A meeting between the ISTC and Sir Keith Joseph was predictably unproductive, although on 21st December the BSC did offer the ISTC and the National Union of Blast-Furnacemen (NUB) an 'extra' three per cent, but this was conditional on the unions agreeing to suspend the guaranteed working week. This system guaranteed BSC employees a minimum of 80 per cent of basic pay for a five day week, even when a shorter period had been worked. In addition to this offer, the Corporation indicated that it would like to begin negotiations on a 39-hour week or a shorter working year, together with the local productivity deals (apparently worth up to 10 per cent) suggested earlier. The ISTC and the NUB both rejected this offer out of hand, on the grounds that no basic 'new' money was available.

At this point, the General Secretary of the TUC, Len Murray, intervened in an attempt at mediation. As a result of his efforts, another ISTC-NUB meeting was arranged with the BSC for 28th December. This time the Corporation came up with yet another variant of its pay/productivity package. On top of the original two per cent, an extra four per cent was now available (i.e. an increase of one per cent on the previous offer), but the conditions were considerably changed. The BSC was now prepared to remove negotiations on the guaranteed working week from national to local level, but also wanted a working party to discuss a deal on flexible working patterns involving changes in practices between craftsmen and manual workers.

The third element was the most contentious, in that it asked the unions to accept the loss of 12,000 jobs as part of the 52,000 already suggested. As with the initial Scholey announcement of a 52,000 reduction in the workforce, it appeared strange that the BSC management should wish to force the unions into publicly accepting specific numbers of job losses, when in the past these reductions had been achieved quickly by means of local negotiation. On the other points, the national union leaders were obviously reluctant to abandon their negotiating power on such a fundamental issue as the guaranteed working week, while mutual suspicion between manual workers (represented chiefly by the ISTC and NUB) and the craftsmen was, as we have seen previously, a way of life in the steel industry. Thus the ISTC and NUB once again rejected the offer immediately.

As a result of the impossible targets placed upon it by the government, the BSC could not offer 'new' money with no strings attached, and so was restricted to putting together various packages linking pay to productivity. For their part, the unions restricted their demands to pay rather than jobs (although the BSC announcement of 52,000 prospective job losses was a prime underlying cause of the strike). Nevertheless the two issues were obviously linked closely, i.e. the Corporation was saying that more pay could only come with fewer jobs. In this sense, the higher the basic wage settlement achieved by the unions the easier the BSC's job in justifying large-scale job losses! Although this dilemma for the unions was barely remarked upon publicly during the strike, the emphasis on pay suggests that there was little to stop the Corporation achieving voluntary redundancies at the local level, and so the unions' best plan was to work in search of maximum benefits for those left in the workforce, while also seeking to maintain a high degree of negotiating power at the centre. With regard to the strike itself, the unions' aim was not only to cease all BSC steel production (an objective quickly secured), but also to halt the steel using industries. In order to achieve this more difficult latter task, several factors were of significance:

(a) the support of other unions, in particular those representing transport workers (the TGWU, the National Union of Railwaymen and the Associated Society of Locomotive Engineers and Firemen), in restricting the movement of steel. Although official support was obtained from all these unions, at the local level many road transport drivers continued to move steel;

(b) production in the private sector. As we will see, this issue was the cause of a great deal of friction. The steel unions had no official quarrel with the private sector, and initially Bill Sirs was opposed to any action against them. Many of his members took a different view, however, and after extensive, and often bitter, picketing of privately owned steelworks, the workers in these plants were officially called out in order to support their fellow trade unionists in the state sector of the industry. This action was unsuccessfully contested in the courts by the private sector companies, but it did provoke a large-scale political debate on the legality of secondary picketing, and the government's plans to severely restrict this practice;

(c) the import of steel. Despite some official support from overseas unions, steel was imported during the strike through a number of smaller ports. Although steelworkers were successful in blacking imports at a number of larger ports, in reality they could not be everywhere at once. In the long term, the strike had the effect of steel users becoming more accustomed to using steel from overseas than had previously been the case, and this 'second sourcing' continued to frustrate the BSC in its attempts to gain a larger share of the home market;

(d) the amount of steel already in stock. Before the strike began, Bill Sirs declared: 'It is important to us to cripple industry as quickly as we can and get it over with as quickly as we can' (*Financial Times*, 29:12:79).

In the event, manufacturing industry proved surprisingly resilient to the effects of the strike. This ability to continue production could partly be attributed to the general economic recession which was then beginning to bite and reduce demand generally. At the same time, and as reports indicated before the strike began, stocks were also at a high level. At the end of December it was said that stocks of finished steel products suitable for transformation to factories stood at five million tonnes, enough to keep industry in production for sixteen to eighteen weeks. It was also reported that the BSC itself had large amounts in stock, and that since the threat of a strike in early December (Villiers' words suggested even before this) large amounts had been moved to the steel users (*Financial Times*, 31:12:79). In addition the stockholders, who supplied 40 per cent of the market, had also built up supplies to a high level. Although there was limited picketing of the stockholders' warehouses, as with the ports this was only of limited effect, and many stockholders encountered little difficulty in transporting steel to their customers.

Evidence leaked to *The Sunday Times* some months after the strike was over revealed that the BSC and CBI had held regular meetings during the dispute, at which it was decided to leak misleading information to the press. Other documents obtained by the paper suggested that the BSC was aware how badly the strike was affecting many areas of industry, but painted a more optimistic picture in its public pronouncements (*The Sunday Times*, 17:8:80). No doubt both the BSC and the CBI were anxious to present an optimistic face for public consumption, and we will never know what the effects on industry would have been if the strike had lasted for longer than three months, but at the same time there is little direct evidence from industry itself to suggest that the strike had any serious after-effects. Stocks of steel at the beginning of the strike were undoubtedly above average, and in any case the recession which was to destroy substantial areas of UK manufacturing industry was already beginning to bite. If there had been a steel shortage, and if UK industry had been enjoying a boom period, then it might have been a different story, but from a strategic point of view the unions chose a bad time to go on strike.

The high stocks of steel at both factories and warehouses increased the suspicion of steel union leaders that their industry had been singled out by government to act as a suitable example to other Nationalised Industries (in particular coal), that strikes seeking to achieve high pay awards and obstruct rationalisation plans were doomed to failure. We will see that if this was the objective of the government (despite the ability of manufacturing industry to continue production) then to some extent it failed, for the unions were eventually successful in obtaining at least a good part of what they had been seeking in terms of a pay rise, while the government was forced to spend enormous sums of public money in order to keep the BSC afloat.

At the same time, the strike was unable to prevent implementation of the rationalisation plan. We suggested earlier that the unions may have calculated that this fight was already lost (it is interesting to note that, in contrast to the steel strike, the coal strike of 1984–85 was about jobs rather than pay, but notwithstanding a stoppage lasting twelve months the unions were unsuccessful in halting Coal Board closure plans).

Of the four factors vital in determining the course of the strike – support from other unions; private sector production; imports and stocks – none appeared to be in the unions' favour as 1979 came to an end, but the die was undoubtedly cast, and on 2nd January 1980 the first national steel strike since 1926 began.

A timetable of the strike

The steel strike, which lasted thirteen weeks until April 1980, can be divided roughly into three phases:

1. The month of January. There were initial hopes that the strike would end swiftly after the BSC improved its offer to the unions. The unions were not in a negotiating mood, however, and after these somewhat perfunctory talks had broken down they concentrated on making the strike effective by restricting the movement of steel. Some ports and stockholders were picketed, but the focus of attention was placed largely on the state of the private sector. Initially, Bill Sirs was firm in his conviction that, as the privately-owned steelworkers were not a part of the dispute with the BSC, they should be left alone. Some of the steelworkers were less inclined to pursue this moderate course, and particularly in Yorkshire there was extensive picketing of private sector works (there were some violent scenes at the Hadfield plant in Sheffield which attracted a great deal of media attention).

   Bill Sirs eventually gave way to pressure from the more militant sections of the ISTC, and the executives of both the ISTC and NUB called out their private sector members. This escalation of the strike was initially declared illegal by the Court of Appeal, but the decision of the lower court was overturned in the House of Lords. Thus by the beginning of February public attention had largely slipped away from the issues which had led to the strike.

2. During February the unions became increasingly aware that their picketing was not achieving the desired effects. Steel was still being imported and also distributed by stockholders, and manufacturing industry was only marginally affected. In addition, the private sector

strike gradually collapsed. For its part, the BSC was becoming increasingly concerned about losses incurred as a result of the strike, and also about the prospects for regaining lost markets after the dispute was over. During this period, BSC management made a further small increase in its offer, but clearly was still restricted by the severe cash limits imposed by the government. The government maintained (at least in public) a non-interventionist approach, although it made little attempt to conceal its dissatisfaction with the BSC's handling of the strike. This criticism culminated in a scathing attack on Sir Charles Villiers by the Secretary of State for Employment, James Prior.

3. From the beginning of March the unions, now fully aware that the strike could take an indefinite period to cripple manufacturing industry, became much more anxious to achieve a settlement. A joint negotiating committee was formed (itself an achievement in the usually disunited steel unions), and negotiations with the BSC resumed. The BSC itself had by-passed the unions' leaders and held a ballot of its workers on their willingness to accept another ballot on the BSC's pay offer. This ballot on a ballot was a tortuous process, but the better than 2–1 'yes' vote strengthened the Corporation's hand. The BSC refused to improve its offer, although it now appeared that both they and the government were looking for a way out of the impasse.

   By this time the government must have been fully aware that it would inevitably be left to pick up the bill for the hefty cost of the strike and its after-affects. Thus there was a high price to pay for an official 'non-intervention' policy. The idea of an independent mediator, initially suggested by Opposition leader James Callaghan, gradually infiltrated the atmosphere, and eventually a three-man inquiry was set up under the auspices of the Arbitration and Conciliation Advisory Service (ACAS). The outcome of this inquiry was inevitably a recommendation which lay between the BSC offer and the amount sought by the unions. All parties (including government) now appeared exhausted by the strike, and the inquiry's recommendations were accepted quickly on all sides. In the short-term, the unions had gained a significant improvement on the BSC's original pay offer, but in the long-term, the strategic plan on output and jobs (hardly discussed during the strike) remained untouched.

The first month

On the 2nd January 1980, 90,000 ISTC and 12,000 NUB members began an indefinite strike. Within a few days they had been joined (in order) by 15,000 TGWU members, 13,000 from the General and Municipal Workers'

Union (GMWU) and 25,000 craftsmen (represented on the TUC Steel Committee by the NCCC). It is important to note that for much of the strike the ISTC and NUB remained somewhat separate from the other two groups (i.e. the general workers and the craftsmen). As unions with members exclusively in the steel industry, the ISTC and NUB made much of the running. In the case of the TGWU and GMWU leaders the position was more delicate, for with members in a number of industries the question of pay differentials could never be far from their minds. In addition, the TGWU bore the responsibility for attempting to secure compliance from their lorry driver members in stopping the movement of steel, a task which we will see proved impossible to implement fully.

With regard to the NCCC, the endemic tension between BSC craftsmen and manual workers made them uneasy companions. Nevertheless, a united union front in terms of strike action and negotiations with the BSC was (more or less) maintained throughout the strike, and indicated that many BSC employees held genuinely strong feelings about the nature of BSC policy. In addition to the unions on strike the Steel Industry Management Association, representing the middle managers, produced a document during the dispute which expressed no confidence in BSC policy. Thus BSC management had few friends amongst its employees during this period.

In the first week of the strike the ISTC and NUB made every effort to control the movement of steel, and superficially the support they received from other unions appeared promising. Both the train drivers' union, the Associated Society of Locomotive Engineers and Firemen (ASLEF), and the National Union of Railwaymen (NUR) quickly gave assurances that their members would not move imported steel, iron ore for BSC, and products of BSC already in rail terminals, and similar assurances with regard to road traffic were quickly obtained from the TGWU (although in their case specific assistance was left to the local level) and GMWU. In addition, the International Metalworkers' Federation ordered a complete union blockage of steel imports to the UK.

In contrast to the strong action on imports, the unions declined to block movements within the UK of raw materials and private sector steel. There was still great sensitivity amongst union leaders about involving the private sector in the dispute, although this position was soon to change. The TUC also gave its support to the striking unions, and told all affiliated unions that they should give maximum support to the dispute. In supporting the strike it could also be said that the TUC, and in particular its General Secretary, Len Murray, was concerned to retain some part of control over events, both with regard to the conduct of negotiations with the BSC, and also the possible escalation of the strike into secondary picketing and strike action by other unions.

An example of the difficulties faced by unions in blocking imports occurred early in the strike, when flying pickets from Corby and Sheffield received assurances from local TGWU officials that steel movements out of the English East Coast ports of Kings Lynn and Boston would be blocked (i.e. prohibited). Despite these assurances, it was reported that, in the two hours after the pickets departed, at least six trailers loaded with steel left the Kings Lynn dock area (*The Times*, 5:1:80). Thus, throughout the strike even the swiftest flying pickets could not be everywhere at once and a steady stream of imported steel passed through the East Coast ports (it was soon evident that assurances on the blocking of imports given by international trade union organisations counted for virtually nothing). At the end of the strike, it was revealed that steel worth about £300 million had been imported into the UK during the three months of its duration.

Despite the already evident limited effects of picketing by steelworkers, a huge amount of media attention was focused on this activity during January and February 1980, but even in the early stages of the strike there was one determined effort to achieve a settlement. The chief instigator of the new negotiations was Len Murray, who on 4th January visited BSC headquarters to discuss the possibility of re-opening negotiations. Murray was obviously concerned that the strike should not force the TUC into giving support to picketing which might become violent in character, and so was prepared to force the pace in the search for a solution. Thus it was Len Murray himself who led the TUC co-ordinating committee which met the BSC for further negotiations on 7th January.

The TUC committee (which included Bill Sirs; the leader of the NUB, Hector Smith; the General Secretary of the TGWU, Moss Evans; and the General Secretary of the GMWU, David Basnett) proposed a new 13 per cent package. This would have consisted of a basic eight per cent, plus five per cent 'on account' as part of the potential productivity payments.

The union leaders appeared to be hopeful that BSC would take up this offer (which was less in percentage terms than the amount eventually awarded to the steelworkers), but the Corporation refused. The BSC was prepared to offer a general increase of eight per cent, but with the promise that the unions accepted a number of conditions on productivity and flexibility. In addition, the Corporation offered a minimum of four per cent on locally negotiated productivity schemes. Thus the differences between the BSC and unions at this time were not based so much on pay increase percentages, but rather on the productivity improvements required to finance them. Sir Charles Villiers said that the Board was not convinced that proposals would be self-financing, while Bill Sirs was adamant that the unions would not accept an offer which contained no 'new' money. Given the parlous state of BSC finances, the only source of this 'new' money was of course the government, but at this time the impression was given that Ministers were standing firm on the existing cash limits.

Politically, the government saw obvious advantages in maintaining an approach which seemed to be one of standing up to union pressure, while also forcing reality on a Nationalised Industry making huge losses. Even at this early date in the strike, however, there was evidence to suggest that despite the impression given in public, the government was preparing itself to pay both the cost of the strike and BSC's ensuing losses. Almost unnoticed, *The Times* reported on 7th January that Parliament would shortly be asked to approve an increase in BSC's borrowing limit from £4,750 million to £5,500 million. This £750 million increase had been allowed for in the 1978 Act, but it required an affirmative resolution in the House of Commons. At the end of 1978–79, BSC's borrowings against the limit were £3,738 million, while for 1979–80 and 1980–81 the proposed respective limits were £700 million and £450 million. If the BSC had kept to these limits, then by the end of 1980–81 its total borrowings would have been £4,888 million, but the government was allowing for another £612 million on top of this sum (in the event BSC borrowings at the end of 1980–81 were just within the £5,500 million limit and stood at £5,406 million, although by this time the Corporation's borrowing limit had been raised to a potential £7,000 million).

The government's decision to raise the borrowing limit suggests strongly that its public show of equanimity at the prospect of a long steel strike, and its refusal to contemplate a higher BSC borrowing limit than £450 million in 1980–81, was not a true picture of political and financial reality. It might be argued that in January 1980 the government was hoping for a quick concession by the unions and/or the collapse of the strike, but its policy on borrowing limits indicated an anxiety to 'keep its powder dry'.

The collapse of the negotiations on 7th January led immediately to an increase in secondary picketing, particularly of private sector plants. At Ductile Steels in the West Midlands, 250 workers were laid off after pickets had halted the transfer of steel, while at Hadfields in Sheffield six pickets were arrested on 8th January. Bill Sirs, who had said that the private sector should be left alone now admitted:

> I cannot control my pickets, I am very worried ... I am trying to control it but the longer it goes on it is impossible ... Some pickets are over enthusiastic, and it is difficult to get them to listen to instructions from headquarters (*Daily Telegraph* 8:1:80).

For the week following Sirs' acknowledgement that control of the strike was beginning to pass out of his hands, there was a confused picture. In addition to Ductile and Hadfield's, the can manufacturers Metal Box told customers that supplies might begin to dry up in a week unless picketing at three of its factories ended (*Financial Times*, 9:1:80). In South Wales, Brian Connolly, the ISTC divisional organiser, expressed determination that his members would picket all private steel plants, stockholders and major users. Stockholders were picketed, together with Hoover and Metal Box factories, while at Port Talbot there were scuffles between pickets and SIMA members arriving for work. In addition, the leaders of the South Wales section of the NUM decided to take a vote on possible strike action, for they claimed that reductions in steelmaking at Llanwern and Port Talbot would lead to closures at the pits which supplied the steelworkers with coking coal. On 9th January picketing at Ductile Steels was called off after assurances that the company would not use BSC steel, but Alloy Steel Rods in Sheffield announced that it would be forced to close, while by 10th January the number of pickets arrested at Hadfield's had risen to twelve.

The increase in secondary picketing in South Wales and Yorkshire led to Bill Sirs becoming much more aggressive in his statements. Sirs had no doubt calculated that such an approach was necessary if he was to maintain control over his union members seeking more radical action, but it made immediate prospects for a settlement remote. In a speech at Middlesbrough Sirs now claimed that, although his members had previously been seeking a 13 per cent pay rise, they would now settle for nothing less than 20 per cent (*The Times*, 12:1:80). This new figure of 20 per cent eventually became something of a millstone round Sirs' neck, for in later discussions with the BSC many ISTC members assumed that this figure had become a basic demand, when it was clear that the leadership would settle for less. Despite his more militant approach in public, Sirs was still obviously anxious at this time to restrict the scope of the dispute, for in a letter to the National Association of Steel Stockholders he told them that it was not the intention of the ISTC to interfere with stockholders or producers who were not involving their companies in the dispute or seeking to defeat the steel unions' action (*The Times*, 14:1:80).

From the middle of January, Sirs' moderate approach was undermined by the actions of his members. At a meeting on the 15th, representatives of the 20,000 ISTC members working in the private sector of the industry voted to obey any instruction to strike in sympathy with their state sector colleagues. The ISTC and NUB leadership decided that this strike would go ahead from 27th January. Some delegates wanted an immediate stoppage, but Sirs secured a deferment in the hope that the mere threat of this strike would lead to a settlement.

The BSC raised the political temperature of the dispute still further, however, when on 17th January it announced its plans for cuts at the South Wales plants. The Corporation set a ten week deadline for the unions to agree to these cuts, otherwise it threatened that either Port Talbot or Llanwern would close. The plan involved a cut of 6,883 jobs at Port Talbot, and 4,454 jobs at Llanwern. This represented a total loss of 11,337 jobs out of a combined workforce of nearly 22,000. With regard to capacity, the combined total would be 2.8 million tonnes per annum, compared with the existing figure of 5 million tonnes per annum. By spelling out these new cuts, at a time when the unions were threatening to expand the strike, the BSC had once again shown little political sensitivity, for Bill Sirs' task of holding back his more militant colleagues had been made much more difficult.

At this delicate time the government finally descended from Olympus and took some positive part in the dispute, although on a strictly limited scale. Initially, Sir Keith Joseph and the Secretary of State for Employment, James Prior, were to meet ISTC and NUB leaders, which would be followed by the same union officials speaking to the Prime Minister herself. At the same time it was deliberately leaked (by James Prior) that a proportion of the BSC's £450 million borrowing limit for 1980–81 could now be used for pay rises, rather than all the money going towards investment and redundancy payments (*The Guardian*, 18:1:80). The involvement of James Prior in the talks was significant, in that he was regarded as having a somewhat different political philosophy to that of Mrs. Thatcher and Sir Keith Joseph. As Secretary of State for Employment, he was then in the midst of piloting through Parliament a Bill which would allow those subject to secondary picketing to take legal action to prevent its occurrence. The BSC steelworkers' threat to spread their strike to the private sector placed Prior at the centre of the debate on this aspect of the strike, and so (at least indirectly) in a position to determine its outcome. Despite attracting antagonism from the unions for his Departmental policy, Prior was generally regarded as being more conciliatory towards union interests than many of his Ministerial colleagues. By his taking a more direct interest in the course of the strike, the steel unions could thus hope that Prior might have some success in persuading the government to be more accommodating to their needs.

When it took place, the ISTC and NUB meeting with Sir Keith Joseph and James Prior, followed by one with the Prime Minister, produced no positive results, but at least the government could now claim that it was not standing completely aloof from the dispute. By this time, both Sir Keith and the Prime Minister had declared that the £450 million borrowing limit for 1980–81 would not be used for pay rises – denials which apparently contradicted the earlier Prior statement. Like Prior, Conservative MPs were also wondering if this £450 million could not be used to settle the strike, but at this stage Mrs. Thatcher appeared content to bide her time, and the government generally was still placing its faith in ACAS to bring the two sides together (as eventually it did).

Despite the government's hopes for the role of ACAS, the unions were still in no mood to resume negotiations, and so decided to go ahead with their plans to bring out members in the private sector. Meanwhile the Wales TUC called a national strike for 29th January. On 24th January the ISTC decided to call out their members in the private sector (numbering about 20,000 out of a total workforce of 65,000) from the 28th of the month. Once again there was evidence of a split between the unions, however, for on the same day the other unions (representing 70,000 BSC workers) involved in the dispute went to ACAS to resume negotiations, and called on the ISTC and NUB to attend. The ISTC and NUB were not sympathetic to this approach (although there was a suggestion that the BSC had dropped some of its conditions for granting pay increases), and were determined to intensify the strike.

For the next few days the strike took on a new political intensity when sixteen private steelmakers sought a High Court injunction to prevent the strike going ahead. The High Court refused an injunction but three judges in the Court of Appeal (presided over by Master of the Rolls Lord Denning) unanimously came down in favour of the private steelmakers and ordered the ISTC to call off the strike. In essence, the ISTC had claimed immunity for the strike under the 1974 Trade Union and Labour Relations Act, on the grounds that this new action was in furtherance of its dispute with the BSC. The judges disagreed, and ruled that the strike would be in furtherance of a political rather than a trade dispute (i.e. it was intended to put pressure on the government rather than the BSC), and was thus illegal. A detailed discussion of the law surrounding secondary picketing and strikes lies beyond the scope of this book (the 1980 Employment Act did have the effect of severely limiting the potential scope of this activity), but in the Appeal Court judgement, Lord Denning expressed the opinion that:

> The second dispute (between the unions and the private steelmakers) could not be regarded as a trade dispute. It was said that it was connected with the first dispute because it concerned 'terms and conditions of employment'. But it was arguable that those further acts were done to bring pressure on the government and not in furtherance of a trade dispute...It was arguable in the present case that there was no immunity for calling out the men in the private sector because it was a dispute with the government, with no immunity (*The Times*, 28:1:80).

At the time it was generally assumed that the unions would enjoy immunity so long as they subjectively believed that their action was in furtherance of a trade dispute. This had been the judgement of the House of Lords in an earlier case (Express Newspapers v. McShane in 1980) but Lord Denning and his colleagues argued that a private sector strike was deliberately intended by the ISTC to influence the government rather than the BSC. Thus the Appeal Court concluded that a strike which set out to change government behaviour was of a political nature and not a trade dispute. Nevertheless, the judges themselves conceded the key role played by the government. One of Lord Denning's colleagues, Lord Justice Lawton, put the position succinctly:

> The history of the past two decades tended to show that whenever there had been a dispute between unions and Nationalised Industry there was a tendency for government intervention, followed fairly quickly by a settlement to the advantage of the strikers. Unfortunately for the union in the present dispute there was no government intervention, and it had become clear that there was not going to be any intervention in the forseeable future. It followed that there would be no quick victory and that the dispute would be one of long attrition (*The Times*, 28:1:80).

Despite this admission that the government possessed the ultimate power to settle a dispute between a union and a Nationalised Industry, Lord Justice Lawton concluded that action taken by unions with the object of coercing government was political rather than furtherance of a trade dispute. Yet it was the government which controlled the BSC, had the power to appoint the Chairman and Board, and also possessed the capacity to provide finance on a large scale. Putting aside the merits of the ISTC's attempt to call out its private sector workers, there was no doubt that the government could not be an entirely passive spectator of the dispute, and was indeed integrally involved in the whole future of the BSC.

The Court of Appeal decision caused much confusion, and also raised the political temperature of the dispute to a new peak. The then President of the Yorkshire miners, Arthur Scargill, advised the steel workers to ignore the Court's ruling. Scargill was speaking after the Yorkshire Area Executive of the NUM had decided to black steel deliveries to collieries and Coal Board workshops (in the event the ISTC received little practical assistance from the miners during the strike, an occurrence which was to work against the miners when they sought help from the steelmen during their own strike of 1984/85). Bill Sirs himself burst into tears at a large rally in Cardiff when he said that if his union executive decided to defy the Appeal Court ruling then he could be sent to prison. He added: 'I am fighting for the working class and I do not care what happens' (*The Times* 29:1:80).

Sirs' response to the Appeal Court decision illustrates the trend of events at that time. There was no doubt that Sirs himself had no wish to defy the Court, but the attitude within the traditionally moderate unions was becoming increasingly militant. Sirs was speaking to a rally held on a day when

100,000 Welsh trade unionists went on strike in support of the steelworkers, while 16,000 ISTC members defied the Appeal Court ruling and also went on strike. This resulted in 43 out of the 44 private steel plants going on strike, with only the Sheerness plant in Essex working normally.

When the ISTC Executive met on 29th January, Sirs was able to persuade the Committee not to defy the Appeal Court. His task was probably made slightly easier by widespread legal opinion that the House of Lords would probably overturn the Appeal Court ruling. Thus the union decided to apply for leave to appeal to the Law Lords, although the Court of Appeal had refused them this facility.

By 31st January 80 per cent of the private sector strikers had returned to work, and the immediate threat of an illegal strike had been removed. Nevertheless, it was now clear to both the BSC and government that they could not hope for an immediate collapse of the strike in the state sector, and that new initiatives were required. Significantly, at the same time that Sirs was speaking to the Cardiff rally, the BSC was involved in fresh negotiations with the TGWU, the GMWU and the craft unions. These unions represented 70,000 BSC workers (compared with 102,000 for the ISTC and NUB). Although money was not discussed, it was reported that significant progress had been made with regard to productivity agreements (*The Guardian* 28:1:80). It appeared that the BSC was now willing to make concessions on the productivity improvements it would require in order to pay out 'new' money, and there was even a hint from Dr. David Grieves, BSC Managing Director of Personnel, that there could be some potential flexibility on the money offer itself. It appeared that the BSC (and also the government) was becoming anxious that the strike might escalate into a more general affair, although they were also hoping to split the notoriously fragmented steel unions. The craft unions in particular were traditionally highly suspicious of the ISTC, and if they could be persuaded to achieve a settlement with the BSC, then this could put pressure on the other unions to follow in the same direction.

The negotiations between the BSC and craft and other unions, while the ISTC was seeking to bring out its private sector workers, suggested that at this stage the strike was a two-tier affair. At the end of the first month, however, there were still no positive signs that the strike as a whole was nearing a conclusion.

One postscript to events at this time concerned an EEC involvement. Henk Vredeling, the Dutch socialist Commissioner for Social Affairs, claimed he was amazed that the UK government had not applied for financial assistance from the EEC to help BSC with its proposed plant closures. The government replied that it had not applied for the £22 million immediately from the social fund because in order to receive the aid it would need to provide another £26 million itself. The Labour Opposition and the unions inevitably made political capital out of Vredeling's comments, but in a wider context this incident hinted at the increasingly interventionist role of the EEC Commission in the affairs of the UK steel industry. The EEC Industry Commissioner, Count Etienne Davignon, had already supported the BSC's proposed cuts, and we will see that in the years to come Davignon's influence over all EEC steel industries would be considerable.

The second month

During the first week in February the strike finally spread to the private sector, but new attempts were also made to achieve a settlement. The ISTC was initially granted leave to appeal to the House of Lords over the Appeal Court decision, and on 1st February five Law Lords decided unanimously that an extension of the strike into the private sector constituted a trade dispute and was not a political strike. The ISTC was now free to spread its strike, although Bill Sirs typically urged caution and asked his Executive for an eight day delay before calling out the private sector. Having achieved a significant legal victory, however, the ISTC leaders generally were in a bouyant mood, and so the Executive voted in favour of an immediate strike.

A private sector strike put greater pressure on steel supplies to customers, but despite the battles in the courts, and the widespread public attention they aroused, the real effects on industry continued to be marginal. It was estimated that only 10 per cent of stockholders had been subject to any kind of picketing, while the CBI had instigated a 'swop shop' between steel users who encountered a shortage in a particular type of steel. For obvious reasons, manufacturers were reluctant to discuss publicly their arrangements for receiving steel, particularly with regard to imports, but it does appear likely that steel was moving more freely than was widely presumed at the time.

Politically, it could be said that during the second month the strike was taking place at three levels. Firstly, there was the highly public aspect of the legality and effects of secondary picketing. This subject received a huge amount of media attention, and indeed the casual observer could easily have assumed that this subject was purely what the strike was about. If the steel unions could have halted the movement of all steel (and even more, if a general strike had ensued) then this perspective would have been correct, but the reality was different.

Thus, the second and more significant level concerned a situation where steel users were still in a relatively comfortable position (an exception was Metal Box who laid off 2,500 workers), and the unions were becoming increasingly desperate. As we will see, towards the end of February the strike in the private sector was to collapse, leading to a final push towards a settlement.

The third political level concerned these negotiations which, notwithstanding all the rhetoric, were continued in some form for nearly all three months of the strike. As we have seen, despite its show of non-intervention, the government was already aware that it could be forced to bail out the BSC at the conclusion of the strike, and so although the steel unions' action was not effective in halting industry, the government could not afford to be sanguine about the prospect of an indefinite strike (there was also the remote prospect that the strike could still be widened into other sectors such as coal).

For a brief period during early February, negotiations took place between the ISTC, the NUB and the BSC. The tentative progress already made in negotiations by the TGWU, the GMWU and the craft unions had placed new pressure on the other two unions. The two sides took advantage of a European Coal and Steel Community meeting in Luxembourg, and Bill Sirs was sufficiently encouraged by the BSC's response to call together his full 60-man negotiating team! By this time Sirs himself seemed prepared to settle for an offer in the region of 13 per cent, but many of the delegates still had in mind the 20 per cent figure mentioned earlier.

When the two sides met, on 8th February, the talks broke up after only twenty minutes amidst much confusion and recrimination about the Corporation's actual offer. In a letter to the unions three days earlier, it had offered basically nine per cent on the table (although seven per cent of this money had some productivity strings attached), plus: 'A four per cent minimum guarantee against which locally determined productivity schemes would be paid'. Perhaps with some justification the ISTC and NUB took this to mean that the 4 per cent would be guaranteed, but the BSC denied this interpretation (*The Times*, 9:2:80). Instead, the Corporation claimed that the four per cent would be conditional upon local productivity schemes being agreed, although it was prepared for this four per cent to be paid unconditionally for 13 weeks in order that the local negotiations could be initiated. That the negotiations had reached this 'fine print' stage did represent some kind of progress, and with hindsight it must be wondered why Sirs broke up the meeting after only 20 minutes. When taken in the context of the time, however, feelings were still running high, with many of his fellow negotiators still insisting on a 20 per cent pay rise, while at the same time hoping that the private sector strike could force the BSC and the government into making more concessions.

For the remainder of February the strike remained in a highly confused state, with little or no progress towards a settlement. Shortly after the ISTC and NUM negotiations had broken down, the BSC made an offer totalling 14.4 per cent to the craft unions. The Corporation once again hoped that this new offer would split the fragile unity of the unions, and when the negotiators for the craftsmen tentatively accepted this offer, they appeared to have succeeded. These hopes were dashed, however, when a full delegate meeting of the craftsmen voted 151–74 to refuse the offer. As with the ISTC and NUB, the 'rank and file' delegates of the unions were proving to be more militant than their leaders.

This point was also seen on the strike front itself when the ISTC employees at the private sector Hadfield's plant in Sheffield voted to return to work, and Bill Sirs recommended that Hadfield's should be made exempt from the strike. The Chairman of Hadfield's, Derek Norton, had become a prominent media figure in the strike by his strong verbal attacks on the picketing of his works, but he also declared that he supported the ISTC's case against the BSC, and these words appeared to persuade Sirs that Hadfield's should be left alone. The South Yorkshire Strike Committee thought otherwise, and the picketing of Hadfield's became even stronger and more violent, with 13 arrested on 12th February, and 22 on 14th February. After this rebuff, Sirs reversed his decision and told the Hadfield's men that they must come out again, an instruction which met with a mixed response. Apart from Hadfield's, only Sheerness, of the privately owned steelworks, decided to defy the strike at this time, but this apparent unity was not to last for long.

At this stage, we should perhaps remind ourselves that the BSC's long-term strategy on jobs and plant closures, surely the most crucial subject of all, had hardly been mentioned throughout the dispute. All sides had adopted entrenched positions, and by now exhaustion was setting in. Yet the dispute about pay percentages was really only the tip of the iceberg. The strike had no doubt slowed down the closures and rationalisation programme, but the BSC was content that this basic strategy had not been questioned. Militant pickets could shout: 'What do we want? – 20 per cent' for months on end, but for many of them their days as BSC employees were numbered.

Meanwhile, the General Secretary of the TUC, Len Murray, was understandably becoming increasingly concerned about the course of the strike. In South Wales, a general strike was still being threatened, and Murray's concern found a significant echo from the BSC and the National Coal Board and then the government. Firstly, the BSC agreed to limit its supplies of imported coking coal to 4 million tonnes per year (out of a total requirement of 9 million tonnes), and forego the extra 1.3 million tonnes it was threatening to bring into the country. The price of coal was a perennial bone of contention between the BSC and the NCB (now British Coal), and the Corporation had estimated that buying the extra 1.3 million tonnes of domestic coal would cost it £33 million more than importing it from the USA and Australia. Thus, in return for the BSC buying more British coal, the

NCB agreed to increase its subsidy of the price of coal supplied to the Corporation from £15 million to £22 million. This off-loading of costs from one Nationalised Industry to another illustrates the interdependency between them, particularly as the NCB was also a prominent customer of the BSC. As we will see in the next chapter, Sir Charles Villiers' successor as BSC Chairman, Ian MacGregor, was to make the cost of coal to the BSC a more prominent political topic.

Although the government had refused BSC coking coal subsidies to both the Corporation and the NCB, we must conclude that Ministers were highly relieved to see a settlement achieved. Its chief political effect was to bring assurances about jobs to the South Wales coalfields, where support for the steelworkers was (together with South Yorkshire) at its strongest. Having adopted a united front with the BSC since the strike began (at least in public), there were now clear signs that the government was becoming increasingly exasperated with the Corporation's failure to achieve a settlement. In a letter to Len Murray, the General Secretary of the TUC (who himself appeared determined to take a more prominent part in negotiations), the Chancellor of the Exchequer, Sir Geoffrey Howe, wrote, with regard to the BSC's closure and redundancy proposals:

> We have been assured by the BSC that they are anxious for further consultations to take place as soon as possible between themselves and the unions about their proposals and the basis for them. The government hopes that these consultations can proceed just as soon as possible (*The Financial Times*, 14:2:80).

Sir Geoffrey also hoped that the government and the TUC could conduct a joint examination on the scope for using more EEC funds to alleviate the problems of the UK coal and steel industries. Although these issues were not what the strike was specifically about, the conciliatory tone of the letter suggests strongly that the government had decided it was time to take more of an initiative in the dispute. From this point on, Ministers made little effort to hide their dissatisfaction with the performance of BSC management, and there was already speculation about who might succeed Sir Charles Villiers when the Chairman's contract ran out later in the year. John Biffen, Chief Secretary to the Treasury, and regarded at that time as a close ally of the Prime Minister (although also noted for a candour not always heard in the words of Ministers), admitted that negotiations had been bungled (*The Guardian*, 16:2:80), but it was James Prior (definitely not regarded as a close ally of the Prime Minister, particularly on economic policy) who was Sir Charles' severest critic.

During an 'off the record' meeting with journalists, Prior suggested that Villiers might leave the BSC before his contract expired, and then could be replaced by someone from overseas (no doubt Ian MacGregor, a US citizen, was already under consideration). Significantly, Prior also considered that amongst Ministers the feeling was one of: 'We cannot go on much longer as we are' (*The Times*, 19:2:80). In speaking to the press in this way, the Secretary of State for Employment was striving to give a new impetus to negotiations between the BSC and unions, by using Sir Charles as something of a sacrificial lamb. By this time the BSC Chairman was in an embattled position, with loss of faith on the part of the government adding to obvious union antagonism, while his management colleagues (as we saw earlier with regard to reorganisation) could not be relied upon for support.

Given Prior's known doubts about government economic and industrial policy, the reference to 'not going on as we are' indicated that he was increasingly dissatisfied with the government's conduct of the strike, and felt that some sort of intervention was necessary. Inevitably, the Prime Minister was not going to take Prior's criticism lightly, and in the Commons the following day expressed every confidence in Sir Charles and BSC management to achieve a settlement with the unions (Hansard, 19:2:80. Cols 236–7). Regardless of her own doubts about the Villiers style of management (he had after all been appointed by a Labour Secretary of State), Mrs. Thatcher could not afford to jettison Sir Charles in the middle of the strike, for to do so would be tantamount to admitting that both BSC and government strategy were wrong. Sir Charles himself defended his position with an uncharacteristically aggressive response. He accused the government of completely bungling its communications, while also managing to muster a pointedly witty remark when asked who he thought had been responsible for the early retirement reports: 'I don't have to name the person. I had no prior warning' (*The Guardian*, 20:2:80).

The differences within the government, and between the government and the BSC, could bring new heart to the unions. In addition, the BSC now acknowledged that the 52,000 job losses would have to be postponed beyond August. This good news for the unions was counterbalanced by the growing awareness that the private sector strike was not having the desired effect. Large-scale picketing at Sheerness failed to close the works down, while by the end of the month steelworkers at Hadfield's and at Round Oak in the Midlands had voted to return. In any case, there was still no concrete evidence that steel-using industries were being adversely affected to a serious degree. A further blow came when an all out strike call by the South Wales NUM delegates was completely rejected by rank and file members. There was thus now a considerable incentive for the unions to seek a settlement, and the ISTC suggested a new package of 15 per cent across the board in return for an agreement on changed working practices, plus five per cent in return for a national commitment to negotiate local productivity deals 'at the earliest possible moment' (all this involved a 15-month agreement from January 1980 to April 1981). Once again, the magic figure of 20 per cent had been produced by the steel unions (giving them parity with the miners), and not surprisingly the BSC rejected this proposal.

The Corporation then decided to increase the pressure on the leadership of the steel unions by means of by-passing them and holding a ballot of its employees on the 14.4 per cent offer. This technique had already been employed with some success by the Chairman of the BL car company, Michael Edwardes. Bill Sirs called this new move 'a dirty trick', while characteristically the Prime Minister was all in favour of the idea. Perhaps more surprisingly the General Secretary of the TUC, Len Murray, refused to express an opinion on the ballot, but by now both the TUC and the Labour Party appeared anxious to find an end to a strike which threatened to continue indefinitely. Memories were still fresh of the public sector 'winter of discontent' strikes during the previous year, which had played a significant part in Labour's defeat in the 1979 General Election. As a result, both the TUC and the Labour Party were anxious about the political effects of an indefinite steel strike. On the penultimate day of February the leader of the Opposition, and former Prime Minister, James Callaghan, suggested that the strike could be settled through the appointment of an independent mediator. This proposal was gradually to infiltrate the political atmosphere during the next month, and eventually bring an end to the strike.

The third month

By the beginning of March 1980 the ISTC itself estimated that 9,000 of its 20,000 private sector members were back at work, with more returning each day (*Financial Times*, 1:3:80). The private sector strike had failed to bring manufacturing industry to a halt, and now the Labour Party (with the encouragement of the TUC) was making a concerted effort to get the steel unions off the hook. Significantly, James Prior agreed with Callaghan's suggestion for an independent mediator, and so for the first time there appeared to be a tripartite consensus between the Conservative government, the Labour Party and the TUC as to how the strike could be settled. Although it might have been expected that, as Secretary of State for Employment, James Prior would be an interested party in seeking an outcome to the strike, it now appeared that he had become the government's chief spokesman on the subject. In contrast, the Minister actually responsible for the BSC, Sir Keith Joseph, had become something of a background figure. Having become identified with the official 'no intervention – no compromise' stance, Sir Keith had perhaps become something of a political liability to a government now anxious to find an accommodation with the unions involved. In contrast, although the TUC was highly suspicious of Prior's proposed legislation on controlling secondary picketing and strikes, they saw him as being someone more likely to listen to their case than the Prime Minister or Sir Keith Joseph.

The TUC's role was now to bring all the steel unions together for new negotiations, and this not inconsiderable feat was achieved on 5th March when delegates from all the unions involved met at the TUC headquarters. Under the guidance of Len Murray, a new pay and productivity formula was drawn up. At the same time, it was also announced that picketing would be stepped up, with lorry drivers instructed not to cross picket lines outside BSC customers' gates. This latter ploy could not hope for immediate success, with steel imports still being moved and the private sector steel strike collapsing, but it was probably the last shot in the unions' locker (it did appear to have limited success at Ford car plants and some engineering companies).

Further pressure was placed on the unions when 58,502 voted 'yes' to the BSC's suggestion that a ballot should be held on the 14.4 per cent offer, while 26,517 voted 'no'. This 'ballot about a ballot' did not of course mean that the workers would automatically accept the offer (in any case less than half the workforce voted at all), but it did represent a limited propaganda victory for the BSC. When the unions and the BSC met on 10th March there were considerable hopes that a breakthrough was imminent, but after three days' negotiations ended yet again without a settlement. Little substantive progress was made, with the BSC sticking to its 14.4 per cent (with some concessions on productivity), while the unions were still looking for a figure approaching 20 per cent (including local productivity deals). The crucial breakthrough came a few days later, however, when the unions finally agreed to accept an independent mediator. On 21st March the unions and the BSC agreed terms of reference for an inquiry, which would be restricted purely to pay and conditions and not include BSC strategy itself.

The unions then went to meet James Prior (but not Sir Keith Joseph) to ask him if he would appoint an independent Chairman for the inquiry. Prior refused to do this, on the grounds that the government would not wish to intervene in this way, and instead referred the unions to ACAS. Despite Prior's predictable reaction, there must have been satisfaction that his support for the Callaghan proposal had borne fruit.

Once ACAS had agreed to appoint a Chairman, the fragile unity of the unions broke up, with the craftsmen deciding to be bound by the results of the inquiry, while the other unions said that they would leave the final decision to their members. Nevertheless, the vital factor was that the inquiry would now take place, with ACAS deciding to appoint Lord Lever as its Chairman (there is evidence to suggest that Lever's name was first suggested by the BSC, and then cleared with Ministers (*Financial Times*, 5:4:80). Despite his background as a former Labour MP and Cabinet Minister, Lever had considerable City and industrial connections, and was likely to adopt a conciliatory approach towards settling the dispute. To assist him, there were two advisers – one nominated by the unions and the other by the BSC. The BSC's nominee was Sir Richard (now Lord) Marsh. Also a former Labour Cabinet Minister, Sir Richard was now a Conservative and Chairman of the British Iron and Steel Consumers' Council. Indeed, Marsh offers a classic example of how the political process in a particular sector can have a cyclical appearance (or to put it more prosaically, how the same figures can pop up at different times and in different jobs), for as Minister of Power in 1967 Marsh had been responsible for setting up the BSC and appointing Lord Melchett as its first Chairman. Only a few days before his appointment to the inquiry Marsh had accused the union leaders of leading their members into a disastrous mess (*The Times*, 19:3:80), and it was of course in the interests of the steel users to see a swift end to the strike. The union nominee was Bill Keys, left-wing General Secretary of the Society of Graphical and Allied Trades, and a man unlikely to concede the steel unions' case easily.

After nearly three long, weary months of the dispute, a mood of great urgency and haste prevailed, and the inquiry took place over a single weekend! The Lever inquiry proposed that the BSC should offer 11 per cent across the board in return for a joint document on changed working practices, job flexibility and local productivity bargaining. In addition, the inquiry recommended that the BSC should pay a guaranteed 'lead in' payment of 4.5 per cent for three months while local productivity deals were being negotiated. This represented a total increase of 15.5 per cent, an increase of 1.1 per cent on the BSC's final offer (Bill Sirs calculated that the proposed increase was worth 17 per cent if holiday schemes and pensions were included). The BSC hoped to recoup the estimated £192 million cost of the offer in increased productivity (although the cost of the strike itself was estimated by the Corporation eventually to have been £200 million over the three months). Politically, the inquiry undoubtedly had the effect of breaking the '14.4 per cent versus 20 per cent' impasse which had existed for several weeks. The unions had obtained only a small increase in the BSC's final offer, but the 15.5 per cent should be compared with the original offer of two per cent (in effect nil). Some ISTC members were opposed to acceptance of the inquiry's recommendation, but on 1st April union delegates voted 41–27 in favour of a return to work. The other unions quickly followed suit and on 3rd April, the strike ended exactly three months after it had begun. At the time, it was the longest national strike in Britain since 1945.

The strike as a political event

In spite of the interminable pay negotiations, the 1980 steel strike can be seen as principally a political event. If we examine the behaviour of the three principal participants, i.e. unions, BSC management and government, then we can gain some understanding as to the meaning of the predominantly confused and messy events.

For the unions, on one level the strike had a symbolic effect. By the end of 1979 it was clear that the BSC intended to implement a plan involving a massive reduction in its capacity and workforce. Although this Corporation strategy was barely mentioned throughout the strike, in the eyes of many observers it will be seen as one last (if unavailing) great effort by British steelworkers to preserve their status and identity, as well as their jobs. We saw with regard to the Shelton campaign that the importance of this type of reaction should never be politically under-estimated. The UK steel industry (as in many of the EEC countries) represented a central pillar of an industrial network which had shaped millions of lives for over a century. The dismantling of this framework was never likely to be a politically costless task.

On another level, however, the strike could be judged a less momentous event for the unions. We saw how Bill Sirs himself was continually seeking to exercise restraint on his more militant members. Throughout the strike, he appeared anxious to reach some kind of accommodation with the BSC, while at the same time dealing with widespread calls for a settlement of nothing less than 20 per cent. Superficially, it was surprising that Sirs, or other union leaders, made little attempt to bring the BSC strategy on jobs and output into negotiations. In reality, it is hard to escape the conclusion that they were already privately resigned to the cuts, and so wished to gain the best deal they could for the workers who remained. They were probably correct in adopting this tactic, for the conclusion must be that many of the BSC workers who went on strike were later quite happy to accept redundancy cheques. Apparently, even the industrial heritage had its price! Finally, the traditional disunity of the unions, exemplified particularly by the mutual suspicions of the ISTC and the craftsmen, prevented a united front for all but the briefest of periods. In the end, the unions did achieve a considerable improvement on the BSC's original offer, but the Corporation's strategy remained intact.

The BSC role in the strike is perhaps best seen as one of bungled opportunities, rather than a change in strategy. The Corporation had set its course in 1978 as a result of *The Road to Viability*. The original two per cent offer can be seen as politically naive, but it was in character with its long-term scheme to achieve a complete decentralisation of pay bargaining.

By 1979 the Corporation was in such a bad financial state that it had little alternative but to push on with its plans, but the pressure put on them by the government appeared to induce a panic amongst BSC management, and the ultimatum of 52,000 job losses in eight months was the result. The 'softly softly' approach had been abandoned, and a confrontation with the unions became inevitable.

The political weakness of BSC management at this time (exemplified by Sir Charles Villiers' perception of himself as a scapegoat) brings us to the role of the government. Many critics of the government suspected that it had chosen the steel industry as a particular example of how to put both the unions and the Nationalised Industries in their place. The government could indeed claim with justification at the end of the strike that the steel unions had failed in their attempts to halt industry, and so had been forced to the negotiating table. It was also true that by setting tight (if unrealistic) borrowing limits for the BSC it had pushed the Corporation into taking drastic measures. As in the case of the unions, the steel strike holds a particular symbolic political importance for Mrs. Thatcher's government, for on one level it can be portrayed as an event which demonstrated a new approach to industrial relations. Strikers were not to be appeased, and inefficient industries such as steel should be taught a salutory lesson about the realities of a changing world.

In a manner similar to the unions, however, the imagery and rhetoric does not match reality. We saw that, even in the early stages of the strike, the government was already contemplating raising the BSC's total borrowing limit by £750 million (all of which would be required within twelve months). In addition, as the strike dragged on, it became evident that Ministers were extremely dissatisfied with the BSC management, and were prepared to take a more direct role in seeking a solution. James Prior in particular voiced public criticism of the BSC, and then put his weight behind the concept of an independent mediator. Indeed, the manner in which the strike was settled – a mediator and his advisors under the auspices of ACAS, could be said to represent the classic UK solution to solving thorny industrial problems. It is surely significant that within days of the inquiry coming into being, the strike was over. Both the unions and the BSC appeared to instinctively accept that this form of arbitration represented a 'proper' way in which the strike could be settled. Thus the government's confrontational approach finally gave way to consensus (although sheer exhaustion on all sides was also a factor).

There is no doubt that the steel strike holds a significant political symbolism for many of the actors involved (and the importance of this factor should never be discounted in political analysis), yet by looking at the events in detail the evidence suggests a more prosaic evaluation. It can be argued that *if* the BSC had made a more substantial offer than the original two per cent (seen as nil by the unions), *if* the government had adopted a more realistic and flexible attitude towards the BSC's financial position, and *if* the Corporation had maintained its 'softly softly' approach, and not issued its ultimatum on 52,000 job losses to the unions, then the strike would probably never have begun. Having begun, it could again be argued that if what might be termed the heavyweight political establishment (i.e. Conservative government, Labour Party and TUC) had placed pressure on the unions and the BSC at an earlier date to accept the idea of an independent mediator, then the strike could well have been settled much earlier. If political symbolism and myth were to be the chief victors of the strike, then they exacted a heavy price, for the BSC was now in an even more serious financial position.

7 Political Facets of the BSC in the Early to Mid-1980s

The MacGregor era, retrenchment and survival

In the years after the end of the 1980 steel strike the BSC more than achieved the reductions in capacity and manpower it had hoped for (and the unions had feared) at that time. The manned capacity was reduced from 21 million tonnes per annum to 14.4 million tonnes per annum, while manpower dropped from 166,400 in 1980 to 53,720 in 1988. Largely as a result of the rationalisation, it was generally agreed that, by the late 1980s, BSC productivity compared favourably with the best in Europe (by the BSC's own figures, 5 man hours worked per tonne of liquid steel in 1988, compared with 14.5 man hours per tonne of liquid steel in 1981). In 1984–85, the Corporation estimated that it would have made an operating profit of £40 million after interest, if it had not been for extraordinary costs incurred (including £180 million as a result of the coal strike), but after this the BSC moved into profit for the first time in a decade, leading to a profit of £410 million in 1987–88. Notwithstanding this progress, the BSC in the early to mid 1980s remained in a delicate political and commercial state, with doubts continuing about the long-term future of the Ravenscraig plant, and the government's hopes for full-scale privatisation still a distant objective.

A highly significant development during the 1980s was greatly increased intervention on the part of the EEC Commission. Overcapacity and low prices led to a 'state of manifest crisis' being declared in 1980. After that time the Commission introduced production quotas, followed by minimum price controls and rationalisation plans for the Community as a whole. In addition, the Commission also conducted negotiations on imports into the Community. It could be said that to a large extent the early to mid 1980s saw the EEC become a politically managed market for steel, with the long-term aim of matching capacity to sales, and thus removing subsidies from the industry.

The BSC itself was more zealous in cutting capacity than many of the other EEC steel companies, although, as we have seen, this was a course which already had been set domestically from the late 1970s onwards. The massive reduction in the workforce from 1980–85 passed off with little political cost to either the government or the BSC, chiefly thanks to redundancy payments which nearly all steelworkers were prepared to accept without a quibble. Thus, the chief cost was financial, with the government writing off massive amounts of the BSC's debt, while also continuing to provide substantial support for the Corporation's current operations. The Chairman of the BSC from 1980–1983, Ian MacGregor, was particularly identified with the cutbacks in manpower. Although, as we have seen, the policy line he pursued was not new, MacGregor's image as a tough American steelman undoubtedly made a strong political impact. The transfer fee agreed with his former employers, the American bankers Lazard Freres, entailed government payments on the basis of performance, and this alone was calculated to stir up political controversy. In addition, MacGregor himself was more likely than previous Chairmen (even including Sir Monty Finniston) to speak his mind on matters concerning the Corporation. At the same time, MacGregor was even more dependent on heavy financial support from the government then previous BSC Chairmen. Without the capital write-offs and generous borrowing limits, it would have been impossible to push through such large reductions in the workforce.

In 1982 the government intervened openly when it took responsibility for the decision to keep all five major plants in operation for at least a further three years. MacGregor made no secret of his opinion that the BSC needed to close at least one of its major plants (probably Ravenscraig), but the government, with at least one eye on the 1983 General Election, disagreed. After this setback to his plans, MacGregor, in a move typical of his character, devised a bold plan whereby steel produced at Ravenscraig would be shipped to the United States to be rolled. After MacGregor's departure from the BSC in 1983 this plan fell through.

After this time the BSC, for the first time in its history, tended to step into the periphery of the political limelight. There are several reasons for this lower political profile. Firstly, having carried through massive cutbacks in output and manpower in the late 1970s and early 1980s, by 1985 there was little of substance that could be achieved in this direction without closure of one of the five major plants. Once the government had made up its mind that Ravenscraig should survive, a decision reaffirmed in 1985, then the scope for further cost-cutting was limited. Hence the halcyon days of the action committees were largely over (although continued doubts concerning the future of Ravenscraig, despite government assurances, led to considerable political activity in Scotland).

Secondly, as a result of the cutbacks, the BSC was of less political significance as an employer. In ten years the BSC workforce fell by nearly three-quarters (220,000 in 1975 to 64,500 in 1985), and so was not able to command the attention of government it could at one time take for granted. A major steelworks closure is always likely to remain an issue of high political salience, but in many areas of the country the BSC no longer occupied its former position as the dominant employer.

Thirdly, from 1983–86 the BSC had a part-time Chairman. After the departure of MacGregor the government appointed Robert (later Sir Robert) Haslam to be the new Chairman. At the same time, Sir Robert continued to occupy his position as Chairman of the Tate and Lyle sugar company. In order to compensate for Haslam's part-time duties, the Chief Executive Bob Scholey was given more day-to-day control of affairs. This new arrangement effectively robbed the BSC of a public figurehead. In their different ways, Melchett, Finniston, Villiers and MacGregor all came to be strongly identified with the BSC as an organisation, but the lack of a full-time Chairman appeared to have bestowed a degree of political anonymity on the Corporation.

Fourthly, a great deal of government time and energy in the 1980s was given over to its privatisation programme. We will see in this chapter that sections of the BSC became more integrated with the private sector, but until the late 1980s complete privatisation was out of the question. In contrast, the government sold majority shareholdings in a number of its public corporations, with the massive sales of British Telecom shares in 1984 and British Gas in 1986 being particular features. This government emphasis on privatisation meant that, in the eyes of Ministers, the BSC had something of a 'Cinderella' image, until its own privatisation in late 1988.

Fifthly, the coal strike of 1984–85 occupied such a central position on the political stage that, for its duration, the BSC was perceived politically as an organisation which might assist in frustrating the aims of the strike (as indeed it did) rather than a group with its own needs and objectives.

Sixthly, there were complex political factors at work with regard to manufacturing industry as a whole. For more than a century, industries such as coal, steel, engineering, shipbuilding and textiles were seen as forming the backbone of the UK economy, and their health was a crucial indicator of the nation's economic well-being. In post-war years all these industries have suffered long-term decline, and latterly acquired the unattractive soubriquet of 'smokestack' – meaning those old industries which had little to offer in the future in terms of employment and wealth creation. The Conservative government itself placed great emphasis on future employment prospects in the areas of information technology and the service industries. Even when demand for steel improved, it was inconceivable that the BSC would ever again become the employer of hundreds of thousands of people. Thus, no government in the future was likely to see steel as an industry likely to contribute in a major way to solving the unemployment problem.

On the other hand, as we argued in Chapter One, it could well be a mistake to see the UK steel industry as being in a state of terminal decline (just as, conversely, the early 1970s was a period of over-optimism). Apart from its obvious strategic importance in the area of defence, steel could benefit if (or more likely when) the interests of manufacturing industry become more politically fashionable. The BSC played its own part in the campaign to develop the infrastructure by becoming a prominent member of an (ultimately unsuccessful) consortium brought together with the aim of building a fixed link across the English Channel. Nevertheless, by 1987 the successful Eurotunnel consortium had begun to dig the tunnel, with obvious potential benefits for the BSC. In addition, a wide variety of interests (including the CBI and TUC) lobbied the government in an effort to obtain more resources for developing and repairing the national infrastructure (e.g. houses, roads, sewers).

The zeal with which the government set about persuading the Japanese Nissan car company to build a plant in the UK (the site eventually chosen was at Washington in the North-east of England) also indicated that car manufacture was another area where Ministers themselves were anxious to encourage a steel consuming industry. As we will see in Chapter Ten, the BSC suffered from UK car manufacturers becoming more reliant on imported steel, but the high politicial salience of the domestic car industry suggests that British Steel (the title of the privatised company) will continue to have a future in this area. Thus it is still possible that British Steel will have a future role as a wealth creator (or at least as an instrument of wealth creation), even if the days of its own central strategic industrial importance as an employer are unlikely to return.

Decentralisation, privatisation and the politics of costs

After briefly looking at the circumstances in which MacGregor took over the Chairmanship of the BSC, we will concentrate in this chapter on examining certain facets of the Corporation's activities which came into particular prominence during the first half of the 1980s. These are re-organisation of the BSC into product based Profit Centres, privatisation prior to 1988, relationships with suppliers, a variety of campaigns to achieve lower raw material and energy costs, and the decentralisation of pay bargaining. The extension of the Profit Centre concept reflected Ian MacGregor's philosophy that greater responsibility for local management would be an incentive for better performance, while it also indicated the government's wish to privatise the BSC, if not all at once, then at least by selling off individual sections.

By its nature, the BSC's search for cheaper raw material and energy prices created a tension with such bodies as the National Coal Board, the Central Electricity Generating Board and the British Gas Corporation. Nevertheless, Ian MacGregor himself became particularly identified with the Corporation's search for cheaper coal and energy prices. Like labour costs and productivity, the cost of domestic coal, together with gas and electricity prices, is greatly dependent on domestic political decisions. The BSC perceived that intense lobbying, if necessary by uniting with other interests, could achieve satisfying results. During the 1980s, activity in this area significantly increased, while the price of scrap became a matter of political salience within the EEC as a whole. In contrast, the price of iron ore, which was by far the largest single cost borne by the BSC, is dependent on currency fluctuations and international market trends, and so lies largely outside the area of UK political debate. As we will see, even when a miscalculation on the BSC's part led to a massive loss on an iron ore investment, its political impact was negligible.

Decentralisation of pay bargaining was a primary cause of the steel strike, but after that time the BSC sought to establish this procedure as the basic method of determining the wage structure, even to the extent of attempting to dismantle the national bargaining machinery altogether. The unions have resisted this trend only spasmodically, and generally failed to halt its progress, although by the time of privatisation British Steel itself appeared content to retain at least some degree of national bargaining.

After examining these aspects of the BSC's activities in this chapter, in Chapter Eight we will turn to the important strategic decisions which the Corporation and the government made during the 1980s. The chief of these was the decision to keep the Ravenscraig plant open. We will also see how domestic political decision-making from 1980–1985 was heavily circumscribed by an increasingly strict regime on prices, output and rationalisation conducted by the EEC as a whole. Thus there was a heightened need for the UK government to muster the maximum possible amount of expertise.

The MacGregor appointment

Once the steel strike was over, the government's attention turned towards appointing Sir Charles Villiers' successor as Chairman. Willing volunteers for the position were not easy to find, but after a lengthy search (forty names were suggested to the 'head hunters' employed by the government) the job was finally taken by 67-year old Ian MacGregor. Originally a Scottish metallurgist (like Sir Monty Finniston), he had lived in the United States for more than 35 years and had become a citizen of that country. The suggestion to appoint MacGregor came from James Prior, who had earlier been so critical of Sir Charles Villiers:

> I knew him [MacGregor] well because my eldest son, David, was working for him in New York. I told Keith Joseph that MacGregor was one of the world's outstanding managers and he should see him (Prior, 1986, 129).

Later, David Prior worked as Personal Assistant to Ian MacGregor at the BSC, and then stayed on at the Corporation after the Chairman's departure.

As Chief Executive of the American Metal Climax Company, Ian Macgregor had won a reputation as a tough but skillful businessman, and his widespread business interests in the US included being a director of Singer, American Cynamid and the Brunswick Corporation. Despite his American background, he had retained contact with the UK, culminating in his becoming part-time non-executive Deputy Chairman of the state-owned BL car company (appointed by the Labour government). He was known to be a keen supporter of Michael Edwardes (Chairman of BL), whose firm-line attitude towards unions included the workforce ballot technique which we saw was copied by the BSC during the steel strike. He thus corresponded exactly with the government's perception of the ideal Chairman for a Nationalised Industry which was considered to be in urgent need of a more hard-headed commercial approach.

Although derogatory remarks inevitably were made about the age of the new Chairman (the same as Sir Charles Villiers on his retirement), MacGregor's subsequent impact on UK industry was considerable, for after leaving the BSC in 1983 he then became Chairman of the National Coal Board (while remaining a member of the BSC board) and led the latter body throughout the traumatic 12-month coal strike of 1984–85. At the BSC, MacGregor immediately ran into a political storm over the size of the 'transfer fee' to be paid to the New York bank Lazard Freres, in which he was a partner. Firstly, the government (not the BSC) was to pay Lazard Freres a basic fee of £675,000 for the services of MacGregor during the three years of his contract. Secondly, Lazard's would receive a fee of up to £1,150,000 depending on the performance of the BSC during and after MacGregor's Chairmanship.

These performance standards were to be judged by a Committee consisting of two members nominated by Lazard Freres, two by the Secretary of State, and an independent Chairman agreeable to both sides. Although MacGregor left the BSC in 1983, the size of the performance payment was not announced until 1985 (it was suggested that the delay was chiefly due to the political embarrassment which might have been caused by announcing the fee during the coal strike). The award should have been made in two parts, but the Committee chaired by John Gardiner, Chief Executive of the Laird Engineering Group, finally settled on a sum of £875,000, out of a possible total of £1,150,000. In other words, the Committee decided that MacGregor had done three-quarters of a good job! The criteria considered by the Committee were profit and loss targets, cost reductions, progression towards privatisation, and an assessment of the BSC's continuing health and strength (*Financial Times*, 14:8:85). The Committee decided to withold a quarter of the money on the grounds that BSC was still making losses, and also that privatisation had only taken place on a small scale.

The government justified the sums paid to Lazard Freres on the grounds that they represented the market rate, but the Labour Party and the trade unions inevitably made political capital out of the affair (some Conservative MPs were also unhappy about an approach which appeared to breach government calls for pay restraint in the public sector). There is no doubt that in many areas BSC performance improved considerably from 1980–85, but the total of £1,550,000 eventually paid to Lazard Freres could perhaps have given Sir Monty Finniston (the last BSC Chairman at that time to report a profit) pause for thought!

A parting shot from Sir Charles

Even when the strike ended there still appeared to be little general appreciation of the depth of the financial trough into which the BSC had sunk. By continuing to emphasise in public that the 1980–81 EFL would not rise above £450 million, the government was able to create an impression that matters were under control, but the opposite was now the case. In 1979–80 the BSC loss before extraordinary items amounted to £545 million. These extraordinary items included the write-down of £1,141 million, and massive redundancy and closure costs of £464 million. During the year, the workforce was reduced by nearly 20,000, from 186,000 to 166,400. The loss obviously included the cost of the strike, which was estimated to be

£200 million (*BSC Annual Report and Accounts*, 1979–80), but even this was not the chief problem for, after the end of the strike, the financial position became still worse.

Perhaps surprisingly, the BSC was well within the 1979–80 EFL of £700 million, at a figure of £579 million, and the unused £121 million was added to the 1980–81 EFL of £450 million. Nevertheless, the government lost little time in raising the Corporation's borrowing limit from £4,750 million to £5,500 million from May 1980 (a move which we outlined earlier had been in the pipeline since at least January 1980).

The full depth of the BSC's troubles finally became clear in June 1980, shortly before the departure of Sir Charles Villiers. As a result of the strike the BSC's domestic market share had been depressed to below the 50 per cent mark (at around 47 per cent). For the first time, Sir Charles' letter of the 1st February to Sir Keith Joseph became public. As we described in the last chapter, this letter warned the Secretary of State that the BSC was likely to require between £300 million and £500 million extra on top of the existing 1980–81 EFL of £450 million. On 6th June, Sir Charles wrote again to Sir Keith explicitly stating that unless drastic action was taken then the BSC would be forced into liquidation. The Chairman considered that an extra £400 million was now required, and offered the somewhat bizarre solution of the BSC selling many of its assets in order to lease them back, and also to sell its debts to factoring companies who would charge a suitable fee.

The contents of Sir Charles' letter were (perhaps not surprisingly) only revealed to the House of Commons on 26th June. Sir Keith Joseph admitted that Sir Charles had previously written to him about the BSC's deteriorating position, but that the Chairman had not asked for more money. It was reported at the time (*The Guardian*, 27:6:80) that Ministers had been surprised by the contents of the June letter, yet the February letter had surely made the position clear. In any case, if the government was really unaware of the BSC's true position, then it made a mockery of the Departmental monitoring process.

Sir Keith's revelations had a traumatic effect on MPs who were generally dumbfounded to hear that the BSC might be on the point of liquidation. For the first time, like the public at large, they were being told the full extent of the BSC's problems. A typical reaction was expressed by Edward du Cann, Chairman of the influential Conservative backbenchers' 1922 Committee. He considered Sir Keith's revelations to be: 'the most grave statement about an industrial matter I have heard in this House, with its frightful economic and family implications' (Hansard, 26:6:80, Col.761).

Sir Keith himself inevitably rejected the leasing and factoring ideas, but then stated that the 1975 Iron and Steel Act made no provision for liquidation. The Secretary of State considerably confused MPs (perhaps an intention of the statement!) by declaring that the government could not enlarge the BSC's EFL until Ian MacGregor had produced new proposals for the Corporation's future. Shortly after this apparent continuation of the hard line approach, however, came the key concession:

> Meanwhile, having considered the Corporation's concern about taking on fresh financial commitments, I have asked the BSC to continue trading as an on-going business. I have told them that in the last resort the government would have to ensure that creditors of the Corporation had their claims met in full (Hansard, 26:6:80, Col. 755).

In other words, the government was finally admitting that it was not feasible to close down the BSC, and so a blank cheque was to be issued in order to cover losses. Even more, the government was claiming that such a closure would be illegal! This argument was not self-evident, for the Secretary of State surely had the power to direct BSC management to close down all steelworks, but this was now an academic debating point. The die had been cast (as in reality it had probably been for the previous twelve months), and the government was now committed to backing the new Chairman for as much as it would cost to take the BSC to the end of the long road to viability. Indeed, the timing of Sir Keith's statement suggests that the new commitment was designed to wipe the slate clean and hence give MacGregor the sense of making a fresh start. As we will see, the 1981 Iron and Steel Act put this government support into more tangible form. At the time of Sir Keith's statement it was reported widely that the government would write off £1,150 million of the BSC's debt to the Treasury (*The Times* and *The Guardian*, 27:6:80) but the eventual figure (£4,509 million) was nearly four times this amount.

On his last day as Chairman Sir Charles Villiers protested that the BSC could have survived by leasing and factoring, and that he did not want more money from the government (*The Guardian*, 28:6:80). However, Sir Keith Joseph had forced him to accept assistance. Sir Charles had presided over a period of great change for the BSC, and also suffered public criticism from Ministers, but his ordeal was now over, although he did remain Chairman of BSC (Industry). In later years Ian MacGregor was to be generally attributed with the BSC's manpower and capacity cuts, but as we have seen much of the spadework had already been done by Villiers. Under his Chairmanship, the BSC workforce was reduced from 208,000 to 166,000, with manned capacity down from 27 million tonnes per annum to 15 million tonnes per annum. Symbolic of Villiers' reign were the major redundancies agreed at Port Talbot and Llanwern shortly before his departure. At Port Talbot, the workforce was to fall by more than 5,000, from 9,353 to 4,899. In addition, another major closure was in the pipeline, and despite a strong fight by the local community, the Consett works in Durham closed in September 1980 with the eventual loss of 3,700 jobs. Consett represented the closure of a complete steelworks, but the policy with regard to Llanwern and Port Talbot indicated that the emphasis was now on reducing numbers at all works.

As we have seen, the outcome of the steel strike could not be described as a defeat for the ISTC, but the dispute had been (at least ostensibly) about money and not about jobs. The ISTC was to continue a rearguard action on the decentralisation of pay bargaining for several years to come, but on the basic output and jobs strategy it remained an 'outsider' body. Nevertheless, the ISTC did produce an alternative BSC strategy during 1980 which set out its case for higher output and the retention of manpower (ISTC, 1980) but neither the BSC nor the government was disposed to admit the steel unions to the inner sanctum of policy-making. On the contrary, Ian MacGregor was much less likely to seek consultations on BSC strategy than any of his predecessors, and the government was in the position where it now had little choice but to back MacGregor's judgement.

Even before the strike, the BSC had encountered little difficulty in securing the redundancies it sought, and the three month stoppage had done nothing to change this position. After the strike, the national leadership of the ISTC was in something of a politically weak position. With pay bargaining being gradually devolved to the local level, and their continued exclusion from the policy-making process, the leadership has not been able to find a truly active role for itself. Consequently (as BSC management no doubt sought) the union has a more fragmented nature than previously. During 1980 the ISTC itself tried to strengthen its political muscle by forming with the National Union of Railwaymen and the National Union of Mineworkers what became known as the Triple Alliance. The chief principle of the Alliance was to be on the lines of 'all for one and one for all', with the hope that each would support either or both of the others in any future strike action. In reality, the interests of the three unions were not sufficiently coincidental to ensure that these aims were achieved. The ISTC leadership was itself somewhat disappointed that the union had not received more help from the NUM during its own strike, and when the miners sought support from the steelmen during the 1984–85 coal strike it was not forthcoming. The ISTC membership feared that support for the miners might lead to more steelworks closures (a particular anxiety at the Ravenscraig plant), and so coal was imported and also moved into BSC works by lorry rather than the usual trains. By these means the Corporation's works remained in operation throughout the strike, but the Triple Alliance was shattered.

Structural reorganisation

As we have already discussed, the chief impact of Ian MacGregor on the BSC was not the concept of capacity cuts and job losses in themselves. Nevertheless, during 1980–81 the size of the workforce was drastically reduced from 166,400 to 120,900 (near to the 52,000 cut proposed in December 1979), although much of this reduction was in the pipeline before the departure of Sir Charles Villiers. In his first corporate plan MacGregor reduced BSC manned capacity only from 15 million tonnes per annum to 14.4 million tonnes per annum, and in 1985 it was still at this figure. Thus although MacGregor was 'hard' on jobs, his approach was comparatively 'soft' on capacity.

In the area of organisational structure there had also been some changes under Villiers, but MacGregor placed great store on making each section of the BSC a Profit Centre, with local management taking greater responsibility for the commercial fortunes of their own areas, while also facilitating a smoother transition to the hoped-for privatisation. MacGregor himself appointed the consultants McKinseys (shades of the Joint Steering Group of the early 1970s) to assess the performance of BSC's management, as well as areas of potential growth, and by September 1980 the new structure was in place. We described in Chapter Five how in 1976 the BSC introduced the dual concepts of Manufacturing Divisions/Product Units and Profit Centres, but this process was now taken much further. In February 1980 BSC Holdings was formed as a management group to comprise the main 'steel products' Profit Centres other than Tubes. Each of these Profit Centres (Tinplate, Stainless, Cumbria, Light Products, together with Forges, Foundries and Engineering) was now to operate as a discrete business, and would no longer report to any of the Manufacturing Divisions.

The complete transition of the BSC to the Profit Centre framework was completed in September 1980. Once more the pendulum swung back to a product based structure, with the two principal operating groups being General Steels and Strip Products. Within each of these groups, however, there were now formed:

> a series of discrete, decentralised, product-based businesses each operating as a Profit Centre. Previously, the commercial responsibility was centralised, but under the new organisation the head of each business was given the commercial as well as the manufacturing responsibility for the product concerned.... The new organisation is designed to assist the achievement of competitive standards of cost, quality and service needed by BSC's customers in a highly competitive world. Each business will look towards separate product markets and will be competing against specific

> domestic and foreign competitors. BSC believes the new organisation will help to motivate employees by enabling them to identify with the progress and success of their particular business (*BSC Annual Report and Accounts*, 1980–81).

BSC Holdings continued to operate as before, although the Tinplate section now became part of the Strip Products Group, BSC Tubes retained its identity as a separate Profit Centre, as did BSC Chemicals and Redpath Dorman Long Ltd (RDL).

The gradual fragmentation of the BSC's empire was to continue during 1981–82, and included the establishment of certain business activities as separate companies under the Companies Act (RDL and BSC (Chemicals) had operated in this way since the early 1970s). The most notable of these businesses to become a limited company was British Steel Service Centres Ltd (the activity of stockholding was assuming an increased importance), but also included several small sections of the Corporation's peripheral activities. The formation of Profit Centres, and the introduction of limited companies, illustrated that, in addition to seeking an improvement in commercial performance, MacGregor was also making every effort to privatise the Corporation. We will see later that, as the decade progressed, the BSC made strenuous efforts to develop a closer relationship with the private sector.

Before looking at privatisation, one further MacGregor feature with regard to organisation concerned a reduction in the size of the main Board. After its enlargement in 1978, MacGregor's policy was to strip it to the bare bones, and he appeared to have strong government support for this process. By 1983, the six worker directors and two civil servants who had been Board members in 1980 had all disappeared from the scene, although a retired civil servant remained on the Board. In addition, one senior official from the DTI, and one from the Treasury, continued to receive Board papers and attend Board meetings (Monopolies and Mergers Commission, 1988, 3.10). Despite the BSC's parlous financial position, MacGregor was not disposed to seek assistance from trade unionists or officials! Given the size of (unrepaid) government loans to the Corporation, it appears strange that Ministers would not want to have had some degree of direct civil service representation on the Board, but they appeared content to leave affairs in the hands of MacGregor and his management team, together with external business expertise. Between 1980–85 the number of executive directors increased from three to four (excluding the Chairman and Chief Executive), while for non-executive directors the number was reduced from five to four (although this category now included Ian MacGregor himself). Officially it was the Secretary of State who appointed members of the Board, but in this case it appears that it was MacGregor himself who made all the running in reducing its size.

Privatisation

One of the criteria by which Ian MacGregor's performance was to be assessed concerned the extent to which he was successful in privatising the Corporation. In this respect, his achievement was strictly limited, for between 1980 and 1985 only 11 per cent of the BSC's assets and 14 per cent of the workforce had moved into the private sector. Even this figure does not tell the whole story, for in many of the hived-off businesses the Corporation has retained a significant shareholding. From the inception of the BSC until the early 1980s the private sector had experienced less financial hardship than the Corporation. Its share of the domestic market (around 25 per cent) remained fairly steady over the years, for the higher quality steels in which many of the firms specialised had not been so susceptible to troughs in demand as had the BSC products.

From the early 1980s onwards, however, several of the larger privately owned works (usually operating as electric arc 'mini' plants) began to suffer increasing hardship through the general slump in demand. Inevitably, in product areas involving an overlap with the BSC, the private sector became increasingly unhappy about competing against what it considered to be unfairly subsidised steel. This unrest in the private sector created a delicate situation for the Conservative government, who were obviously keen to assist the private sector, but equally could not afford to see the BSC collapse. As we will see, the government eventually provided direct financial assistance to the private sector in order to assist the rationalisation process, but the other element in the policy consisted of creating joint venture companies between the BSC and the private sector. The resultant companies officially became part of the private sector, and both the BSC and government could claim that a privatisation programme was taking place, but in reality the outcomes were not so clear cut. Rationalisation often involved major closures in the private sector, while the BSC was left as the major beneficiary. Politically, these developments led some members of the British Independent Steel Producers' Association (BISPA) to become highly suspicious of what they considered to be BSC infiltration of their sector.

The first joint venture, known as Phoenix One, involved negotiations lasting nearly two years, but these were successfully completed in February 1981. The project involved a merger between the wire rod and bar manufacturing interests of BSC and the large engineering company Guest, Keen and Nettlefold (GKN). The new company, known as Allied Steel and Wire, had estimated net assets of £130 million, a turnover of £200 million, and a workforce of 4,800. BSC and GKN each held 50 per cent of the joint share capital, but the majority of the assets for the new company had previously been held by GKN, and the latter provided five out of six of the executive directors. The company was set up in 1981, with the BSC (or it could be said indirectly the government) providing initial working capital of up to £50 million. In 1987 a financial reconstruction took place, in which GKN withdrew from the company altogether and the BSC reduced its shareholding to 20 per cent. A group of City institutions took control, although the company management put £700,000 of their own money into the £181 million deal. The transaction gave the BSC £44m in cash and a continuing significant involvement in a now profitable company which relied on the Corporation for steel supplies. In 1988 ASW Holdings (as it had become known) was floated on the Stock Exchange.

After the success of Phoenix One, the tortuous series of negotiations on Phoenix Two must have come as a considerable disappointment to the government. Phoenix Two was devised as a means of reducing overcapacity in the engineering steels sector, and involved considerable rationalisation in the private sector. Firstly, the Duport company withdrew from steelmaking by closing one of its works and selling off a rolling mill to the BSC. Secondly, Hadfield's of Sheffield (which had been a prime centre of picketing during the steel strike) closed one works and introduced severe cutbacks at the other. Later, Lonrho, the owners of Hadfield's, formed a company with the BSC and GKN known as Hadfield's (Holdings) Ltd which eventually closed down the other Sheffield works. Thirdly, Tube Investments sold out to BSC their 50 per cent share in Round Oak steelworks jointly owned with the Corporation (Round Oak itself was shut down in 1982). The recession thus brought about rationalisation without the help of Phoenix Two, but unhappily for the government's privatisation plans it involved closures in the private sector and a relative increase in the BSC's empire.

After this rationalisation, BSC and GKN were (as in Phoenix One) left as the only Phoenix Two participants, but in February 1982 negotiations between the two parties broke down. It was intended that about two million tonnes of BSC's capacity per annum (out of a total of 14.4 million tonnes) would be integrated into Phoenix Two, while GKN would contribute only 400,000 tonnes. GKN feared that BSC wanted the implementation of Phoenix Two to consist mainly of the closure of the former's Brymbo works in North Wales (which it had purchased from the BSC in 1973) and at that time the company considered an improvement in demand was imminent. This optimism was not justified, however, and in 1983 negotiations were resurrected. Eventually, joint plans were submitted to the government, but despite their enthusiasm for privatisation, Ministers were reluctant to give their support to the new venture, for it was likely to involve them in providing substantial sums of new capital to the operation. Nevertheless, the government did eventually agree to inject £55 million into the new company, which began operating in April 1986 under the name of United Engineering

Steels (UES). The shareholding was split 50–50 between the BSC and GKN, even though the Corporation contributed 60 per cent of the assets. The creation of UES represented the largest piece of BSC privatisation to date, with 6,500 workers being transferred from the Corporation to the private sector.

Phoenix Three came to fruition more quickly and took place in November 1982 when Firth Brown, the steelmaking division of Johnson and Firth Brown, joined forces with BSC's River Don works in Sheffield to form the new company of Sheffield Forgemasters. The intention was to create a stronger base for the British forgings and special steels sector, but again this involved considerable rationalisation, with more than 1,000 of the 5,000 new jobs inherited by the new company disappearing. BSC had a 50 per cent share of Forgemasters, and a succinct comment on Ian MacGregor's bargaining expertise was provided by Johnson and Firth Brown's Finance Director: 'MacGregor knew that once the deal was announced [on the creation of Forgemasters] we were committed. But he could always walk away. He was very subtle. He knew exactly how hard to push' (*Financial Times*, 29:3:83).

From the outset, Forgemasters encountered considerable financial hardship. Despite rationalisation and help from the BSC (detailed below), the company lost £48 million in its first four years of trading. Finally, in 1988, Forgemasters was sold to its management for £26.2 million. After a redemption of loan stock, the BSC and Johnson and Firth Brown were left with just over £11.5 million between them (*Independent*, 17:5:88). Initially, the BSC had put in assets worth £41 million, including £17 million in cash, and so Forgemasters acted as a salutory lesson that privatisation was no guarantee of future success.

One further joint venture during the MacGregor Chairmanship of the BSC involved the creation of British Bright Bar Ltd with GKN and Brymill (the BSC took a 40 per cent stake). Again this involved the closure of two private sector works. Thus these joint ventures represented privatisation of a peculiar type. Assets were transferred to the private sector, but the BSC continued to be integrally involved in their activities, and it was indeed private sector works which often appeared to be in greatest jeopardy of closure as a result of these arrangements.

Perhaps the best example of the evolving relationship between the BSC and the private sector is represented in a plan which failed to reach fruition. In November 1982 Elkem, the Norwegian parent company of Manchester Steel (with two works in that area) was offered £16 million to close down by a consortium of companies operating in the same sector (wire drawing and reinforcing). The consortium consisted of Templeborough Rolling Mills and Allied Steel and Wire, in both of which the BSC owned 50 per cent of the equity, and Sheerness Steel. The presence of Sheerness Steel in this consortium was ironic, for it had long prided itself on its independent status (being the only private sector company to continue operations throughout the steel strike). In 1981 it had withdrawn from BISPA because it objected to the Association's campaign for government assistance, and also wanted a firmer stand against government aid to the BSC. By late 1982, however, Sheerness was operating at a heavy loss, and consequently was prepared to join forces with its traditional competitor in order to ensure survival.

Twenty-five per cent of the £16 million could have come from the government as part of its drive to rationalise the private sector steel companies, a factor which was seized upon by Manchester MPs who were angered that government money could be used to create redundancies in an area already suffering from high unemployment. In the event Elkem turned down the consortium's offer, but the threat to Manchester Steel by a joint public/private sector effort indicated that rationalisation of the UK steel industry could be a politically confused and messy business. (The fate of Manchester Steel was finally sealed in 1985 when Elkem sold its two plants to Allied Steel and Wire, which promptly closed them down!)

Sheerness Steel's ambivalent attitude towards the BSC was mirrored in other parts of the private sector, and led to a split within BISPA itself. In December 1982 one of the Association's largest members, Glynwed, followed Sheerness in resigning from BISPA. In addition to the objections formerly raised by Sheerness, Glynwed complained that, because of the BSC's large shareholding in a number of private steel companies, BISPA no longer represented only the private sector. The President of BISPA expressed sadness at Glynwed's resignation, but said that the Association's policy was to accept that there were more and more hybrid companies in the industry.

Politically, BISPA could do no other than take this line, for the private sector was already receiving government money to assist in its rationalisation process, and the government itself was keen to see more joint venture companies created. An intensive lobbying campaign by BISPA for government aid culminated in December 1981 when its President accused Ministers of giving extensive financial aid to the BSC while ignoring the needs of the private sector. Later that same month, the government announced that £22 million was to be made available for schemes which promoted rationalisation and restructuring of the industry in order to reduce overcapacity (in 1982 a further £12 million was made available to the scheme). The government also allocated a further £3–4 million in grants to promote restructuring and elimination of overcapacity in the steel castings industry. After a slow start all these sums of money were eventually taken up, with the consequent closure of several plants. The recession thus led BSC and BISPA towards a closer relationship with each other. The joint ventures fitted in well with the government's privatisation policy, but force of circumstance was also a prime mover in securing their implementation.

After the departure of Ian MacGregor several more relatively small-scale joint venture companies were formed (progress on privatisation was also given by the government to Robert Haslam and his successor Bob Scholey as a term of reference for the Chairman's job). BSC's shareholdings in these companies ranged from 74.5 per cent (a merger with Tube Investments known as Seamless Tubes Ltd) to 22 per cent (BSC's refractory interests merged in G.R. Stein Refractories Ltd). In addition, the BSC had also disposed of a miscellaneous group of largely peripheral interests. One of the largest of these occurred during April 1982 when the BSC's wholly owned construction subsidiary Redpath Dorman Long Ltd (RDL) was sold to Trafalgar House for £10 million. Trafalgar House already owned RDL's principal competitor, although RDL's workforce were anxious to purchase the company for themselves.

One sale which did involve considerable employee participation concerned Victaulic, a company making pipe joints and seals. The distinctive feature of the sale (for a sum of £15 million) was that at that time it marked the second largest employee purchase arranged in the UK (the largest being the sale of the National Freight Corporation). The Victaulic employees contributed a total of £838,000 to take a 40 per cent stake in the company, although the BSC itself purchased 30 per cent of the shares. In 1988, Victaulic was floated on the Stock Exchange.

Between 1980 and 1985 the BSC disposed of assets with a book value of £349 million, excluding land and property sales, which realised £74 million (*BSC Annual Report and Accounts* 1984–85). The completion of Phoenix Two in 1986 added a significant sum to that amount, but the BSC's close involvement in many of the newly privatised enterprises, and the Corporation's predominant position within the UK steel industry, suggested a blurring of the distinctions between the state and private sectors, rather than an outright switch from one to the other.

Phoenix and the DTI

Further light was thrown on the Phoenix projects in 1985, when the House of Commons Public Accounts Committee (PAC) produced a report on the control and monitoring of the BSC's investment in these curiously hybrid organisations (House of Commons Committee of Public Accounts 1984–85, HC 307). In order to assist them in their task, the members of the PAC had commissioned a Report from the National Audit Office. As a result of this Report, and oral evidence given by Department of Trade and Industry (DTI) officials, the PAC concluded that the assets put into Allied Steel and Wire by

GKN were probably less than the net book value suggested. In setting up Allied Steel and Wire the BSC had undertaken to provide cash for the company at particular intervals, and the PAC concluded that this money (the actual amount was not revealed) had been injected into the company in advance of its needs, thus considerably enhancing the trading results. The DTI justified this action on the grounds that the government intended Allied Steel and Wire to be a free-standing company, and to have rearranged the BSC payment would have entailed a degree of monitoring inconsistent with this intention.

On Phoenix Two, the BSC had spent £71 million in acquiring Duport Steel Assets and Round Oak, but the Corporation calculated that these acquisitions had contributed £32.5 million to its results by March 1984. The PAC was doubtful about this figure, and also disappointed that the DTI had not done more to check its validity.

Phoenix Three, Sheffield Forgemasters, had encountered great difficulty in remaining viable, and by the end of 1984 was in danger of collapse. At this point the BSC stepped in by providing £5 million in trading facilities and a £5 million loan for the purpose of further restructuring. The DTI conceded that this £10 million lay well outside the original guidelines for Forgemasters, but that Ministers had always intended these guidelines to be *desiderata*, and not firm rules.

The PAC Report indicated the extent to which the Phoenix projects had placed the government on the horns of a dilemma. Despite the considerable BSC financial input into Phoenix projects (and its 50 per cent shareholding in both Allied Steel and Wire and Forgemasters), the government wished to give the impression that these new companies were free-standing undertakings which required only the minimum amount of attention from the DTI, yet from 1980–85 the BSC itself was receiving massive sums of money from the government, at least some of which was likely to be set aside for the Phoenix companies.

The PAC was understandably concerned about the DTI's monitoring of the BSC-Phoenix relationship. When questioned on this subject, Sir Brian Hayes, Joint Permanent Secretary at the DTI, replied that the BSC Board received a quarterly report on Phoenix projects (passed on to the DTI), apart from the fact that it had directors on the boards of Phoenix companies (House of Commons PAC 1984–85, Q. 1604). When pressed further on this matter Mr. Murray, Head of the Minerals and Metals Division at the DTI, conceded that these directors had no separate reporting function to the BSC, as some of them were also members of the BSC board (HC 307, Q 1606). Strangely, in later supplementary evidence to the PAC, the DTI claimed that since 1984 the BSC had revised its monitoring of Phoenix companies. BSC Directors also on Phoenix Boards reported on a formal basis to the Corporation's Managing

Director, Finance. As a result of this monitoring the BSC Board now received a summary report each month on major holdings, together with any smaller holdings where there was a significant adverse change in performance. These reports were passed on to the DTI (it should be noted that by 1985 all civil servants had been removed from the BSC Board) (House of Commons PAC 1984–85, 27).

Thus the questioning from the PAC had apparently allowed the senior DTI officials to discover for the first time exactly how they managed to monitor the BSC-Phoenix relationship! The Phoenix affair once again raised the vexed question of the extent to which the DTI effectively monitored the activities of the BSC. In Chapter Eleven we will be returning to this subject, for it is one which continues to be of crucial general importance in determining how massive amounts of public money are spent.

Raw material and energy costs

In the BSC's search to cut costs, the lion's share of political attention was focused inevitably (and rightly) on the human element. A plant closure or a general reduction in the workforce has severe social repercussions which require urgent political attention. The BSC itself laid great emphasis on improving labour productivity, and in the 1980s achieved considerable success in this field. It can be argued that reducing the size of its workforce was one area where the BSC, in co-operation with the government, could act in a relatively autonomous way. The unions were basically by-passed in offering generous redundancy terms to steelworkers, and so the BSC's radical slimming operation was achieved with little significant political opposition.

In contrast, reductions in raw materials and energy costs are not so simply achieved. In the case of the most basic raw material, iron ore, the BSC's decision from the early 1970s to base its plant configurations on imported ore meant that it was highly dependent on international market forces and currency movements. The same was true with imported coal, while domestically the BSC's relationship with British Coal, formerly the National Coal Board (NCB), was frequently tense on the subject of coal prices. In the energy field, the BSC became increasingly aggressive with regard to the price of electricity, gas and fuel oil. Thus the Corporation found itself in direct opposition to three other Nationalised Industries – British Coal, the Central Electricity Generating Board and the (eventually privatised) British Gas Corporation. The BSC was able to wring some concessions out of these public (and former public) corporations, but the campaign continued up to and beyond privatisation.

In its own domestic policy community the BSC was clearly the most powerful interest, but when it came to dealing with interests, either at home or abroad, who possessed equal or even greater power than the Corporation, then objectives were not so easily achieved. In this wider world the Corporation, particularly with its financial problems of the 1980s, could appear to be a politically vulnerable organisation, e.g. the political power of the miners and the economic strength of the gas and electricity corporations allowed all three industries at times to pass on costs to the BSC. 'An integrated policy on energy pricing' is a favourite slogan of governments, but in reality, to paraphrase an old saying of Harold Wilson: 'One public corporation's price increase is another one's cost increase'. In some matters the Nationalised Industries learned to talk with a united political voice (the formation of the Nationalised Industries Chairman's group in 1978 is evidence of the industries perceived need for this united lobbying power), but in other areas, such as pricing, there was endemic rivalry.

For the BSC, success in cutting raw material and energy costs was vital, for potentially they would make or break the Corporation's viability. Notwithstanding the political attention given to labour costs, the BSC was not a labour-intensive organisation. In 1980–81, when at the end of the year the BSC workforce stood at 120,900, employment costs represented 30.7 per cent of total expenditure. On the other hand, for the same period materials represented 41 per cent of expenditure, and fuel costs 11 per cent (*BSC Annual Report and Accounts*, 1980–81). By 1984–85, when the BSC workforce at the end of the year had fallen to 64,500, employment costs had also fallen to 23.5 per cent of total expenditure. Materials had now jumped to 50 per cent of total expenditure, while fuel and utilities stood at 17.4 per cent (*BSC Annual Report and Accounts*, 1984–85).

In this section we will be examining briefly how these costs arose, and also the BSC's push, at least in some areas, to achieve savings. Tension between the BSC and its suppliers was of course endemic, and therefore of long duration, but after MacGregor took over as Chairman there was undoubtedly an increased amount of political activity, particularly with regard to coal and energy prices. In these cases domestic political pressure bore fruit, but we begin with iron ore, where it is international forces which predominate.

Iron ore

After 1945 the UK steel industry became increasingly dependent on imported iron ore. This process was considerably strengthened by the creation of the British Iron and Steel Control, which operated as a central buying agency to the companies as a whole. Heal outlines the situation thus:

> The role of the BISC (Ore) which had been established during the War was to obtain all imported ores for the British industry and to average the costs to all users, subject to minor price differences or inefficiencies. An almost inevitable consequence of this cost-levelling procedure was that it removed the impetus to perfect the local operations, because the financial gains would not be fully realised locally but would be spread throughout the industry. Capital investment in ore docks was, therefore, less attractive than investment within the works themselves, and would remain so until such time as the state of the ore supply system threatened to undermine the whole financial health of the industry. There was the added difficulty that improvements to the ore docks involved co-operation with outside bodies, negotiations with which were apt to be inconclusive, and certainly protracted (Heal, 129).

BISC (Ore) purchased a number of ore carriers, although these tended to be smaller and therefore less efficient than those operated by other steelmaking countries. It took a long time for 60–65,000 tonne carriers to be accepted (until the mid-1960s the companies generally thought that 30,000 tonne carriers were adequate) and even longer for the docks to be built which could accommodate these larger carriers. After several years of discussion, approval was finally given in 1967 for construction of the Port Talbot ore terminal. This development was completed in 1970 at a cost of £20 million, with the terminal able to take vessels carrying 100,000 tonnes (see Heal, 127–134). We also saw earlier that the £100 million Hunterston development in Scotland only came into use in 1979 after several delays.

Despite this slow development, the generally cheaper cost of foreign ores compared with those produced domestically constituted an irreversible change, and was obviously a major factor in determining the development strategy of the early 1970s. From 1964, consumption of imported ore outweighed that for home mined ore, and since then the gap has increasingly widened. As a result, in 1984 home mined iron ore amounted to only a trickle at 403,000 tonnes (chiefly in Lincoln and Northants, a relic of what had once led to the development of Corby), whereas the figure for imported ore stood at 13,525,000 tonnes (although this total could be contrasted with the 19,039,000 tonnes of imported ore consumed in 1976, when 4,992,000 tonnes of home ore was also used) (UK Iron and Steel Statistics Bureau).

The UK's reliance on imported ore means that it must bargain in a world market. In the 1980s, as supply began to outstrip demand, prices tended to slacken, but during 1984 some improvement in trade for the steelworks led to increased demand for iron ore, and as the BSC commented in its *Annual Report* for 1984–85: 'frustrated attempts by consumers to achieve major dollar price reductions to offset adverse currency effects'. These adverse currency effects referred to the decline in the value of the pound against the dollar – a development of considerable importance to the BSC given that, as the 1984–85 Annual Report also pointed out, approximately 50 per cent of the Corporation's supplies and transport expenditures were linked to the US dollar. Forward currency cover arrangements delayed the worst of the dollar's strength, but the importance of exchange rates to the BSC was clear. In giving evidence to the Trade and Industry Select Committee in 1984, BSC Chairman Sir Robert Haslam thought it fair to say that: 'for every cent the dollar hardens against the pound we lose about £4 million on our profit and loss account' (House of Commons Trade and Industry Select Committee, 1983– 84, Q13). This currency movement was obviously beyond the BSC's control, yet when the value of the pound against the dollar can vary by over 40 per cent, as it did during 1984–85, then Sir Robert's own calculation meant that the plus or minus effect on BSC finances would be potentially considerable.

In order to offset adverse currency movements, the UK must attempt to strike the best bargain it can for imported iron ore. This bargaining can lead to wide fluctuations in the amounts of ore imported from different countries. In 1984 Canada remained the UK's largest supplier of iron ore, but the amount decreased to 3,645,200 tonnes from a figure of 5,227,500 tonnes in 1983. In contrast, Australia, which aggressively marketed its ore throughout the world in the 1980s, increased its ore imports into the UK from 1,093,300 tonnes in 1981 to 3,149,200 tonnes in 1984 (ISSB). By 1985, the Australian share of the European market for iron ore was said to have increased to 14 per cent from 11 per cent the previous year.

Yet notwithstanding this increased penetration, the largest Australian iron ore companies, Hamersley and Mt. Newman, were able to obtain significantly higher percentage price increases in Europe than in Japan. For Japan, the percentage increases in 1985 compared with 1984 were 1.42 per cent for iron ore fines and 2.2 per cent for lump ore. In Europe, the comparable figures were 4.25 per cent and 6.4 per cent (*Financial Times*, 17:7:85). Thus, as with currency movements, a three or four per cent difference in ore price could make a difference of several million pounds to BSC accounts (the value c.i.t. of UK iron ore imports totalling 14,159,600 tonnes in 1984 was set at £225,623,000, ISSB). Nevertheless, in the 1980s the advantage in price lay with the steelmakers rather than the ore producers. This advantage is likely to continue, for Brazil, the world's largest iron ore exporter (although only third in the UK table behind Canada and Australia), is continuing to expand its output, with the massive Carajas project (deposits of 18 billion tonnes) beginning production in 1986, and then rapidly building up to a production level of 35 million tonnes a year (*Financial Times*, 21:5:84).

The predominance of international market forces in the iron ore market meant that little was heard of this important component of BSC costs within the domestic political debate. Unfortunately, one attempt by the BSC to achieve a more stable environment for its iron ore supplies ended in a disastrous failure. In 1975 BSC joined the Quebec government-owned steel group, Sidbec, in a joint project to develop an ore mine on the north shore of the St. Lawrence river. A US Steel Company subsiduary (QCM) undertook to develop the mine and operate it, and so Sidbec (50.1 per cent), BSC (41.67 per cent) and QCM (8.23 per cent) became joint shareholders in the mine. The attraction for the BSC was that Sidbec-Normines had a relatively low capital cost, while the ore was to be purchased at the Great Lakes price (fixed by US and Canadian producers in the area), which at that time was below world prices generally. Unfortunately, by the time the mine came into use, world ore prices had fallen well below the Great Lakes price.

The catch for the BSC was that it had undertaken to buy ore from Sidbec-Normines in proportion to its shareholding, and so in purchasing 2.5 million tonnes of ore a year was making a loss of £20 million a year. At the same time the Corporation was understandably wary about paying closure costs, not to mention repaying its original investment. Finally, in October 1984 the BSC did write off its Sidbec-Normines investment, but at a cost of £103 million. As an epitaph to this disastrous undertaking a BSC official said: 'What seems a good idea at the time ain't necessarily so' (*Financial Times*, 12:12:84). This comment seems something of an understatement, yet the £103 million write-off, not to mention the millions lost over the years in operating the mine, attracted negligible domestic political attention. These Sidbec-Normines losses could be contrasted with the £180 million which the BSC estimated was the cost to its finances of the 1984–85 miners' strike. This is not to minimise the effects of that strike on the BSC, but it does illustrate how seemingly obscure (if highly expensive) events in far-off places could be written off by a public corporation as 'just one of those things' while political attention was engaged elsewhere.

Scrap

In contrast to iron ore, imported scrap plays only a negligible part in UK steelmaking. Indeed, the industry itself produces nearly half its own needs through the nature of the steelmaking and milling process in which steel scrap may be recycled (in 1984 scrap arising from the steel industry itself amounted to 3,736,000 tonnes, out of a total consumption of 7,886,200 tonnes (ISSB). The price of scrap has tended to fluctuate widely over the years, and is notoriously difficult to predict. We saw in the case of Shelton that the cost of scrap was one of the primary bones of contention between the Action Committee and the BSC, over the potential cost of the steel produced by the arc furnace route. In the UK, arc furnace steel which is produced entirely from scrap, was principally the domain of the private sector. For this reason the price of scrap was of more importance to the private steelmakers than the BSC. Nevertheless, in the late 1980s the BSC operated 15 of the 39 arc furnaces at steelworks in the UK (chiefly in special steels around the Sheffield area) and these in total represented about 10 per cent of the Corporation's output (some scrap is also used in oxygen steelmaking).

During the 1980s the already delicate relationship between the steelmakers and the British Scrap Federation became even more fraught over the question of exports. As the UK demand for steel slumped in the late 1970s, the British Scrap Federation pressed for the official restrictions on exports to be lifted. The new Conservative government granted this request in 1979, after which time the level of scrap exports more than trebled in five years (from 1,338,700 tonnes in 1979 to 4, 315,500 tonnes in 1984, ISSB). By far the chief destination for these exports was Spain, which took 2,222,700 tonnes of scrap in 1984, compared with only 646,000 tonnes in 1979, ISSB). The enormous increase in scrap exports, worth more than £302 million in 1984, inevitably heightened the tension between the scrap merchants and the steel producers. Scrap prices more than doubled between 1983 and 1985, leading to pressure for the re-introduction of export controls from both the BSC and BISPA.

They were joined in this campaign by steel producers from several other countries, in particular Belgium and Italy, and the EEC Commission came under great pressure to restrict exports outside the Community generally. The problem for governments was to balance the needs of steel producers, particularly those using arc furnaces, against the increasing value of scrap as an export business. From the UK point of view the admission of Spain into the EEC in 1986 made the imposition of controls more difficult, for restricting exports to a fellow EEC member was likely to be a politically more difficult process than limiting trade with a country outside the Community. Nevertheless, in contrast to iron ore, scrap is a commodity which is susceptible to political management. In the UK, the almost total reliance on scrap arising domestically means that the government is forced into making a political adjudication between rival interests.

Coal

In the case of iron ore domestic political activity is largely irrelevant, while with regard to scrap the issue of exports tended to become a more politically delicate problem after the departure of MacGregor from the BSC. In contrast, on the subject of coal, MacGregor made a major political issue of the relationship between the BSC and the National Coal Board (NCB) almost immediately after taking over the Chairmanship of the Corporation. He summed up one aspect of his argument when giving evidence to the House of Commons Industry and Trade Select Committee in 1981:

> For years the British steel industry ran on the fact that we had relatively low costs of labour and we were therefore using our domestic coals to produce iron in quite inefficient ways – a very high consumption of coke, maybe as much as 1,500 lbs per tonne at one stage. Nowadays we cannot do that because we no longer have a low-cost wage structure. During the inflationary period of the 1970s our unit costs rose substantially faster than those of our competitors (House of Commons Industry and Trade Select Committee, 1980–81, Q101).

MacGregor was here making the point that the BSC had not always made the best use of its coking ovens, but he was also greatly concerned about the cost of coal. As coke is a basic ingredient in the production of pig iron, then it was inevitable that the iron and steel industry would initially be developed in close proximity to the coalfields. Many of the steel companies were also pit owners, although we saw in Chapter Four that the more labour intensive nature of coal mining made that industry more susceptible to labour unrest than the steel industry. When the iron and steel plant configuration moved towards the development of coastal plants, the obvious implication was that the BSC would obtain a higher percentage of its coal needs (and not just iron ore) from abroad. The rundown of the domestic iron ore industry was implemented with little political debate, but coal is a different matter altogether.

In the 1980s pit closures became a highly emotive subject, culminating in the year-long miners' strike of 1984–85. We will be examining the implications of that event for the BSC in the next chapter, but from the beginning of the 1980s the BSC sought to loosen its reliance on domestic coal still further, while also receiving a subsidy for maintaining some degree of home supplies. To a large extent it succeeded in achieving both these objectives, although the Corporation remained the NCB's second largest customer after the Central Electricity Generating Board. The government itself performed something of a U-turn on this subject, for in 1979 it removed the separate coking coal subsidy which had been in operation for five years. By 1981, however, it was prepared to encourage the NCB to sell coking coal to the BSC at the subsidised price. Pryke sums up the situation in 1979 thus:

> Home produced coal was a third more expensive, its use involved considerable cost penalties and it seemed likely that foreign coal's price advantage would widen. Moreover the costs of imports would have had to rise about 50 per cent in order to make British coking coal competitive. Also it must be remembered that much of the coal which the Corporation received from the NCB was being produced at a large loss. As the world steel industry was so depressed there would be no problem in obtaining supplies from abroad and the only obstacle to their use – apart from the miners and the Coal Board – was the handling problems that would arise at the ports if imports were increased beyond 6 million tonnes (Pryke, 207).

Ian MacGregor saw a rapid increase in imported coal as a primary means of reducing BSC costs, although quality was also a factor. A mixture of coals is necessary in order to obtain a satisfactory blast, and much of the best quality coking coal comes from abroad. Shortly after his appointment, MacGregor proposed to Sir Keith Joseph that henceforth the BSC should import all its coal. The Chairman estimated that this switch in orders would result in a saving to the Corporation of £135 million a year, and it was reported at the time that Sir Keith had expressed support for this statement to the Cabinet itself (*Now*, 8:8:80). Not surprisingly, many Ministers were alarmed at this suggestion, for the government was extremely sensitive about forcing a confrontation with the Mineworkers. Memories of 1974, when a united NUM strike had led to a General Election and the downfall of the previous Conservative government, were still relatively fresh at this time, and the government was not yet ready to risk a repeat performance.

We described in the last chapter how the BSC's commitment to buy at least 5 million tonnes of its coal from the NCB during 1980 had cost the NCB a subsidy of £22 million to the Corporation during that year. This transfer of costs from one loss-making Nationalised Industry to another was the economic price the government was prepared to pay in order to maintain peace at both BSC headquarters and the coalfields (Sir Keith Joseph had acknowledged that the government ultimately was responsible for the losses of the state-owned industries).

In addition to subsidising the BSC, a further problem for the NCB was the Corporation's decreasing demand for coal brought about by the slump in steel demand. By 1980–81, BSC coke consumption at its blast furnaces was reduced to 4,984,300 tonnes, from a figure of 6,808,900 tonnes only two years previously in 1978–79 (BSC Annual Statistics 1980–81). Thus the Corporation could stay within its import restrictions while decreasing its use of NCB coal. In 1979–80 the BSC purchased 6.9 million tonnes of coal from the NCB, compared with 8.5 million tonnes the previous year. On the other hand, BSC imports of coal from 1977–78 to 1979–80 (chiefly from the

USA, Australia and Poland) had more than doubled from 1 million to over 2 million tonnes a year. Even with its reduced use of NCB coal, the BSC had a sizeable coke surplus during 1980, and was reselling it at prices which undercut those of the Coal Board! (*Financial Times*, 22:8:80).

MacGregor, having laid down extremely aggressive markers for the Corporation's negotiations with the NCB, was not prepared to make easy concessions, but in September 1980 agreement was achieved on domestic BSC coal purchased for 1981. Despite its threat to purchase only from abroad, the Corporation agreed to buy just over 4.5 million tonnes of coal from the NCB, but for its part the Coal Board was prepared to subsidise the price of this coal up to a sum of £40 million. The NCB intended to claw back over half this subsidy by buying larger amounts of imported steel.

The continued interdependence between the BSC and the NCB was thus largely a political construction, and reflected the government's wish that neither of these major Nationalised Industries should desert the other in its (and the government's) hour of need. Unfortunately for the government, unrest in the coalfields continued to grow, until by February 1981 there was the real prospect of another major national strike by the miners. Faced with this confrontation the government retreated, and in addition to withdrawing a list of proposed pit closures, Ministers also promised action on restricting imports. This decision was taken hurriedly, and apparently without consulting the two chief customers of the NCB, the CEGB and the BSC. Thus despite its September 1980 agreement with the Coal Board, the BSC was now faced with the prospect of renewed government pressure to increase its purchase of domestically produced coal.

Characteristically, Ian MacGregor was initially resistant to the prospect of reducing BSC coal imports and stated that: ' it would be tragic if the British Steel Corporation was asked to take a step backwards from its attempt to make itself reasonably efficient' (House of Commons Industry and Trade Select Committee 1980–81, Q 100).

The first major test for the new arrangement came in April 1981, when a contract to supply Port Talbot with 450,000 tonnes of coal a year came up for renewal. This coal had previously come from the United States, but the NCB won the new three-year contract. The NCB price was subsidised to comply with world prices, at an estimated cost of £4.5 million to the Coal Board (*Financial Times*, 24:4:81). The government played a prominent part in ensuring that the NCB won this contract, for it could not afford to renege on its February commitment to the NUM, whose leaders were now demanding that coal imports be set at an 'irreducible minimum'.

After 1981, the BSC and NCB signed contracts on a similar basis. Thus in October 1983 they concluded a three-year agreement whereby 4 million tonnes a year of NCB coal would be provided to the BSC at prices related to those of imports. The miners' strike of 1984–85 inevitably interrupted this arrangement, but there were obvious political pressures to ensure the agreements continued. Figures for coal supplies for the years 1980–85 are given below, and indicate that, until the miners' strike year 1984–85, the ratio between NCB and imported coal remained relatively steady.

Table 1 : BSC coal supplies 1980–85 (million tonnes)

	NCB	IMPORTS
1980–81	4.6	2.7
1981–82	5.5	3.1
1982–83	3.8	2.0
1983–84	4.2	2.6
1984–85	0.3	6.9

Source: BSC

The source of the imports was generally restricted to three countries – the USA, Australia and Poland. In 1983–84 these imports comprised 42 per cent from the USA, 42 per cent from Australia and 16 per cent from Poland, a pattern similar to earlier years. Thus, it appeared that the BSC had settled into a pattern in which it purchased about 34.5 to 38 per cent of its coal from abroad.

Apart from steel demand, coal usage by the BSC would also decrease as a result of increased efficiency in the coke ovens. Sometimes the new technology could lead to problems, as at Redcar in 1982 when the Corporation was forced to rebuild ovens which were only three years old after deterioration of the brickwork. This setback led the BSC to increase its purchase of coke – a commodity in which it would normally have hoped to be virtually self-sufficient.

Table 2 : BSC purchased coke 1982–84 (million tonnes)

1982	1.1	(43% imported)
1983	1.2	(61% imported)
1984	1.4	(81% imported)

Source: BSC

The UK has lagged behind many other EEC countries with regard to coke consumed per tonne of iron (see Darnell *et al.*, 1984, 30) and in 1985 the BSC estimated that there could be a deficit equivalent to 20 per cent of coke consumption. Thus the economics and efficiency of coke ovens has become an area where significant investment is thought to be needed. In 1985 the BSC announced that it would spend up to £10 million on equipment to permit direct injection of pulverised coal into the Ravenscraig blast furnaces. It was intended that this new process would displace some of the higher-cost coke at the blast furnaces. In this case the use of lower quality coke could possibly have helped the NCB to increase its supply of coal to the BSC, although on the whole the BSC did not wish to become greatly dependent on domestic coal supplies. The BSC also sought to extend the lives of the coke ovens by means of silica welding, and also increased the use of scrap in BOS vessels.

An interesting incident occurred towards the end of 1985, when the NCB considered development of a £90 million coking coal pit at Margam near Port Talbot. Ian MacGregor wished to shelve the plan, apparently on the grounds that the NCB would not be obliged to receive regional aid (the government denied this assistance to the NCB on the grounds that it was receiving enough subsidy already). Confusion reigned, however, when Energy Secretary Peter Walker (the Minister who, more than twelve years earlier, had given the green light to the BSC for its massive development plan) declared that his Department would be ready to discuss how the government could help in furthering the project. MacGregor then claimed that the plan was being abandoned because of the weak international demand for coking coal (*Financial Times*, 12:12:85). Apart from illustrating the lack of communication between the NCB Chairman and the Energy Secretary, the incident also indicated that the government might be forced to backtrack once again and grant the Coal Board a subsidy in order to prevent the BSC from obtaining still more of its coking coal needs from abroad.

In the wake of the 1984–85 miners' strike, the number of South Wales pits was halved, and although doubts remained about the price of coking coal, British Coal (as the NCB had become) decided in 1987 to approve the £90 million Margam development. It was intended that the new pit would employ about 800, and produce, for use by the BSC, 1.2 million tons a year of high quality coking coal. Politically, Margam offered an opportunity for the government to build new bridges in an area which had become highly alienated as a result of the miners' strike.

In addition, Margam was also used as a lever for BC (with government backing) to introduce new working practices. Government approval of the investment was made conditional on the workforce accepting a six day week, instead of the traditionally accepted five day period. The national leadership of the NUM refused to accept the six day scheme, but locally the South Wales miners voted in favour, and Peter Walker (although refusing BC's request for a regional development grant) gave formal permission for the investment to go ahead. The BSC had also voiced its approval of the six day week and claimed that without it the Corporation would probably be compelled to purchase cheaper coking coal from abroad, rather than take supplies from Margam. In this case, the government was able to use inherent competition between state owned industries in order to win concessions from the miners, and also split the NUM still further. In the final chapter we will expand on this important theme of how government power *between* industries can be more effective than government power *within* an industry (even when state owned).

The campaign for lower energy costs

The BSC's push for lower energy costs was one of the most politically visible aspects of its activities during the 1980s. When the BSC talked about energy it was chiefly referring to electricity, gas, fuel oil and oxygen, for the Corporation classed coal as a raw material rather than an energy source. The chief emphasis was on obtaining lower prices and greater flexibility for gas and electricity contract prices, and a reduction in the tax on fuel oil. All these are domestic political matters, and it is surely no coincidence that the BSC chose to make a major political campaign out of the issue, i.e. the Corporation realised that it is an area where pressure on the UK government itself (as in the case of coal from the NCB) can reap direct financial dividends.

In order to pursue its energy costs campaign, the BSC chose to work closely not only with BISPA, but also with a number of other trade associations, the Confederation of British Industry (CBI), the National Economic Development Council (NEDC), and the NEDC's executive arm the National Economic Development Office (NEDO). Thus although the campaign (which continues after steel privatisation) is a domestic one, a variety of interest groups have perceived that to take on the economically and politically powerful energy producers over this issue requires the creation of a strong countervailing alliance. It could be said that, even more than the energy industries themselves, the chief target for the alliance is the government. Electricity and gas (until its privatisation in 1986) were two of the few Nationalised Industries in the 1980s to make a profit, and thus made a positive contribution to government finances. This contribution made the government extremely wary about authorising lower energy costs, for to do so would have put a nasty dent in the government's annual income, which then needed to be made up elsewhere (in higher borrowing or taxes) if public expenditure was not to be reduced accordingly (although it could also be argued that lower energy costs would have reduced the losses of other state owned organisations such as the BSC). A further complicating factor was the government's intention to privatise the British Gas Corporation (BGC). In order to obtain as high a price as possible, the government could not afford to deflate gas prices below a level which made the BGC a less attractive proposition for prospective purchasers. After privatisation of

British Gas, energy consumers switched their pressure to seeking a tighter regulatory framework for the industry, but with only limited success.

For these reasons, although the energy consumers' alliance achieved significant results, the BSC continued to fight an uphill battle on energy costs. In 1979–80, purchased fuels represented 9.2 per cent of BSC costs, whereas in 1984–85 the figure for external charges (which includes fuel and utilities) was 17.4 per cent. The government's own economic and political need for a high rate of return from the energy industries (high profits for BG could be taken as an indicator of successful privatisation) inevitably made further progress on this front a difficult task for the energy using interests.

The campaign gained its first impetus during 1980 when the BSC and BISPA joined forces to attack steep rises in electricity costs. The most prominent case occurred in the BSC's Yorkshire and Humberside Division, where the Yorkshire Electricity Board sought a price increase of 30 per cent. The private sector was particularly worried about electricity prices, given its use of electric arc furnaces, and in July 1980 joint BSC-BISPA submissions were sent to Sir Keith Joseph and the Secretary of State for Energy, David Howell. The chief argument put forward by the steel industry was that UK energy costs were much higher than in nearly every other EEC country, and so acted as a considerable handicap in its efforts to be truly competitive. Specifically, the BSC and BISPA called for an immediate official investigation into the UK pricing and supply policies of the oil companies, the immediate removal of heavy fuel oil excise duty, a reduction in the effective cost of electricity, and a drastic change in the way electricity prices were determined for bulk customers. This last request referred to the industry's wish to negotiate direct with the Central Electricity Generating Board, and so achieve some reasonable stability on prices.

In August, representatives from BISPA met Sir Francis Tombs, Chairman of the Electricity Council, who agreed to examine possible changes in the tariff structure. BISPA's hopes were soon dashed, however, when the Electricity Council announced that it was sceptical about the steelmakers' case and claimed that electricity costs were broadly comparable with the rest of Europe.

By this time Ian MacGregor had become involved in the campaign, and he commissioned the BSC's auditors, Coopers and Lybrand, to revise the Corporation's cost measurements (*The Times*, 20:10:80). MacGregor wanted to obtain increased efficiency in the BSC's own energy use in addition to reducing prices, and linked this with the campaign for lower coal prices. In November and December 1980 the strands of the energy campaign began to come together when, firstly, the CBI produced its own case for lower energy prices, and then a study from the consultants Industrial Market Research (IMR), commissioned by the BSC and BISPA under the aegis of the National Economic Development Office's Iron and Steel Sector Working Party, claimed that the UK steel industry's bill for gas, electricity and fuel oil of over £600 million pounds a year could be cut by £100 million if prices were more in line with those found in the rest of the EEC.

The BSC, as a member of the CBI, could join with BISPA and other private sector trade associations and companies in a joint campaign for lower energy costs. The problem for the energy users with regard to the CBI was that the Confederation also included the major oil companies amongst its members. The BSC and others considered that a campaign run solely within the CBI would inevitably involve a degree of compromise, and so eventually the energy users were to split off into a group of their own.

The NEDO Iron and Steel Sector Working Party (SWP) was a tripartite forum involving the government. The NEDC had been set up in 1962 as a forum for discussion, whereby the principal industrial and governmental interests could come together in an attempt to seek remedies for the UK's economic decline (for a history of the NEDC see Middlemas, 1983). Economic Development Committees (EDCs) and SWPs were created over the years to develop the work of the NEDC at the micro level. The Iron and Steel SWP (later an EDC) enabled representatives of BSC management and the unions to get together in a more informal and unobtrusive manner than was usually the case, and was thus of some use in healing scars after the steel strike (see Middlemas, 229).

Nevertheless, the Iron and Steel SWP, which met around once every three months, was not a forum where the BSC and union officials could hammer out an agreement on the BSC's Corporate Plan. An EDC or SWP could conduct research on topics relevant to the industry, which it may be hoped would filter through to the grassroots level, and it was in this vital (if relatively unspectacular) 'nuts and bolts' area that they had their greatest effect. Little Neddys – the term by which EDCs and SWPs have often been known, were reconstituted every two years. In 1985, the Iron and Steel EDC consisted of an independent Chairman, three representatives from the BSC, two from BISPA, three from the trade unions, two officials from the Department of Trade and Industry, and two from NEDO. During its two year term an EDC would often choose to concentrate its attention on a particular aspect of the iron and steel business, e.g. in 1979 the SWP produced reports on international comparisons of steelworks efficiency, but in early 1981 it was decided to turn the spotlight on energy prices. Apart from the intrinsic importance of the subject to the industry, it was also one on which the BSC, BISPA and the unions could all unite – an important consideration when memories of the steel strike were still fresh.

In January 1981 joint pressure from EDCs and SWPs representing high energy users such as iron and steel, chemicals, foundries and paper and board, succeeded in putting energy prices on the NEDC agenda itself, with the result that a special tripartite Task Force was set up to examine the problem. The creation of this Task Force signified that energy prices had now become a politically salient topic to which the government would be forced to direct its attention. The Task Force produced its first report in March 1981, and this claimed that UK electricity prices were 20–35 per cent more expensive than those found in France, and 25 per cent more expensive than West Germany, while on gas the disparity with the EEC was claimed to be 10–15 per cent. With regard to fuel oil, the Task Force sought a reduction in excise duty in the forthcoming Budget.

In the event the energy users were to be disappointed on the subject of the fuel oil tax, as they were to be in succeeding Budgets (although the tax remained static, leading to a decrease in the real price), but there were concessions on gas and electricity. The major policy change was on gas, where in 1980 the government had decided that each year for the next three years gas prices would be 10 per cent above the rate of inflation. Industrial and commercial users of gas were now given a price freeze which would last until the end of the year. In the event, apart from a rise of 1p per therm in March 1982, these prices remained frozen for a further three years, until April 1984. The concessions on electricity were less generous, and amounted to around 3 per cent provided users agreed that supply could be interrupted at 15 minutes notice for up to 365 hours a year. A limited number of arc furnace steelmakers took up this offer during 1981, but in the 1982 Budget there were further potential reductions of up to 16 per cent for large consumers of electricity. In addition, in November 1982 the Electricity Council announced a price freeze for the whole of 1983, while after 1984, rises in the price of electricity were much lower than formerly.

This campaign for lower energy prices can thus be claimed to have been a success, although the heavy energy users had still not succeeded in their aim to negotiate prices directly with the Central Electricity Generating Board. In addition, in August 1982 the government refused to make any specific concessions on energy prices to the BSC itself, despite the Corporation's claim that these costs were placing it at a disadvantage of around £40 million compared with France and West Germany.

The NEDC produced one further Energy Task Force report in 1981, but after that time the intensive energy users settled down, from their initial 'fire brigade' type of activity, into a more permanent (but less politically visible) lobby. The Iron and Steel EDC continued to do work on energy costs, while the CBI basically took over the Task Force role and produced annual reports comparing UK industrial energy prices with those to be found in other EEC countries. The most interesting development in this area was the creation of the Energy Intensive Users' Group (EIUG). This is basically an offshoot of the CBI trade associations who initially campaigned for lower energy costs, but nevertheless wish (for reasons we described earlier) to remain a separate entity from the CBI itself. The original members (known colloquially as the 'Energetic Seven') were the BSC, BISPA, the Chemical Industries Association, the National Federation of Clay Industries and the British Ceramic Manufacturers Federation, the British Man-Made Fibres Federation, the Glass Manufacturers Federation, and the British Paper and Board Industry Federation. They were later joined by the Cement Makers Federation. The importance attached to the EIUG by its members is illustrated by the fact that the Glass Manufacturers Federation subsequently resigned from the CBI, but remained a member of the group.

The economic importance of the trade associations involved in the EIUG, together with the continuing close involvement of the NEDC and the CBI, ensured that the energy users' alliance (of which the BSC was a prominent member) had a firm 'insider' status, whose interests could not be ignored by government. On the other hand, as we have seen, the government had its own reasons for maintaining energy prices and taxes at a certain level, and so the BSC was not likely to achieve any easy victories in this area.

Privatisation and regulation

In 1986 British Gas was privatised, while after its victory in the 1987 General Election the government announced its intention of privatising the electricity industry within the lifetime of the new Parliament. For the EIUG (and the BSC itself) these changes did not entail any lessening of the campaign for lower energy prices, but instead caused a switch in emphasis, particularly with regard to gas. The privatisation of BG as a single unit, and the lack of any competition in supply, pushed the industrial users toward seeking a tighter regulatory framework for gas prices, while also endeavouring to win a more direct influence over the affairs of BG. This new campaign led, in 1987, to one significant failure and one notable success.

The fall in oil prices during 1986, a commodity to which the price of gas is linked, did allow some customers to obtain lower prices, but the EIUG complained that BG had not adjusted its prices to the extent of other European suppliers. For the government privatisation of BG took precedence, and Ministers were not sympathetic to any changes which entailed a deflation of gas profits. After privatisation, gas prices to the domestic consumer were tied to a relatively strict regulatory framework linked with the rate of inflation, but for industrial consumers the government allowed BG to continue its existing contracts. The lack of regulatory control disturbed the EIUG, and led the troubled steel company Forgemasters to propose Sir Ian MacGregor (now retired from both the BSC and BC) as a

Director of BG. It was intended that Sir Ian would campaign specifically for lower gas prices to industry. Not surprisingly, the appointment of Sir Ian was vehemently opposed by the BG Chairman, Sir Denis Rooke, himself a formidably aggressive political operator. At the first AGM of BG in August 1987, shareholders voted by a large majority against the appointment of Sir Ian to the Board.

The setback for the EIUG was more than compensated for later in 1987, when the group won a significant victory on gas prices. During the year the EIUG lobbied the Office of Fair Trading, the Monopolies and Mergers Commission (MMC) and the EEC Commission, over what they considered to be BG monopoly profits won at the expense of industrial consumers. The 1973 Fair Trading Act provided for the Director General of Fair Trading to order the MMC to conduct an investigation if he considered that sufficient evidence existed to throw doubt on the current level of gas prices. BG protested that its contract pricing policy had remained unchanged for twenty years, with the full approval of successive governments, but in November 1987 the Director General of Fair Trading invoked the Act and ordered an MMC investigation. In October 1988 the MMC published its Report, and concluded that there was extensive discrimination by BG in the pricing and supply of gas to contract customers. Consequently, amongst its recommendations, the MMC considered that BG should be required to publish a schedule of prices at which it would be prepared to supply firm and interruptible gas to contract customers and not to discriminate in pricing or supply (Monopolies and Mergers Commission 1988, para. 1.6). The conclusions of the MMC Report represented a considerable triumph for the EIUG, and in 1989 BG did begin to publish a schedule of prices for its industrial customers.

This success for the EIUG was also an embarrassment to the government, who must have hoped that privatisation of gas would lead to a more tranquil political environment on energy. Instead, the new regulatory system allowed the head of a quango to take a less sanguine view of industrial gas prices than successive Ministers, and consequently gave the subject more political prominence than it had enjoyed under nationalisation. Thus privatisation did not remove the gas industry from the political agenda. Indeed, a regulatory system one step removed from government could compel Ministers to make decisions on industrial gas prices which had been avoided during the years of nationalisation (under the 1973 Fair Trading Act a Secretary of State can direct an organisation to implement MMC recommendations).

In the final chapter we will discuss briefly the manner in which privatisation and a regulatory framework may confer more potential power on government as a mediator between interests, than would have been supposed generally when organisations such as BG left the public sector.

With regard to electricity, large consumers such as the BSC won concessions from the Electricity Council during 1986, but the prospect of privatisation created a climate where government wished to maximise the profits of the electricity industry.

The BSC's electricity prices inevitably were dependent to a large extent on the relationship between BC and the CEGB. In the wake of the miners' strike, the CEGB won considerable price concessions from BC. It was calculated by the CEGB that the major agreement on coal supply and prices struck between themselves and BC during 1986 would save them £600 million over two years, and consequently prevent price rises to industrial consumers of 7–8 per cent. The EIUG was still not satisfied, and so the CEGB proposed that larger scale consumers should buy 4 million tonnes of coal a year from BC at a reduced price. Thus the electricity produced from this coal would be for the exclusive use of energy intensive customers. Despite internal opposition to this plan within the Electricity Council itself, it was approved in September 1986 (*Financial Times* 19:9:86). The scheme was estimated to cost £50 million and, given the continued financial difficulties of BC, largely represented an indirect government subsidy to large consumers such as the BSC.

After this success, events swung against the consumers towards the end of 1987 when the government announced that electricity prices would have to rise by an average of 8–9 per cent in 1988, with a further 6 per cent increase in 1989. The reason given for these increases was that the industry needed to fund a £40 billion investment programme, although there was widespread suspicion that the decision was also linked to the proposed privatisation of the industry. Regardless of the motivation, the BSC and its fellow consumers once more found themselves ranged against an enormously powerful contervailing interest, with the government largely holding the balance of power between the two sides.

Decentralisation of pay negotiations

After the 1980 strike the BSC pushed on successfully with its intention to decentralise pay bargaining. Union opposition to this trend was not strong, as there was no inclination to force another long confrontation with the management of the Corporation. The ISTC in particular put up spasmodic resistance to the BSC's proposals on pay bargaining, but on the whole achieved few concessions.

After the 1980 strike, there was considerable speculation concerning the next round of wage bargaining. As part of Ian MacGregor's first corporate plan for the BSC, it was proposed that a pay freeze should operate between

January and July 1981, after which a national self-financing increase of 7 per cent would be awarded . Repeating its tactic of twelve months earlier, the BSC held a ballot of all its workers on the corporate plan as a whole. Only 65 per cent of the Corporation's employees voted, but of these 78 per cent said 'yes' to the plan.

This time the ISTC was better prepared for the Corporation's tactic, and at the same time held a ballot of its own. The ISTC ballot asked separate questions on the corporate plan's proposal for further job losses (over 20,000) and on the pay offer. Only 51 per cent of these ISTC ballot forms were returned, but of these 18,392 accepted the job losses, while 8,442 rejected them. On the question of pay 15,525 voted against the BSC Plan while 11,558 voted in favour. Thus both BSC and the ISTC could claim at least some kind of victory, an apparently confusing situation but one which in reality had little effect on subsequent events. The contradictory element in the ballots tended to devalue both of them, whilst the relatively low level of returns indicated that it was a device with diminishing political value. After the ballot Bill Sirs immediately made it clear that the ISTC would not take any strike action, and conceded that the corporate plan would be pushed through regardless. The craft and general unions had already accepted the plan, and so the ISTC was already out on a limb, although the union's own ballot probably persuaded the BSC that this particular technique had served its purpose.

The BSC began paying the 7 per cent extra in July 1981, regardless of the ISTC's opposition, and for the next year the BSC reverted to its pre-strike stance and proposed no national pay increase, with all bargaining being devolved to the local level. For a time the ISTC threatened an overtime ban, but by February 1982 all the unions had accepted the BSC's offer. Later in the year 2.5 per cent of the lump sum local bonuses were transferred into basic rates, but after two years of relative quiet all the steel unions united early in 1983 to seek a national pay review. The more resolute intent of the unions was demonstrated in February when six ISTC branch officials at Port Talbot were stripped of office by the national leadership for signing a local pay and productivity agreement with the Corporation. Bill Sirs said that the unions were not going to be 'conned again' by the BSC offering large local bonuses which failed to materialise, and at a joint meeting of the 14 steel unions it was agreed to seek a national pay increase of between 5 and 9 per cent.

At this stage it might have appeared superficially that a repeat of 1980 was in the offing, particularly as 10,000 workers in BSC's South Yorkshire Special Steels Division went on strike in April over redundancies at the Aldwarke and Templeborough mills. This strike was settled after eight days, however, and early in May the unions conceded the BSC's case by agreeing to forego any sort of national pay deal for that year. In addition to pay, the Corporation also began to introduce the 39-hour week within the context of local deals.

Despite an initial hard-line approach, the leadership of the steel unions had decided that a national strike over pay was no longer a policy option, and by the end of 1983 all the unions had agreed to a third year of local deals only (although some of the local bonuses were later consolidated into national rates). After this, the management of the Corporation was further emboldened to seek a complete dismantling of the national bargaining machinery, but then surprised the unions in December 1984 by suddenly offering a national pay deal for the first time since 1981. This U-turn by the Corporation could be attributed to the effects of the miners' strike. As we will see in the next chapter, the steelworkers had continued to operate plants and assist in unloading raw materials, despite extensive picketing and pleas for help from the NUM. This national pay offer could therefore be seen not only as a reward to the steelmen for their loyalty, but also a 'sweetener' to ensure continued operation of BSC plants. The ISTC even went so far as to agree a 15-month deal with the BSC, but the Corporation still intended to try and push through its plans for dismantling the national bargaining machinery altogether.

The union leaders were understandably reluctant to make this concession, and so progress was slow, but apart from commercial considerations there was an obvious political element in the BSC's push for complete and permanent decentralisation of pay bargaining. Firstly, a large part of the *raison d'etre* would be taken away from the national union leadership, who had failed over the years to become part of the Corporation's strategic planning process. Secondly, the chances of a united national strike over pay would be greatly reduced, and thirdly, privatisation of the BSC (which was still the government's long-term aim) would be more easily facilitated if each Profit Centre (which might be sold off as a separate entity) had its own pay bargaining machinery.

In the event, British Steel was privatised as an entity, and in the later years of the 1980s the Corporation was more content to retain a mix of national and local pay bargaining. As the days of large-scale cuts in manpower receded into history, then much of the steam went out of pay negotiations and also the BSC's desire to dismantle the national structure. In 1988, prior to privatisation, the BSC was able to conclude a two-year deal with the ISTC which, in addition to annual national rises, allowed for a consolidation of bonus pay into basic rates in each of the years. By this time, membership of the ISTC within the BSC workforce had shrunk to 26,000, which still represented 52 per cent of the total, but was a far cry from the 90,000 ISTC members who had gone on strike in January, 1980.

8 The EEC and a State of Manifest Crisis

The BSC and strategic planning in the 1980s

If the UK steel industry was employed as a political football for more than twenty years after World War Two, and if the 1970s was the era when the BSC's vision of ever-expanding steel production melted away like a desert mirage, then it could be said that the period 1980–85 was characterised by a steady increase in the political authority and influence over the UK industry of the European Economic Community (EEC) Commission. The chronic collapse in demand during the early 1980s, and resultant over-production within the EEC generally, led governments and national steel industries to cede power to the Commission, in the hope that a better balance might be achieved between supply and demand. Controls on output and prices, together with intervention on restructuring and subsidies, made the EEC a prominent actor in the affairs of the BSC, with the long-serving Industry Commissioner (1977–85), Count Etienne Davignon, giving his name to the measures as a whole. Thus the Davignon Plan has become a shorthand reference which gives instant identification to a traumatic era in the history of the European Coal and Steel Community. Although it had been hoped that EEC Commission controls would be abolished by the end of 1985, the steel sector was still in too fragile a state to allow this to happen, and so at least some of the regime remained intact. The regime constructed by Davignon was only dismantled finally in 1988.

Despite his own free trade philosophy, Ian MacGregor was a staunch advocate of the EEC intervention. He no doubt recognised that the BSC was not in a position to withstand a long price-cutting war, and so sought a more stable environment in order that the Corporation could win some much needed breathing space. For the UK, the Davignon Plan proved a useful ally, although as we have seen the BSC's initial restructuring plans were introduced as a result of BSC discussions between the Corporation and the Labour government in 1977, and were not the result of pressure from the Commission. Ironically, by the mid-1980s the BSC's own performance had improved to such an extent, that there was general frustration at the manner in which the Davignon Plan had ossified trade within the Community at a level which reflected the Corporation's weaker position in the early 1980s. The Commission could argue, with some justification, that the BSC should not expect to be able to have its cake and eat it, but there was inevitably domestic tension at what some saw to be deliberate checks being put on the UK steel industry's capacity to revive itself.

The chief domestic political interest during the 1980s centred on the future of the Ravenscraig plant in Scotland. As we have described in Chapter Two, the development of Ravenscraig in the 1950s came about largely as a result of political decision-making by the Macmillan government. It had been hoped that the customers for Ravenscraig steel within Scotland itself would prosper and grow, but by the early 1980s these dreams were turning sour. A particular blow was the closure of the Linwood car factory in 1981, which took large amounts of Ravenscraig steel. Despite these setbacks, the Conservative government in both 1982 and 1985 took direct responsibility for keeping the plant open, notwithstanding opposition from the BSC itself. There was obviously a strong political element in both decisions, for the prospect of losing Ravenscraig created a great deal of opposition from a wide spectrum of Scottish interests, and support for the Conservatives in Scotland appeared to be slipping as a result. At the same time, there were also economic arguments which could be put forward in favour of retaining steelmaking at the plant. Significantly the BSC, which in the early 1980s was pressing the case for closure (MacGregor himself was in favour of closing two strip mills and not just one), began putting forward some of these points itself once it became clear that the government was not prepared to budge on the subject.

Some supporters of the campaign to keep Ravenscraig open suspected that the hand of the EEC Commission lay at the root of the threat to the plant. At one point the EEC Competition Commissioner did say that Ravenscraig should close, and the Commission was undoubtedly anxious to cut back EEC steel capacity to the maximum extent possible. At the same time the government, at least on this occasion, chose to stand firm, and so got its own way.

The other major political event for the BSC during this period was the miners' strike of 1984–85. From April 1984 the NUM and the rail unions attempted to deprive BSC works of their raw materials. Despite extensive picketing, the BSC managed to maintain these supplies by using lorries instead of trains. Although the ISTC initially offered limited support for the NUM, the steelmen eventually supported the BSC management in their efforts to keep the plant open. This complete alienation between the ISTC and the NUM illustrated the extent to which the Triple Alliance rested on extremely flimsy foundations. The leadership of the ISTC had always considered that they received insufficient support from the NUM during the 1980 steel strike, and the steelmen were not prepared to risk their own jobs to support a union which itself was divided about the justification for a strike.

The BSC judged that without the miners' strike (estimated to have cost £180 million) the Corporation would have made an operating profit of £40 million in 1984–85 (it should nevertheless be noted that there were exceptional costs of £264 million, which included the Sidbec-Normines debacle). By 1986 the BSC was operating at a profit, and so the departure of Sir Robert Haslam to follow Ian MacGregor as Chairman of the NCB (a MacGregor-Haslam succession was clearly a habit the government found difficult to break!) marked a return to some type of viability for the first time in more than a decade.

The Benn-Finniston debate, which politically marked the beginning of the crisis (and the end of the profitable years) back in 1975, had attempted to clarify the government-Nationalised Industry relationship. During the succeeding ten years governments became the BSC's paymasters in order to keep the Corporation afloat. Even a Conservative government strongly committed to a non-interventionist strategy was in reality drawn into periodic political intervention. In the final Chapter we will be examining how the patterns of power within the policy system came to shape the government-BSC relationship.

The development of the European Coal and Steel Community

The European Coal and Steel Community (ECSC) came into being as a result of the Treaty of Paris, signed in 1951. The perceived strategic importance of the coal and steel industries, particularly on the part of France and West Germany, led to six nations coming together as the trail blazers for greater European integration and industrial harmonisation (Belgium, Holland, Luxemburg and Italy also signed the Treaty). The initiative for the ECSC came from Robert Schumann, the French Foreign Minister, and massive political elements behind its creation were of course the memories and devastating effects of two World Wars. None of the governments which signed the Treaty wanted a repetition of these terrible events, and this factor alone was motivation enough to secure agreement. On a more prosaic level, however, and particularly with regard to steel, it was hoped that trade liberalisation would produce a more modern and efficient steel industry within Europe as a whole.

The powers given to the ECSC's High Authority, as Tsoukalis and Strauss point out, nevertheless suggested that the member countries feared (with considerable justification) that in certain circumstances competition and price-cutting would lead to a state of crisis. Quoting Hallstein, they describe the system set up by the Treaty of Paris as one of 'regulated competition', with provisions made for central crisis management (Tsoukalis and Strauss, 1987, 189). Thus the High Authority was given powers to raise levies on steel production, extend loans to firms for capital investment, finance research and development, and re-train redundant workers, in addition to vetoing investments which ran counter to its own economic forecasts (Tsoukalis and Strauss, 189). Even more, Article 58 of the Treaty stated that if 'a state of manifest crisis' was declared, then the High Authority had the power to set production quotas and minimum prices. It was this article which allowed the EEC Commission, nearly thirty years after the Treaty was signed, to greatly enhance its control over ECSC steel industries.

The UK was not an original member of the ECSC, neither was the UK a signatory to the Treaty of Rome, which created the EEC in 1957. Indeed, in common with the majority in the UK, the steel companies appeared to view developments on the Continent of Europe with a mixture of distrust and disdain. Vaizey suggests that, immediately after the War, steel masters hoped that Britain would take Germany's predominant place in the European steel cartel which had helped to support the Nazi war effort (Vaizey, 167). If so, then it might be said that the British government chose to remain in a damagingly aloof position. It appeared that both the government and the industry feared that by joining the ECSC Britain would place itself under what they considered to be the excessive power of the High Authority, and so would not be able to exploit the UK's supposedly stronger position. It was generally thought, quite mistakenly, that the British iron and steel industry was in a considerably more robust state than any within the ECSC, and so as a result of this complacency the UK lost any opportunity it might have had to guide the Community through the vital early post-war years (Vaizey, 168).

On the question of British membership of the ECSC and EEC there was, as Moon illustrates, an extremely low political salience within the UK political establishment generally:

> This is in marked contrast to the salience which the issue achieved within the Six. Whilst the events among the participating countries were far-reaching, these were not recognised within the British political system. British membership of the EEC was widely assumed to be out of the question, and for the most part not even discussed. Although the Liberal Party and a few individual backbench MPs viewed the issue in a different light, they were unable to alter the muted status of the issue in the face of the Government, Opposition, business, trade union and press consensus (Moon, 1983, 39).

By the 1960s a complete U-turn had taken place, and the British government was eagerly seeking admission to the EEC. By this time, however, Britain faced formidable opposition to its membership in the shape of President De Gaulle of France, and it was not until 1973 that the UK finally joined the Community. Although the ECSC and the EEC had merged administratively, the Treaty of Paris remained in place. Britain's entry into the EEC coincided with the decision to push ahead with the BSC's massive capital development programme. There is an obvious connection which can be made between the two events, for with the BSC now more open to competition from the rest of Europe, the need to modernise was perceived as being still more urgent. By this time the BSC was fighting a losing battle to keep imports at bay, for imported finished steel deliveries as a percentage of the UK home market had risen from 5.5 per cent in 1970 to 12.7 per cent in 1972 (Pryke, 194). This trend accelerated after 1973, so that by the late 1970s imports were capturing over one fifth of the UK market. Much of this new penetration came from other EEC countries, but the BSC itself was unable to significantly increase its exports to the Community. Thus, as Pryke puts it:

> It may be replied [to the argument that the BSC had become less competitive in the 1970s] that imports still account for a smaller proportion of domestic consumption than in any of the other Common Market countries and that they, like us, have been flooded by cheap steel from the Far East. However, every other major steel producing nation in the Common Market had higher net exports in 1978 than in 1965. Our net exports of steel slumped from 2,125,000 product tonnes to under 550,000, largely because of a surge in imports from the Six. Imports from the Far East have played a minor part (Pryke, 195).

From the late 1970s until the late 1980s the position with regard to both imports and exports remained relatively stable, but the damage had already been done. Thus in 1984 imports of finished steel products into the UK totalled 3,553,900 tonnes, with 2,631,800 tonnes coming from other EEC countries. These figures can be compared with a 1979 total for imported steel industry products of 3,784,000 tonnes, of which 2,357,700 tonnes came from the rest of the EEC. On exports, the respective totals for steel products in 1979 and 1984 were 4,398,600 tonnes and 3,966,200 tonnes. The totals for exports to the EEC are remarkably similar, with 1,542,000 tonnes in 1979, and 1,526,000 tonnes in 1984 (these export figures are for the UK industry as a whole) (ISSB). Since the late 1970s the EEC Commission has been signing trade agreements with third countries, and to confirm Pryke's point about the Far East, imports from Japan declined from 168,000 tonnes in 1979 to only 73,700 tonnes in 1984 (ISSB). Trade with the EEC, both with regard to exports and imports, was thus of increasing importance for the UK steel industry.

EEC crisis in the 1970s

1975 marked the beginning of crisis not only for the UK steel industry, but for the EEC as a whole. The French asked for a state of manifest crisis to be declared, but this plea was opposed by West Germany. As the traditionally largest and strongest of the EEC steel industries, West Germany has been highly suspicious of what it considers to be unjustified subsidies given to rival national industries by their governments, which it fears can create unfair competition. Thus throughout the 1970s both the West German government and the (predominantly privately owned) steel companies only acceded to greater Commission control with reluctance. Eventually, even the West German steel industry was to feel the cold draught of recession, but in 1975 the crisis was not sufficiently severe to secure the required unanimity.

The French request for a state of manifest crisis was also complicated by the UK referendum on continuing membership of the EEC. The anti-marketeers portrayed the Commission as exercising an undue influence over the UK industry, while the pro-marketeers argued that the Commission could only act after a unanimous vote by the Council of Ministers. If the Commission took on greater powers while the referendum campaign was in progress, then this would have been a political present to the opponents of UK membership, who could claim that their worst forebodings were justified. In the event the referendum resulted in an overwhelming vote in favour of continued membership (67.2 per cent in favour to 32.8 per cent against). The Labour government itself took pains to play down any potential Commission powers, and in any case at that time the full implications of the steel crisis did not appear to have permeated the UK political community.

A problem for the Commission with regard to the UK industry was the BSC's continuing massive investment programme. By the end of 1975 it was becoming clear that EEC steel capacity as a whole was outstripping demand by many millions of tonnes, yet the BSC was relentlessly pressing on with its expansion plans. By 1976 the EEC had itself loaned the UK £400 million, chiefly to finance the development programme, with a further £160 million in loans coming from the European Investment Bank. The Commission might have wished to curtail this investment strategy, but was reluctant to deny the BSC its opportunity to 'catch up' with the rest of the EEC.

The UK's political ambivalence towards the EEC in the mid-1970s mirrored the general situation within the Community's steel industries as a whole. Although founded on the principles of greater European integration and general liberalisation of the market, at the first sign of bad times there was a tendency towards unofficial cartelisation. At the same time, as we described earlier, there was a mutual distrust between West Germany and France. The Commission thus faced the Herculean task of constructing some sort of official cartel, which could restrict output and hold prices while a rationalisation of the industry took place, together with greater control over trade. Between 1975 and 1980 the Commission attempted to perform this feat without genuinely coercive control over the various national industries. Eventually, it had to admit defeat, and seek greater control under Article 58 of the Treaty of Paris.

The split within the Community was demonstrated early in 1976 when a private cartel, known as the Denelux Grouping, was formed between West Germany, Holland, Luxembourg and the Flemish steel company Sidmar (a subsidiary of Arbed, the Luxembourg steelmaker). The companies involved in this cartel were generally the most profitable within the EEC, and its formation threatened a schism between them and the French, British and Italian companies (together with the state-owned Cockerill in Belgium). The Commission was already producing quarterly forecasts of steel production per country, in the hope that the companies would operate voluntary controls, but with only mixed success. As a result of the recession the majority of companies were reducing output as a matter of course, but were still reluctant to accept direct advice from the Commission.

Towards the end of 1976 it had become clear that the worsening steel crisis was not of the usual cyclical nature, but represented something more fundamental. The ECSC Consultative Committee, consisting of steelmakers, unions and consumers, called for output cuts and also for severe restrictions on third-country imports into the Community. By the end of the year the Belgian EEC Industry Commissioner, Henri Simonet, had produced a new plan, which included the setting of voluntary production or delivery quotas for individual companies in six categories, and also stand-by power to publish minimum reference prices. The introduction of the Simonet plan coincided with the creation of a new EEC steelmakers' club known as Eurofer. The Denelux grouping had obviously galvanised the other EEC steel industries into action, and the existence of Eurofer put in place an organisation which would be able to operate in tandem with the Commission, if and when a tighter regulatory framework was to be required.

After introducing his plan in January 1977, Simonet was replaced as Industry Commissioner by his fellow Belgian, Count Etienne Davignon. It quickly became evident to both Davignon and the steelmakers that the Simonet plan was not working as it should, and pressure was put on the Commission by Jacques Ferry, the influential French President of Eurofer, to declare a state of manifest crisis. Ferry himself sought the ultimate regulatory control of a World Steel Pact, but for the moment the EEC steelmakers still lacked the unity to introduce a state of manifest crisis. Davignon was not to be put off, however, and was soon demonstrating the political skills which were to serve him well in his eight years as Industry Commissioner. His first concrete achievement was to introduce compulsory minimum prices for reinforcing bars. The voluntary prices agreement had been persistently breached by Italian private steelmakers known as the Bresciani. Davignon also tightened up on licences for steel brought into the EEC. Although the Commission had concluded a voluntary trade agreement with Japan during 1976, the EEC steelmakers were seeking substantial reductions in imports which they considered were likely to underprice the domestic product.

Agreement was achieved at the end of 1977 on minimum prices for imports into the EEC. Any steel landed in the Community below that price would be subjected to a countervailing duty for three months. It was intended that this trigger price mechanism would have the effect of encouraging importers to conclude bilateral agreements with the EEC, whereby at least some of the steel could avoid the new tariffs. This system proved to be effective, for during 1978 the EEC concluded bilateral agreements with nearly all the principal importers. As a result, the Community has generally been able to restrict imported steel to about 10 per cent of consumption.

Although the restriction on third country imports probably prevented a further deterioration in the BSC's market position, we have seen previously that the chief threat to the UK came from other EEC countries, rather than from outside the Community. During the 1970s the BSC's share of the domestic market fell from 75 per cent to just over 50 per cent, and the bilateral agreements had no effect on reversing this trend. A serious problem for the BSC was the failure to enforce the guideline prices set out in the Davignon plan. During 1978–79 the Corporation attempted to introduce price rises, but was persistently undercut by cheap imports. There was of course no unity on this matter within the UK steel community itself, for the British Iron and Steel Consumers' Council (still an influential body despite losing its statutory status on Britain's accession to the EEC in 1973) had no wish to pay higher steel prices if it could be avoided, while the members of the National Association of Steel Stockholders produced their own price lists, which were usually below those recommended by the Commission.

In the EEC generally, the price system proved difficult to enforce, to the extent that the Commission took more than 20 companies to the European Court of Justice for breaches of the mandatory price code on reinforcing bars, merchant bars and coils (other steel products had only voluntary price codes attached to them) (Tsoukalis and Strauss, 202). The problems caused by the failure of the Davignon price code were compounded by an increasingly protectionist approach in the United States. The US itself introduced a trigger price mechanism during 1978, and as a result served six anti-dumping suits on the BSC. It was ironic that the BSC, which persistently complained about what it considered to be dumping by other EEC countries in the UK, should itself be accused of a similar practice on the other side of the Atlantic. The unwelcome reality for the BSC was that it faced a deteriorating market position in which it was ill equipped to compete. Even for the steel it sold it could not charge the prices to cover costs, and it was in this atmosphere that the Corporation made its ill-judged offer to the unions at the end of 1979, and which led directly to the steel strike.

In the EEC generally, there was a slight improvement in demand during 1979, but the position worsened again during 1980, so that by the autumn of that year it was clear that the first Davignon plan had failed. Undaunted, Davignon sought a new set of measures which would be capable of producing a more fundamental change in the EEC steel industry than had hitherto been the case.

A state of manifest crisis

The BSC was to the forefront of those seeking greater Commission control over the steel industry. Ian MacGregor went to the extent of threatening that the Corporation would start an all-out price-cutting war if tougher measures were not taken quickly. Paraphrasing the words of Al Jolson, more than fifty years earlier, he declared: You ain't seen nothing yet – there will be mayhem (*The Times*, 6:10:80).

The Chairman of the BSC, and later the NCB, MacGregor on many occasions proclaimed the virtues of the free market, but clearly circumstances altered cases. The EEC steel industry was about to enter a process which would eventually lead to mandatory controls on output and prices, linked to rationalisation of capacity and (at least in theory) a phasing out of state subsidies.

The US had suspended its trigger price mechanism earlier in 1980, but in October this device was reintroduced. Although the American steelmakers dropped anti-dumping suits against EEC producers, it was clear that one of Europe's principal markets would in future be more difficult to penetrate. The French and British governments were both strong supporters of a state of manifest crisis being introduced, but the West German government was initially defiant in its opposition. Instead, the West German steel companies proposed a continuation of voluntary controls, for they feared that an EEC industry controlled by the Commission might jeopardise their more efficient and profitable operations. By this time, however, Davignon had set his mind on compulsory controls, and accused the West Germans of creating considerable uncertainty and causing further damage to the steel industry (*Financial Times,* 24:10:80).

The West Germans' particular concern was the proposed inclusion of special steels in the new EEC regime. Special steels tend to be one of the more profitable steel products, and their exclusion from output restrictions would have allowed the West German steel companies to maintain higher returns. In order for the state of manifest crisis to become official, it was necessary to achieve unanimity within the Council of Ministers. Control of the EEC steel market had now become overtly political, for it was national governments rather than the steel companies which bore the responsibility for negotiating the terms on which the Commission would assume power. Actual implementation of the policy would depend largely on maintaining some sort of unity within Eurofer, while the framework of controls would be monitored by the Commission's inspectors. Enforcing the rules would not be easy, but eight out of the nine EEC governments had now agreed to cede some control of their national steel industries to the Commission.

After much haggling and brinksmanship within meetings of the Council of Ministers during October 1980, the other eight governments finally conceded to the West Germans that small firms, tinplate steel, steel rail, small tubes and certain categories of high alloy special steels should be excluded from controls. These products amounted to about 20 per cent of EEC output, and were chiefly in high profit areas, where the West German industry was strong. Many of the smaller special steel companies in the UK were upset that their products were to be excluded from controls, for they feared that the domestic industry could not carry out an orderly rationalisation without protection. After much lobbying of the Commission by these companies in 1981, it was agreed to draw a 'ring fence' around them in terms of minimum prices and imports. This type of arrangement indicated the extent to which the UK steel industry was to look towards the Commission as a protector of its interests.

In return for exemptions from controls, the West German government agreed to compulsory production curbs in four main categories – coils and rolled products, sheet steel and other flat products, heavy sections and lighter long products. The Commission then faced the enormous task of communicating the quotas arranged for each of the 350 steelmaking companies now subject to this control. The quotas were based on a producers' best 12 months in the 36 months up to June 1980. It was intended originally that these quotas would continue only until June 1981, but in reality they were to become a feature for most of the 1980s. Fines were to be imposed on those companies found to be breaking these quota controls. In the event, the Commission encountered difficulty in actually collecting the fines, the most notable example being the large West German company Kloeckner-Werke. From the outset, Kloeckner refused to recognise its output quota, and by the beginning of 1984 had accumulated unpaid fines of over £40 million. At this point it finally agreed to join the regime, but the fines remained unpaid.

Kloeckner's behaviour raises serious questions about the legitimacy of the Commission's controls. Some would argue that the whole cartelisation of the EEC steel market represented an unfair trading practice which penalised consumers while featherbedding producers (see Jones, 1983). On the other hand, the measures taken did have the potential to give some sort of breathing space to organisations such as the BSC, which by this time were suffering almost uncontrollable losses. The Commission's power during the 1980s was derived not so much from a strict interpretation of the Treaty of Paris, but rather as a result of the steel companies themselves accepting that the gravity of the situation demanded a central arbiter. As the Kloeckner-Werke example illustrates, however, neither the Commission nor national governments could ensure that their will would be carried out. In this sense, the EEC steel community remains a network of interests rather than a hierarchy of command and control.

The Commission did not attempt to impose new mandatory prices in 1980, and indeed it was to be 1984 before these were introduced. From 1980–83, even where quotas were successfully imposed, price discounting was rife, leaving the BSC finances in a continuing precarious state.

The 1980 Corporate Plan and the 1981 Iron and Steel Act

The BSC losses for 1979–80 amounted to £545 million (not including extraordinary items which amounted to the amazing sum of £1.24 billion), of which £200 million were attributed to the steel strike. The strike had compelled the BSC to abandon its timescale of 50,000 job losses by August

1980, but Ian MacGregor was determined to continue radical change in this area. Immediately after the strike the BSC's domestic market share slumped to just over 46 per cent, although by the end of 1980 it had been lifted to 52 per cent. Nevertheless, steel consumers and stockholders were now even more wedded to the principle of 'second sourcing', and succeeding years were to see little change in this pattern. By the end of September 1980 steelmaking had ceased at both Shotton and Corby, while the integrated Consett steelworks in Durham had closed completely. The excess demand throughout the EEC kept prices at a low level, and the BSC itself complained of what it considered to be dumping on the UK market by its European competitors.

In this gloomy environment BSC losses accelerated as 1980 proceeded. In September the government began to implement its 'blank cheque' policy by raising the Corporation's EFL for 1980–81 by a further £400 million. Together with the unused £120 million from 1979–80, it brought the BSC's new EFL to £971 million, more than double the £450 million which only a few months earlier the government had claimed was not negotiable. After this immediate help the next task for the government was to put together a financial reconstruction of the Corporation which would dissolve a significant proportion of its massive debt commitments.

By the end of 1980 the BSC was losing £2 million a day and its output was declining even before the introduction of compulsory quotas as part of the new Davignon plan. Given this extremely depressing environment MacGregor's first corporate plan, produced in December 1980, was not as radical on closures as some had feared. Nevertheless, it did include closures and cutbacks at a variety of works – the major closures being the Normanby Park plant and Appleby-Frodingham rod mill – involving a further reduction in the workforce of 20,000 (i.e. on top of the 52,000 job losses announced a year earlier). By the end of 1980 the BSC workforce had been reduced to about 140,000, a loss of 26,000 since the end of the steel strike. On output the plan proposed a further reduction from 15 million tonnes per annum to 14.4 million tonnes per annum (presumably in line with EEC production quotas), but it was indicated that none of the five main steelmaking plants was in immediate danger of closure. The commercial objectives of the plan were modest: to regain the pre-strike domestic market share of 54 per cent and a small (but undefined) improvement in export sales.

As we described in the previous chapter, the BSC once more employed the workforce ballot technique on acceptance of the corporate plan, but this time with a rival in the form of an ISTC ballot. Although the Corporation won widespread acceptance for its Plan (78 per cent to 22 per cent), while the ISTC vote went against the 7 per cent pay offer linked to the Plan, both ballots were probably unnecessary. The steel strike had exhausted the unions, and there was little significant opposition to implementation of the corporate plan. The chief political attention now turned to the government, which had at last produced a capital reconstruction for the BSC.

The 1981 Iron and Steel Act wrote off £3 billion from the BSC's 'Section 18' debt, with provision for a further reduction of £1 billion. In addition, the Corporation's debt of £509 million to the National Loans Fund was entirely extinguished. Early in 1981 the BSC's borrowing limit had been lifted to £6 billion, but as a result of the write-offs this was now reduced to £3.5 billion. The provision for the further £1 billion write-off was taken up during 1982, and the borrowing limit reduced to £3 billion. Thus as a result of the 1981 Act £4.509 billion effectively vanished into thin air. This huge weight taken off the BSC's shoulders reduced its borrowings from £5.406 billion in 1981 to £2.361 billion by 1983. Given that £4 billion of the write-off was in the form of 'Section 18' capital, which the BSC did not have to repay on an annual basis, then the borrowing reduction made only a small difference to the Corporation's real world finances. It nevertheless had the effect of removing the extremely distant possibility that any of this money might be repaid, while the government could hope that relief from this debt would eventually make the Corporation a more attractive candidate for privatisation.

This privatisation goal was also evident in the other significant part of the 1981 Iron and Steel Act. We described early in Chapter Three how the 1967 Iron and Steel Act set out certain statutory duties for the BSC. These duties (restated in the 1975 Iron and Steel Act) included the Corporation being given the responsibility of producing steel to satisfy the needs of UK manufacturing industry and also to 'further the public interest in all respects'. The Corporation was also given the task of promoting exports, but not to discriminate against customers in the supply and price of steel. These statutory duties were now abolished, with the result that the Corporation appeared to have no legal basis for its continued existence. Three strands of political motive can be identified on the part of the government for introducing this apparently radical policy change.

Firstly, by taking away the BSC's statutory protection from Ministerial interference, it allowed the Secretary of State more potential control over the officers of the Corporation. This was particularly relevant with regard to privatisation, where the government was anxious to create more Companies Act companies within the BSC, prior to selling them off. In fact, as we have seen, the previous Conservative government had created two Companies Act companies within the BSC in 1971, and so the concept was not new. If Companies Act companies were capable of being created under the old legislation, it could be argued that the statutory change was unnecessary for this purpose. In the event, the government's immediate privatisation plans were not fulfilled to the extent that they had hoped, while even the Phoenix projects could not be described as privatisation in the true sense.

Thus privatisation in itself does not appear to have been a sufficient motive for the government to abolish the statutory duties. The second strand of government thinking is perhaps of more importance in this respect. Sir Keith Joseph had been greatly embarrassed during 1980 to admit that the government was forced to issue a 'blank cheque' to the BSC in order to keep the UK steel industry afloat. Sir Keith claimed that existing legislation did not allow the Corporation to be put into liquidation, although Section 5(b) of the 1975 Act stated that none of the BSC statutory duties should be taken as: 'precluding the Corporation from disposing of an asset used, or capable of use, for the purpose of any of these activities or from disposing of any interest in a company in which any such asset is rested'. Although this wording appears to be explicit, the government claimed that some of its legal advisers considered that the sub-section did not allow for the complete rundown of the Corporation. Even if this section of the Act did not go far enough, we argued in the previous chapter that the Secretary of State also had the power to direct the Corporation on matters of strategic policy, which presumably could include an order to shut down all state-owned steelworks.

Given the government's political sensitivity to financing a Nationalised Industry on a huge scale, it is perhaps understandable that it should wish to demonstrate that the BSC was not to be featherbedded eternally. The government claimed that the new Act, which allowed for the creation of Companies Act companies, would allow the BSC to be placed into liquidation if necessary. In reality, this argument appears to be somewhat academic, for many BSC works were closed down without any chance of them becoming Companies Act companies. In addition, the government at that time displayed no inclination to make any of the five major works a Companies Act company, and indeed had taken direct responsibility for keeping all of them in operation. Nevertheless, by publicly declaring that the BSC could henceforth be placed into liquidation, Sir Keith Joseph might claim that he was not compromising his free market principles. In this respect, the abolition of the BSC's statutory duties could be described as being of more symbolic than real importance.

The third strand concerned the nature of the BSC's business. The statutory duties, and a strong argument given by the Labour government in defence of nationalisation, stressed the manner in which the BSC should serve the needs of UK industry. The expansionist philosophy of the early 1970s embodied the hope that a state-owned steel corporation would be a powerful component part of a new dynamic growth for UK industry as a whole. The publication of *The Road to Viability* brought this phase to an end, and henceforth the BSC was to be treated as an organisation seeking to attain free-standing commercial credibility. Apart from the privatisation goals, the Conservative government, for obvious financial reasons, was desperately anxious to see the commercial objective achieved. Abolition of statutory duties therefore emphasised the extent to which the Corporation had moved away from being an instrument of national industrial policy, into an organisation seeking to find business and make money when and where it could.

All three strands of political motivation behind the 1981 Iron and Steel Act could thus be described as being of symbolic rather than real significance. As Foster argues, the statutory framework supporting a public corporation can be open to many interpretations (Foster, 23–25). It could not be said that abolition of the BSC's statutory duties led to real policy changes, and in this respect, despite appearances, the 1981 Act was not a radical piece of legislation. By choosing to interpret the 1975 Act in a particular way, the government could nevertheless claim that the new Act was setting the BSC on a fresh policy course. The political symbolism of the new Act (as with the steel strike) conveyed a message which was important for a government anxious to display a radical exterior, even if its actions were relatively conventional.

The abolition of statutory duties was a somewhat esoteric element of the 1981 Iron and Steel Act. Notwithstanding the more prosaic part of the Act (writing off £4.509 million of debt), the financial year 1980–81 represented the nadir of the BSC's fortunes. The loss, before extraordinary items, amounted to £668 million, even more than the £545 million loss made in the 1979–80 strike year. Only 11.9 million tonnes of steel was produced, compared with 14.9 million tonnes in 1979–80, which also included the three month strike. These losses were made despite a huge decrease in the size of the workforce from 166,400 to 120,900. Indeed, in addition to the loss itself, closure and redundancy costs amounted to £164 million, while a further write-off of asset values accounted for £188 million. Given these figures, it was little wonder that the government was forced in February 1981 to lift the BSC's EFL for the year by a further £150 million up to £1,121 million, of which £1,119 million was taken up by the Corporation. Despite the Davignon regime, prices generally remained low, while the high value of the pound and the decline in domestic demand for steel were other factors accounting for the deterioration.

Corporate plans as acts of faith

The delicate path trodden by the BSC in the early 1980s was recognised by the House of Commons Select Committee on Industry and Trade in May 1981. The Committee remained sceptical about the BSC's plans, but admitted that it could see no practical alternative to the Corporation's interim short-term survival plan, which had been submitted to the government in December 1980. The Select Committee published its Report in an attempt to influence the deliberations between the Department of Industry and the BSC, as the latter produced its new corporate plan in the Autumn of 1981.

The Committee report provides an apt summary of the state of British steel policy in the early 1980s. Thus the Committee noted that both the Secretary of State for Industry and Ian MacGregor had emphasised that the industry was at 'the brink of a precipice' (House of Commons Industry and Trade Select Committee, 1980–81, para 2). It also noted that the 'financial reconstruction' of the BSC finances, which the government had arranged, was a euphemism for writing off enormous sums of outstanding debt. The survival plan, which MacGregor had himself agreed was 'optimistic', had left the Committee with doubts about its realism. These doubts were encouraged by the fact that:

> In December 1979 the BSC's Business Plan had proposed, *inter alia*, an annual crude capacity of 15 million tonnes, and in the financial year deliveries of 11.7 million tonnes and a net profit of £45 million. In the event, deliveries were put at only 9.4 million tonnes and the loss was £660 million. This catastrophic shortfall of performance against a target was another in a line of successive failures to meet targets set in good faith (House of Commons Industry and Trade Select Committee 1980–81, para 16).

The Select Committee was critical that the 1980 corporate plan was no more than a 'survival plan', and both the government and the BSC refused to look beyond 1982. It must be said, however, that long term forecasting had never been the BSC's strongest point, and it probably made sense not to make extravagant forecasts at a time when the Corporation was only existing through the continuation of the government's 'blank cheque' policy. As so often, the continuing mood of crisis provoked a great deal of secrecy on the part of both the BSC and government, and the members of the Select Committee only obtained confidential copies of the Plan after considerable difficulty and delay. As for the TUC Steel Committee, they were once more left out in the cold altogether, and were never given the opportunity to study the plan at all. In a manner similar to Villiers and Varley in 1977, MacGregor and Joseph in 1981 were not anxious that the public should see just how far the BSC had still to travel on the road to viability.

The principal objectives of the 1980 corporate plan (made public in summary form) had been:

1. To regain pre-strike domestic market share of 54 per cent;
2. A modest increase in export tonnages;
3. Improved productivity as a result of substantial reductions in personnel;
4. Reduced and rationalised capacity;

5. Reduction of energy costs;
6. Restrictions on capital expenditure;
7. Reduction of the BSC interest burden.

By the mid 1980s all these objectives had been more or less achieved, and the Corporation was making a small operating profit. Despite this eventual improvement, 1981–82 proved to be something of a false dawn for the BSC. The loss before extraordinary items of £358 million was over £300 million less than the previous year's total of £668 million. Borrowing totalled £694 million, within the limit of £730 million set by the government, but additional commitments of £72 million (including the setting up of Allied Steel and Wire) in reality took it above the EFL maximum. The Corporation's output improved to 14 million tonnes, from 11.9 million tonnes in 1980–81, yet UK steel consumption was down by 4 per cent, and the improvement was thus due almost entirely to building up stocks which had sunk to an exceedingly low level during 1980.

As with all governments, Ministers were anxious to convey the optimistic impression that a significant upturn in demand for steel was in sight, and the new Secretary of State for Industry, Patrick Jenkin, predicted that the BSC could be in profit by the end of the financial year 1982–83. Compared with Sir Keith Joseph, Jenkin was generally regarded as a Minister less likely to take a hardline approach with the BSC, and some comparisons could be made with Eric Varley taking over from Tony Benn as Labour Industry Secretary in 1978. These comparisons did not perhaps bear too close an examination, for as we have seen, Sir Keith was prepared to assist the BSC to the tune of billions of pounds when the Corporation was threatened with complete collapse, while Patrick Jenkin was to ensure the preservation of Ravenscraig when Ian MacGregor was strongly pressing for closure.

The Davignon regime falters

There was further irony in a government, apparently wedded to free market principles, strongly supporting the Davignon plan of mandatory output and voluntary price controls. The stance was of some political embarrassment to the government, and so Ministers were happy to support a strong push by the West German government in the Council of Ministers to set a deadline for government aid to national steel industries. It was not that the UK government had any wish to see a rapid end to these subsidies, for the BSC was in no position to withstand such a deprivation, but a commitment to phase out aid could demonstrate a political will to eventually return to a freer market, while also putting greater pressure on the Corporation to improve its financial position. For their part, the West Germans continued to be upset at what they considered to be unfair competition within the EEC, and at one point threatened to ban imports from other Community countries if a commitment to end subsidies was not given. Agreement was finally achieved in June 1981 to cease subsidies within the EEC by the end of 1985, with operational aids being terminated by the end of 1984. This was a date sufficiently in the future to at least postpone any political crisis within the EEC, but it did put some new pressure on those EEC industries, such as the UK, France and Italy, which were being heavily subsidised.

Ironically, despite its ideological commitment to ending subsidies, the UK government was put in a sticky position by the EEC Commission when the latter withheld its permission for £360 million of the government's aid to the BSC, and even hinted that the Corporation should cut its manned capacity to 13.9 million tonnes. The Commission took the view that the BSC still held more than 6 million tonnes of capacity in reserve, and wanted this plant removed altogether. The BSC and UK government countered by arguing that only about 3 million tonnes of unmanned capacity remained, and that in any case the UK had reduced capacity at a greater rate than any other EEC country. There was undoubtedly some validity in this argument, although it could also be said that, as the least successful steel industry within the ECSC, the UK had little choice but to implement these cuts! The UK government was also upset that other EEC countries, chiefly France and Italy, were not 'playing the game' and notifying the Commission of all subsidies.

The Commission eventually gave the go-ahead for the £360 million, but only on condition that BSC steel production in 1981–82 did not exceed 13.9 million tonnes, that the government must submit the BSC's next three year plan for joint examination before London could take a decision about it, and that the Commission would also investigate the British government's policy towards the BSC over the next three years (*The Guardian*, 13:11:81). These conditions left little doubt as to the sovereignty over their domestic steel industries which the EEC governments had ceded to the Commission, even if the Commission (as in the case of Kloechner-Werke) did not necessarily have the power to enforce its decisions.

9 The Long Road to Viability and Privatisation

Ravenscraig reprieved

After a relatively quiet period in the financial year 1981–82, the BSC's position began to deteriorate rapidly in the latter part of 1982. Destocking within the UK began once more, but even worse, despite the Davignon Plan, prices tumbled throughout the EEC. The improvement in demand predicted by the government had not materialised, and in 1982–83, BSC production tumbled to an all-time low of 11.7 million tonnes.

One crisis averted in 1982 concerned exports to the US. In response to the deteriorating position of its own steel companies, the US government was adopting an increasingly protectionist position, and in February 1982 the International Trade Commission decided to proceed with 38 of the original 93 cases submitted by the companies. The BSC calculated that if the suits were successful, about 100,000 tonnes of its anticipated exports to the US would be blocked. In June the US Commerce Department declared judgement on 28 of the cases, and found that 3.9 million tonnes of US steel imports were subsidised illegally by foreign governments. The BSC was singled out for special treatment, with the US government alleging that the Corporation was receiving aid worth 40.362 per cent of selling prices (a highly political calculation later reduced drastically to 20.33 per cent). The UK government protested that the public money granted to the BSC formed part of an essential restructuring programme, but the Corporation now had to post bonds on US steel exports equal to the estimated subsidy.

After much threatening and haggling during the latter half of 1982, agreement was reached in October between the EEC Commission (which negotiated on behalf of its members) and the US government to restrain the exports of ten EEC product categories to just under 5 per cent of the market until the end of 1985. A further, more vaguely worded, deal pegged the EEC exports of tubes and pipes to 5.9 per cent of US consumption, unless American pipe-producers were unable to meet an increase in domestic demand (*Financial Times*, 22:10:82). The allocation of the quotas was decided on licensing arrangements operated by ECSC member states. Exports of UK steel industry products to the US had shot up from 213,800 tonnes in 1980 to 563,600 tonnes in 1981, but in 1982 these exports had slumped to 388,300 tonnes. By 1984 UK steel product exports to the US had risen once more to 482,400 tonnes, and indicated that even within the agreement there was scope for growth (ISSB).

Protectionism in the US, combined with the various measures which comprised the Davignon Plan, indicated the manner in which the BSC was increasingly becoming part of a politically managed market for steel. The excess of output over demand had led to the construction of a European cartel, in which producers were relying on the restraint and co-operation of their competitors in order to survive. As the actual terms of the Davigon Plan were thrashed out within the Council of Ministers, matters inevitably took on a more overtly political hue, with the result that the steelmakers themselves were placed in a somewhat subordinate position. Nevertheless, through Eurofer the producers still had a highly influential voice. Thus although the EEC Commission set the broad national output quotas, it was Eurofer which decided how these various quotas should be distributed between each particular company (production quotas were at times bought, sold and exchanged within Eurofer, an operation which was little publicised, but illustrates the considerable power of the cartel).

By 1982 the chief focus of attention within the ECSC was on maintenance of prices and co-ordination of rationalisation plans. During the year prices fell severely, and this contributed to a serious decline in the BSC's financial position. Losses shot up from £3 million a week from April to June 1982 to more than £9 million a week from October to December. The Corporation complained that many countries were dumping surplus steel in the UK at rock-bottom prices, with the result that BSC capacity had fallen from an operational use of 91 per cent to one of 72 per cent. MacGregor himself claimed that, as a result of these imports, BSC's market share was down to 45 per cent of the domestic market (although this figure soon improved again). Steel imports did rise from 3,314,700 tonnes in 1981 to 3,836,300 tonnes in 1982, although it should be noted that the 1982 figure was only slightly above the 1979 imports figure of 3,784,000 tonnes (ISSB). Probably the most significant figure was that UK steel consumption in 1982–83 fell by 6 per cent from 1981–82. The high value of the pound and high interest rates, combined with the recession internationally, left the major part of UK manufacturing industry in a depressed state, and the BSC suffered accordingly.

By the autumn of 1982 the BSC had abandoned its hopes of making a profit during 1983, and thoughts now turned to speeding up the rationalisation plans. Indeed, since June 1982 there had been rumours that the Corporation was reappraising its plans, and that the closure of one of the five major steelmaking plants was in the offing. In the 1970s Llanwern had generally been thought of as the most vulnerable of the 'big five', but by 1982 attention had switched to Ravenscraig. MacGregor himself heaped praise on Llanwern, where manpower had been halved over two years to a figure of 4,446. Greater flexibility in working practices had also been introduced into the plant, and consequently in April 1982 Llanwern's productivity had reached a record of 4.08 man hours per tonne, compared with 5.7 m.p.t. at Port Talbot and 6 m.p.t. at Ravenscraig. These three works represented the BSC's manufacturers of strip steel (usually referred to as strip mills), a significant part of which was used in the motor vehicle industry.

We saw in Chapter One how Ravenscraig largely owed its existence as a major steelmaking plant to a political decision by the Macmillan government in the late 1950s. The development of Ravenscraig was linked to the hope that Scottish manufacturing industry would also enjoy an expanding future. In the vanguard of this new era was the Rootes Group, which opened a car plant at Linwood, near Glasgow, in 1959. Unfortunately, the Linwood plant was never able to operate with any great success, and after two changes of ownership (the American company Chrysler and the French firm Peugeot-Talbot) it finally closed in 1981, with the loss of nearly 5,000 jobs. The loss of Linwood robbed Ravenscraig of one of its principal customers, and the distance between the Scottish works and the principal car manufacturing plants in the Midlands of England highlighted the weaknesses of its relatively remote position. In addition, since the closure of steelmaking at Shotton, Ravenscraig had become the chief supplier of steel to that works. The BSC might well have hoped that it could confine its strip steel Division entirely to the confines of Wales, by means of increasing the manned capacity of Port Talbot and Llanwern. Scottish confidence was not enhanced when one of Ravenscraig's three blast furnaces was closed down in August 1982.

A further political factor was the issue of privatisation. Although it could be possible to run Port Talbot, Llanwern and Ravenscraig together at an overall profit, a potential buyer might be expected to prefer a compact strip mills Division with one or two plants operating at or near capacity. The emphasis placed on privatisation by the government and Ian MacGregor led many in the Scottish steel industry to believe that the plant might be sacrificed. Countering this argument, however, there were two other political factors which would eventually lead the government to decide that Ravenscraig should remain open, at least in the immediate future.

Firstly, there was increasing unease at Westminster generally that other ECSC steel industries were not rationalising at the same rate as the UK. Although Britain had started from a position of relative disadvantage, by the end of 1982 the BSC's manned capacity had nearly halved since the mid-1970s, while in the same period the workforce had been reduced by over two-thirds. Thus many MPs, including a good number of Conservatives, were expressing the opinion that the government should insist on more urgency on the part of such countries as France and Italy to rationalise their steel industries, rather than close yet another UK works.

Secondly, by the autumn and winter of 1982 a General Election was in the offing. The Conservative government was, with good reason, nervous about its likely performance in Scotland, where its popularity appeared to be in severe decline. At the first whiff of a Ravenscraig closure a concerted campaign was mounted in Scotland, in which a wide cross-section of the community was to join. Apart from the Linwood car factory, Scotland had recently experienced a series of closures, including the Singer sewing machine works in Clydebank, the pulp mill at Fort William, and the Invergordon aluminium smelter. To lose Ravenscraig on top of these was considered by the Scots to be a blow of catastrophic proportions, from which their manufacturing industry could possibly never recover.

During the latter part of 1982 the government faced severe pressure from three powerful interests: the BSC, the EEC Commission, and virtually the whole of Scotland. With regard to the BSC, the government asked Ian MacGregor to prepare a detailed list of options for dealing with the Corporation's increasingly heavy losses. At the same time, Industry Secretary Patrick Jenkin made it clear that the government would make the final decision on major closures. As with the Labour government's Beswick review, Conservative Ministers were not prepared to leave such matters of high politics purely in the hands of BSC management. The revised corporate plan was submitted to the government by MacGregor towards the end of November, and included the closure of Ravenscraig as an option. Although MacGregor was officially non-committal at that time about the future of Ravenscraig, early in 1983 he made it clear that he would have been quite happy to see both Ravenscraig and Llanwern close (House of Commons Industry and Trade Select Committee 1982–83, para 363). Bill Sirs held the opinion that MacGregor had set his mind on closing Ravenscraig (Sirs,

117–118), such was the Chairman's zeal for pushing ahead towards privatisation. MacGregor thus clearly favoured the closure of Ravenscraig, although uncharacteristically he did seem quite prepared to leave the matter in the hands of the government.

It could also be said that the Ravenscraig campaign diverted attention from the closures, cutbacks and redundancies which did take place during 1982–83. In August 1982 the BSC announced closure of the Clydebridge plate mill, with the loss of 575 jobs, and the Parkhead Forge in Glasgow with 347 jobs lost. In September, 700 redundancies were announced at Scunthorpe, bringing the number employed at the works down to 7,600, compared with 16,000 three years earlier. At the time, there was also considerable speculation about the future of the nearby Redcar blast furnace. This modern furnace, costing £400 million, had failed to work efficiently since its construction three years earlier, and was operating at only two-thirds of its capacity. In the event the blast furnace stayed in operation, although its size remained at a testament to the over-ambitious investment strategy of the 1970s.

The next works to fall was Round Oak, a major engineering steel producer in the West Midlands, with the loss of 1,286 jobs. This was closely followed by an announcement that the bar-rolling section for special steel at the Craigneuk works in Motherwell (the site of Ravenscraig) would close, with the loss of 427 jobs. At the beginning of December, 1700 job losses were announced at special steel plants in South Yorkshire. This was quickly followed by another 2,200 losses in Wales and on Teesside. A cynic might conclude that the Ravenscraig issue performed a useful function for the BSC in diverting attention from all these other cuts, although by this time the steady reduction of the BSC workforce had apparently become a fairly routine matter, albeit at a high financial cost. During 1982–83 the workforce was reduced from 103,700 to 81,000 (this figure included 1,500 joining the private sector as a result of the creation of Sheffield Forgemasters), with redundancy and related costs amounting to a massive £483 million.

While the BSC was implementing these cuts, the EEC Commission was putting new pressure on ECSC countries to reduce capacity. Despite the production quotas of the Davignon Plan, the voluntary minimum prices mechanism was beginning to fail, and capacity was still well above demand. When EEC Industry Ministers met in June 1982 they agreed in principle on the need for a 30 million tonnes a year reduction in capacity over the following three years. Agreement in principle was one thing, but to actually implement these cuts throughout the ECSC was another matter entirely.

At the same meeting it was also agreed to renew the compulsory quota regime for a further twelve months. The inclusion of wire rod meant that about 80 per cent of the ECSC's output was now covered by these quotas.

These controlled steel goods were known as Treaty of Paris products, after the Treaty which set up the ECSC, and included these goods within its ambit. The uncontrolled goods were known as Treaty of Rome products. Although there were voluntary quotas on most of the Treaty of Rome products, there was still some controversy, on special steels in particular, about their exclusion from mandatory quotas. As we saw earlier, however, the West German government wished to maintain a free market for the high value special steels, and so pressure from Patrick Jenkin to bring them within the mandatory regime came to nothing.

A further feature of this June meeting was a plea put forward by the Italian government for an extra quota of 720,000 tonnes. After discussions within Eurofer, it secured 510,000 tonnes of this amount. The political significance of Italy's case was that it demonstrated the difficulty of creating in that country a consensus of opinion that its steel industry should be severely shrunk. The Italian government had a strong political commitment to maintaining a large-scale industry, and feared the political consequences of implementing a radical rationalisation programme. It was this type of problem which inevitably produced a counter-reaction in the UK, where many politicians from both right and left of the political spectrum argued strongly that the UK should not put through any more major closures until countries such as Italy at least began to address themselves to this issue.

When ECSC Industry Ministers met at Elsinore in Denmark in November 1982, Davignon was particularly outspoken, and claimed that on present trends there would be 48 million tonnes of excess finished steel in the ECSC by 1985 (150 million tonnes per annum capacity against 102 million tonnes per annum demand). Despite this projected 48 million tonne gap, Davignon, no doubt recognising the political difficulties, was seeking a cut of 'only' 30–35 million tonnes. Davignon's calculation was that planned reductions in output capacity at that time amounted to 14.5 million tonnes, leaving another 16–20 million tonnes to come. Patrick Jenkin apparently found sympathy among Commission officials when he proposed that, as the UK had already implemented an extensive rationalisation plan, it should suffer less in the future than other ECSC countries. On the other hand there was an ominous remark from the Italian Minister of Nationalised Industries that the proposed cuts would be politically and socially dangerous (*Financial Times*, 19:11:82).

A further ominous note for all the ECSC governments was struck by the EEC's Competition Commissioner, Frans Andriessen. The role of a Competition Commissioner in the politically controlled market that steel had become was inevitably somewhat incongruous. Where Davignon was the shrewd and patient political operator, Andriessen on several occasions antagonised national steel interests (e.g. he caused controversy in the UK when he stated openly in October 1983 that Ravenscraig should close).

Andriessen now warned the Ministers that the competition authorities had set 1 July 1983 as the deadline by which the competition authorities must vet twenty or more steel industry rationalisation projects worth an estimated £10 billion, plus further state aids (the UK accounted for almost £2 billion of this money). If the steel restructuring proposals did not match the aid then the Commission might block some of this financial assistance. Andriessen did not go on to enlarge by what means the Commission would stop national governments assisting their steel industries. In this respect, there was considerable disparity between the theoretical and actual powers of the Commission. Although the Commission could take a country before the European Court of Justice, in reality the EEC officials were dependent on a politically negotiated consensus amongst the governments that rationalisation was necessary, and that responsibility for this immensely difficult political task should be shared by all.

By December 1982 the UK government faced pressure from both the BSC and the EEC Commission to sanction further rationalisation. At the same time, the political pressure from Scotland, and the steel unions generally, was approaching its climax. On 23 October the ISTC held a one-day strike as a protest against further cuts, but a more serious situation developed towards the end of November when pickets were sent to blockade Japanese and South African imports at Immingham on Humberside. This action, reminiscent of the steel strike, was called off when the pickets became convinced that the steel had a definite customer, but the incident illustrated the existence of a new groundswell of opinion against imports, and their implication for the future of the BSC.

Meanwhile the Ravenscraig campaign found probably its most powerful ally when the Secretary of State for Scotland, George Younger, threatened to resign if the plant closed. Such a departure would obviously have been a considerable embarrassment to the government within a few months of the General Election, and it was therefore an enormous boost to Scottish hopes that the plant would remain open. On 21 December the government officially came down on the side of the Scottish Secretary when Patrick Jenkin announced to the House of Commons that the government had instructed the BSC to keep all of the five main integrated plants open for another three years. The Secretary of State stressed that it was now the turn of other ECSC countries to close steelworks and cut capacity (Hansard, 21:12:82, Cols. 672–74). At the same time, Ian MacGregor released a statement which warned that Mr. Jenkin's decision would put great strain on the Corporation. Thus the government had satisfied Scotland, but at the expense of antagonising MacGregor, and also setting itself up for a hard ride within the ECSC.

The Fairless plan and a new chairman

After originally setting an EFL for the BSC in 1982–83 of £365 million, the government was once more compelled to pay out on its blank cheque, and raise the limit to £575 million, of which the Corporation claimed £569 million. Although losses by March 1983 had been reduced to £7 million a week from £9 million a week in December 1982, there was a distinct danger that the Corporation would be an albatross around the neck of government finances for the indefinite future. The BSC itself produced a Corporate Plan in February 1983, which was submitted to the Commission as demonstrating the Corporation's ability to achieve viability by 1986. The Commission rejected the Plan, however, on the grounds that it was over-dependent on gains from exchange rate movements (MMC, 1988, 27). In this climate, despite the government's assurances on the future of Ravenscraig, speculation about the future of the plant continued to abound. As we described earlier, Ian MacGregor had made it plain to the Industry and Trade Select Committee that he thought the BSC could make do with just one strip mill, presumably implying that both Llanwern and Ravenscraig should be shut down.

MacGregor, never a man prone to passivity, quickly put behind him the setback to his plans imposed by the government and by the spring of 1983 had hatched a new idea (this section taken from Dudley, 1984). This involved a radical tie-up between Ravenscraig and US Steel's Fairless plant in Pennsylvania. The idea was that Ravenscraig would produce semi-finished slab steel, which would be shipped over to America and rolled at the Fairless mills. The plan would thus involve the closure of Ravenscraig's finishing facilities, and also steelmaking at Fairless. Despite being the largest of the US steel producers, US Steel had not kept sufficiently abreast of modern steelmaking developments, and so was quite happy to face the prospect of obtaining steel from abroad (despite the obvious transport costs).

For his part, MacGregor had retained close contact with the US steel companies, and saw this liaison as a means of killing several birds with one stone. Firstly, the retention of steelmaking at Ravenscraig would give any British government at least half of what it wanted. Secondly, a UK steelmaker would penetrate the US market, for the MacGregor plan could by-pass the EEC-America quota agreement of 1982. Thirdly, by exporting steel to the US, the UK industry could become less dependent on the EEC market. Finally, it was intended that Ravenscraig would become a company separate from the BSC, and so would help to fulfil the government's ambitions to privatise the Corporation.

Although details of the MacGregor plan were not made public officially, it was suggested widely that its implementation would halve Ravenscraig's workforce of 4,000, while 3,000 of the 8,000 jobs at Fairless would disappear. Political opposition to the scheme was widespread in Scotland, but perhaps even more significantly there was also great hostility to it in the US from both unions and Congressmen. Thus, although the Prime Minister herself was known to be a strong supporter of the plan, by the autumn of 1983 it was obvious that the political obstacles were too high to clear. The final blow was probably MacGregor's departure from the BSC in September 1983, for the projected deal had borne his personal imprint. BSC management generally appeared to be dubious about the insistence by US Steel that the Corporation sink £400 million into the project, and so by the end of 1983 the scheme had fallen through. In any case, the BSC was for the moment less keen to see the demise of Ravenscraig, for during the year work had begun on a major £171 million strip mill development at Port Talbot, and so for some time to come Ravenscraig's facilities would be required to stand in for those temporarily out of commission at the Welsh plant.

At the time of MacGregor's departure the BSC's finances were slowly improving, but it could not be said that he had wrought any kind of personal innovative revolution at the Corporation. As we observed previously, the radical rationalisation plan had been set in motion by Villiers, although under MacGregor manpower had been reduced from about 166,000 to around 80,000. To halve the workforce in three years was indeed a radical execution process, but it is difficult not to conclude that a similar programme would have taken place under another Chairman. Compared with other costs such as raw materials and energy, labour was one which could be relatively easily reduced by the simple expedient of offering generous redundancy terms. Given the government's willingness to provide the cash for this process, then it did not present any insuperable obstacles. Although MacGregor did have some definite success with regard to coal and energy prices, we saw that there were many powerful interests stacked against him, and so the room for manoeuvre was limited.

Similarly, despite his free market values, MacGregor had no choice but to accept the controls exerted by the Davignon plan, while on privatisation only one Phoenix actually rose from the ashes during his reign, together with a number of peripheral BSC interests. The Fairless plan indicated a more individualistic side to MacGregor's personality, but taking the three years of his Chairmanship as a whole, this aspect of his character was little in evidence. Overall, like the Prime Minister and her government, MacGregor was a political prisoner of the BSC's chronic financial position. No doubt both the Chairman and Mrs. Thatcher would have liked to privatise the whole Corporation in the early 1980s, but this policy presupposed that there was a profitable organisation to sell. Ironically the government and MacGregor had to accept considerable control and regulation in the shape of the Davignon Plan, in the hope that their eventual objectives of profitability and privatisation could be achieved.

Not surprisingly, the government encountered difficulty in finding a successor to Ian MacGregor. Characteristically, MacGregor himself expressed a wish to combine the Chairmanships of the BSC and the NCB, but the government did not appear to be keen on the idea (watching MacGregor negotiate with himself on the price of NCB coal to be sold to the BSC would have made an interesting spectacle). The initial front runner for the job was Sir Alistair Frame, Deputy Chairman and Chief Executive of the metals company Rio Tinto Zinc. Sir Alistair was upset when his name was leaked, however, and after the BSC's own long-serving Deputy Chairman, Bob Scholey, was passed over, the choice finally fell on Robert Haslam. Formerly Deputy Chairman of ICI (he had left the company after being pipped for the Chairmanship by John Harvey Jones), Haslam was now Chairman of the sugar company Tate and Lyle. By the time of Haslam's appointment, in June 1983, the Conservatives had won a resounding victory in the General Election held earlier that month, and so the new Chairman could at least be confident that he would not lose his job merely through a change of government.

Haslam's salary was fixed at £55,000 a year for a part-time appointment. To compensate the BSC for the loss of a full-time Chairman, Bob Scholey was given greater authority and the title of Chief Executive, to go with his existing post of Deputy Chairman. (During MacGregor's Chairmanship Scholey had been given the title of Chief Operating Officer, but he now regained the title of Chief Executive which had been taken by MacGregor in 1980). As we remarked in Chapter Seven, the loss of a full-time Chairman inevitably had the effect of downgrading the political importance of the BSC as an organisation. This status was restored in 1986 when Scholey, after more than a decade as number two at BSC headquarters, became the new Chairman when Haslam left to take charge of the NCB (by now British Coal).

The Davignon plan tightens

After introducing quotas and price guidelines, by the summer of 1983 the EEC Commission was ready to push harder for rationalisation of ECSC industries. Franz Andriessen, the Competition Commissioner responsible for steel restructuring, had become increasingly impatient at the manner in which he considered that too many ECSC governments were dragging their feet with regard to cutting steel capacity. So far, the Commission had only received commitments to capacity cuts totalling 18.4 million tonnes per annum on the 1980 total EEC capacity of 168.6 million tonnes per annum (the figure of 18.4 million tonnes represented a commitment to cuts, not cuts themselves). Andriessen wanted to see at least another 9 million tonnes go by the end of 1985, and after much heated discussion between ECSC Industry Ministers it was finally agreed to remove 26.7 million tonnes (i.e. 8.3 million tonnes on top of that which had been committed already). It could be said that this compromise represented a defeat for Davignon and Andriessen, who had both been seeking total cuts of 30–35 million tonnes, but at least the figure had moved towards that target.

Cecil Parkinson, destined only briefly to be Secretary of State for Trade and Industry (the two Departments had once more been merged), argued strongly that a major UK steelworks closure was not politically negotiable, as the country had nearly fulfilled its obligation of a 4.2 million tonnes per annum cut. In contrast many other countries, and in particular Italy, had barely begun to implement their cuts programmes. Parkinson's arguments were sympathetically received by the Commission, and consequently the UK was expected now to contribute only a further 500,000 tonnes. Italy, on the other hand, was required to cut a further 3,460,000 tonnes, Belgium 1,400,000 tonnes and West Germany 1,200,000 tonnes.

There was inevitably considerable uncertainty about the prospects for achieving all the required cuts, for nearly all the ECSC countries faced formidable implementation problems. In West Germany, a Commission appointed by the steel industry to investigate the possibility of rationalising the steel companies, advised the creation of two large groups. One involved the merger of Thyssen (the ECSC's largest steelmaker) and Krupp, and was to be known as the 'Rhine group'. The other would have brought together Hoesch, Kloeckner, and the state-owned Salzgitter, and would have been known as the 'Rhur group'. The merger plans were backed by the Christian Democrat government, but the companies themselves were not enthusiastic. Hoesch refused to consider a merger with Kloeckner, while in October 1983 talks between Thyssen and Krupp collapsed when the former refused to help absorb the latter's DM 1.5 billion of borrowings.

Despite the failure of these merger talks, the West German government was prepared to approve a £750 million aid package for its steel industry. It was intended that the money would assist in cutting capacity amounting to 11 million tonnes by 1985. Although the West German steel industry is by far the largest and strongest within the ECSC (the price of steel is usually settled in deutschmarks, thus making the relative value of the pound against the deutschmark of considerable importance in determining the BSC's trading position), by 1983 nearly all its steel companies were experiencing financial difficulty. The companies were particularly upset at what they considered to be unfair competition in the form of subsidised steel imports, and despite the aid that they themselves were now receiving, were putting pressure on their government to introduce some form of border tax. In this atmosphere, it was not easy for the West German government to persuade the steelmakers that large-scale rationalisation was required.

In France, the Socialist government had nationalised the state-owned Sacilor and Usinor in 1981. By the end of 1983 the French steel industry was estimated to be losing £870 million a year. President Mitterrand was attempting to walk a delicate political tightrope in seeking to implement cuts while not alienating his natural supporters. Early in 1984 the President announced that a further 21,000–27,000 jobs would have to go, and this immediately provoked riots and a general strike in the traditionally militant region of Lorraine. This resistance indicated that achieving the targets given by the Commission would not be an easy task for the government. Significantly, the French President himself was now suggesting that the ECSC abandonment of subsidies should be put back a further two years, until the end of 1987.

A similar situation prevailed in Belgium, where the state-owned Cockerill-Sambre was losing money at an accelerating rate, yet the resistance to cuts in areas such as Wallonia was formidable.

In Italy, where by early 1984 the closures programme still barely had begun, the problems were even more acute. The state-owned Finsider, largest of the Italian steel companies, itself lost about £800 million in 1983. The Italian government committed itself to implementing the cuts imposed by the Commission, but this decision inevitably caused great controversy in a country with a traditionally strong commitment to growth in its steel industry.

These examples illustrate that, notwithstanding the commitment of national governments to implementing EEC cuts, there was no guarantee that the policy would be fulfilled. In Britain an alliance of sorts between government, steelmakers and unions had produced the desired results, but throughout the ECSC as a whole achievement of planned cuts was still exceedingly problematic, and made the complete cessation of subsidies by the end of 1985 a doubtful prospect.

The five elements of the Davignon plan: import controls, output quotas, price guidelines, a timetable to end subsidies and a firm rationalisation programme, were now in place, but 1983 was not a good year for ECSC steelmen (in addition to domestic difficulties, the US imposed tariffs on EEC imports of special and stainless steels) and towards the end of it, prices were once more being heavily discounted. Thus Davignon was forced to step in yet again. In November 1983 he proposed mandatory minimum prices for a range of products amounting to 50 per cent of steel tonnage subject to quotas (the quotas themselves controlled about 85 per cent of total ECSC output). In addition producers were to pay a deposit (eventually agreed at £8.55) on every tonne of steel shipped, which would be forfeited if price controls were breached. Controlled steel was also to be accompanied by a certificate, so that its origins could be traced.

These new conditions were basically accepted by all the ECSC governments except West Germany. The Germans were still upset about subsidies in other countries, and also increased steel imports into their own country. After some haggling the other governments conceded to the West Germans that, at least for the moment, trade patterns should be ossified. Trade flows were to be monitored, and allied to restraints on imports it was hoped to keep steel movements at the levels obtaining in the second half of 1981, and the first half of 1982. This was not good news for the BSC, which might have hoped to gain new export markets as a result of its new-found efficiency. Nevertheless, if the mandatory minimum prices could hold, then it would at least give the Corporation the opportunity of a more stable income.

In 1983–84 the BSC made a loss of £174 million after interest, compared with £383 million the previous year. Even more significantly, rationalisation and other exceptional costs were down from a massive £483 million in 1982–83 to £79 million in 1983–84 (the workforce had only fallen by 10,000 to 71,100 during the year). Liquid steel production was up from 11.7 million tonnes to 13.4 million tonnes, reflecting a modest increase in demand of 4 per cent. It could be said that the Corporation was at last beginning to benefit from the radical improvements in productivity it had wrought since the early 1980s. Most of the manpower cuts and major closures had now been completed, and for the first time in nearly a decade it was possible for the BSC to look forward to the future with at least some degree of confidence. Life had rarely been placid for the BSC throughout its history, however, and by the beginning of 1984–85 a cloud was appearing on the horizon in the form of the miners' strike.

The miners' strike and the end of the Triple Alliance

Any hopes the BSC might have entertained that 1984 would see it enjoying a period of quiet recovery were quickly dispelled when a strike by large sections of the NUM began in March. When the NCB made a decision to close Cortonwood colliery in Yorkshire, a strike in that county quickly spread to South Wales, Scotland and Kent. Other areas, notably Nottinghamshire, were less compliant, and the refusal of the NUM Executive to call a national strike ballot led to a fundamental split within the union.

The President of the NUM, Arthur Scargill, wished to see the other two members of the Triple Alliance, rail and steel, give full support to the miners, but his hopes were not to be fulfilled. The relationship between Bill Sirs and

Arthur Scargill had been strained since the steel strike of 1980, when Sirs blamed Scargill for starting the ritual cry of 'what do we want? – twenty per cent' (Sirs, 4). Sirs was also upset about Scargill's claim that during 1980 the miners had given full support to the steelmen, when in reality the steel unions had received little tangible support from other industries. When the NUM decided that the steel industry was to be one of the main tools in achieving their chief objective of halting the NCB's closure plans, conflict was bound to follow. At Port Talbot and Teesside, all their coal needs were supplied from abroad, and the coal unloaded at their own docks. Despite some sporadic NUM picketing at Port Talbot, coal supplies to these two works were not impeded. Attention was therefore focused on the BSC's three other major works of Ravenscraig, Scunthorpe and Llanwern, where a high proportion of their coal was not only domestically produced, but had to travel some distance to arrive at each works. During the spring and summer of 1984 the NUM made repeated attempts to stop coal supplies to these works but failed to halt production (a detailed, inside account of this period is given in Sirs, 7–42). The ISTC was ready to give the NUM some support, but was not prepared to risk the complete closure of works such as Ravenscraig and Llanwern. Sirs summed up the situation when he declared that: 'I will not see the steel industry crucified on someone else's altar' (*Financial Times*, 31:3:84).

It appeared that much of the NUM picketing of steelworks was in breach of the 1980 Employment Act, which allowed employers to seek an injunction against unions using secondary picketing as a strike tool. Ironically, the 1980 Act had been created in the emotional climate of the 1980 steel strike, when the ISTC had employed secondary picketing against private sector steel companies. The BSC declined to take the NUM to court during 1984, however, apparently judging that such a move might only inflame the situation still further.

The first crisis occurred at Ravenscraig in April, where the ISTC wished to see two train loads of imported coal a day transported from Hunterston to the plant, whereas the NUM wanted only one load, the minimum required to protect the blast furnaces and coke ovens. After a meeting of the Triple Alliance, the NUM agreed to let two train loads of coal a day into Ravenscraig, but by the end of the month supplies had declined to one train load. Earlier in the month, BSC management had started bringing coal in by road, and now (with the ISTC's approval) they restarted this practice. After this, the railmen stopped even the one train a day, while the TGWU threatened to starve diesel lorry firms carrying coal to Ravenscraig. This was an ineffective threat, however, and the BSC stepped up its supplies by road. For the first time, the NUM now asked the railmen to halt supplies of iron ore to Ravenscraig, but the immediate crisis was resolved on 11 May when at a Triple Alliance meeting in Edinburgh it was agreed that Ravenscraig would receive about 18,000 tonnes of coal (representing 13 train loads) a week.

Meanwhile, by May the coal stocks at Llanwern had also run down, and supplies promised by the NUM had failed to materialise. The miners eventually agreed to an emergency supply of several train loads, and an agreement was then struck to allow two dozen train loads of coal and coke a week into the works, only for this arrangement to also fall through in June.

At Scunthorpe the most serious situation of all developed when a Triple Alliance agreement broke down on 24 May. The BSC immediately began to transport by lorry coke from the Corporation's own ovens at Orgreave near Sheffield. Publicly, BSC management at Scunthorpe expressed serious concern about the condition of blast furnaces at the works, and was prepared to risk crossing miners' picket lines at a coke manufacturing plant which lay in the heartland of the NUM strike. Nevertheless, in addition to the Orgreave coke, 6,000 tonnes of Polish coal was also moved into Scunthorpe by lorry from the Flixborough wharf five miles away. Ian MacGregor himself claims that the NCB was hoping to tie up pickets at Orgreave, as it prevented them travelling to picket the Nottinghamshire pits. In addition, once the Polish coal had arrived at Scunthorpe, he claimed that the BSC had no special need for Orgreave coke (MacGregor, 1986, 207).

Matters came to a head on 18 June, when the NUM gave the ISTC until midnight to seek a national agreement with the miners' National Co-ordinating Committee or face a complete blockade of coal and coke. On the same day, NUM picketing at Orgreave reached its peak, and there then occurred the most violent scenes of the whole twelve month miners' strike. Sixty police and pickets were injured, one hundred miners arrested, and Arthur Scargill himself was taken to hospital suffering from concussion. After this confrontation the BSC diplomatically announced that Scunthorpe now had sufficient coal and coke supplies for the moment, and halted coke deliveries from Orgreave (the Corporation began moving coke from Orgreave again on 9 July, but this time with only a few pickets present).

The ISTC, which at the time was holding its Annual Conference in Scarborough, was extremely angry at the miners' ultimatum, and, despite giving a standing ovation to an NUM representative, the Conference rejected any proposal to halt steel production. The blockade began, but road deliveries of coal were still made to Ravenscraig and Scunthorpe. Up to this point the NUM had not attempted to stop supplies of iron ore, but on 26 June four out of five trains scheduled to deliver iron ore to Llanwern were cancelled when the drivers refused to cross miners' picket lines. Far from a solidifying of the Triple Alliance, the dispute had now developed into a trial of strength between the NUM and the ISTC, and after a meeting between the two unions on 29 June had failed to reach agreement, the ISTC agreed to the BSC moving iron ore by road into Ravenscraig and Llanwern.

The chief centre of attention now became the Scunthorpe works and its supply routes. All supplies by rail ceased on 3 July, and on the following day 1,200 miners attempted to converge on River Trent wharfs where coal and iron ore were being imported, but they were headed off by police. The TGWU was upset that lorries were now being used to keep the BSC plants in operation, and on 10 July called a national dock strike when 700 dockers at Grimsby and Immingham walked out over a dispute concerning the statutory National Dock Labour Scheme. The dockers claimed that the BSC was moving ore without using registered labour, and so was in breach of the Scheme. The national strike received only patchy support, however, and after only four days a local agreement was reached between the BSC and the TGWU. The Corporation agreed to train the local dockers to use the equipment needed to load iron ore into lorries (although iron ore supplies to Scunthorpe were not resumed until 6 September). Apparently, only outside contract labour had hitherto been capable of performing this feat! By now, the steel unions had become determined to back the BSC management in maintaining steel production at any cost, and it appeared that the TGWU, the largest union in the country, was not able to deliver all-out support to the miners.

There was to be one more concerted attempt to halt supplies to a steel plant, however, when in August a dispute arose at Hunterston, which led to a further national dock strike. It will be recalled that the Hunterston terminal, which delivers imported iron ore and coal to Ravenscraig, was only opened after a six month delay in 1979 when the Conservative government agreed to it becoming a National Dock Labour Scheme port. This piece of pragmatism now rebounded on the government somewhat when a dispute broke out between the TGWU dockers at Hunterston, the BSC, and the Clyde Port Authority (the dockers' employers). The registered dockers refused to unload the Ostia, a Liberian registered vessel carrying 95,000 tonnes of coking coal, because of their support for the miners. There was considerable confusion over the exact terms of the Scheme agreement at Hunterston, but by the end of August the BSC claimed that coke supplies at Ravenscraig were becoming dangerously low, and so unloaded the Ostia without the help of registered labour.

Once again the TGWU called a national dock strike, and once again the support was patchy. Over half the ports remained open, and what could have been the biggest test of the government's resolve to resist the miners' strike failed to achieve its objective. Ostensible support for the miners was given in resolutions passed at the TUC Conference at Brighton in September, but in reality the unions lacked both sufficient unity (including the NUM itself) and grassroots support to exert truly damaging pressure on the NCB and the government. On 16 September the dispute at Hunterston was settled without the BSC agreeing to use registered labour in future, while the BSC also won its demand that Ravenscraig should receive 22,500 tonnes of coal a week. Shortly afterwards the three-week old national dock strike came to an end.

Although the miners' strike was to last for nearly another six months, the NUM's challenge to BSC supplies was not resumed. In the end, the steel unions had chosen to support BSC management and preserve steel production, rather than prolong the life of the Triple Alliance. The steel unions had also scored a significant political point in that it would be extremely difficult for the government to sanction closure of a plant such as Ravenscraig, when its workers had stood up to severe pressure from the miners and maintained production throughout the year long strike.

The BSC estimated that the cost of the miners' strike to the Corporation totalled £180 million, involving approximately 8 million tonnes of material being switched from rail to road transport. The Corporation actually made an operating profit of £40 million for 1984–85, but after the cost of the miners' strike and exceptional items (including the Sidbec-Normines write-off), the loss became one of £409 million. Thus although the EFL for the year was eventually set at £343 million, the Corporation breached this amount by a large margin, the final figure being £523 million.

Another life for Ravenscraig

Because of the miners' strike, the government was not prepared to submit a BSC corporate plan for viability to the EEC Commission. When the strike came to an end in March 1985, the government was once more faced with the Ravenscraig dilemma. The three-year guarantee to the five major works was due to be terminated at the end of the year, and the government faced pressure from both BSC management and the EEC Commission to shut the plant down. On the other hand, the Ravenscraig workforce had played a large part in securing a continuation of steelmaking during the miners' strike, while the Secretary of State for Scotland, George Younger, was still worried about the effect of a Ravenscraig closure on the fortunes of the Conservative Party north of the border. In addition, there were definite signs that the commercial fortunes of the BSC were on the upturn, with a small operating profit being made in the first months of 1985–86. This new optimism was reflected in the BSC's Corporate Plan of August 1985. Profit projections in the range of £150–£275 million for 1987–88 were considered to meet the Commission's viability criteria, and so justified the release of the remaining financial aid due to the Corporation. Consultants appointed jointly by the Commission and the DTI concluded that, if anything, the BSC's profit projections were on the low side, although they also considered that the Corporations's sales forecasts stretched credibility to the limit (MMC, 1988, 27–29). In the event, the BSC's forecasts on profits and sales proved to be somewhat on the conservative side!

As expected, the government eventually came down in favour of keeping the five major plants in operation for a further three years, but this time the decision was not so clear cut as in 1983. Although Ravenscraig was to survive, the government gave its consent to the closure of the Gartcosh cold strip rolling mill in March 1986. The BSC decided that the further investment needed to modernise Gartcosh was not justified, and that the Gartcosh steel supplied by Ravenscraig should be rerouted to Shotton. The Gartcosh closure was to cost 710 jobs, and the government no doubt hoped that the political penalty of this would be minimal. As Gartcosh took one third of Ravenscraig's hot mill steel, however, many of the interests which had fought to save the major Scottish plant now feared that the loss of Gartcosh was the thin end of the wedge which would eventually see off Ravenscraig.

The government was nevertheless content that it had secured the immediate future of the five major works, particularly in the face of pressure from the EEC Commission. When it was pointed out to the Secretary of State for Trade and Industry, Norman Tebbit, that the Commission was looking for a reduction of over 24 million tonnes in EEC capacity over the next five years, he replied that Britain would only be contributing in a minor way: 'When you are looking for big amounts they are not really here any more' (*Financial Times*, 8:8:85).

Any pressure from the Commission over the retention of Ravenscraig was somewhat eased by the BSC's purchase of the privately owned Alphasteel strip mill in South Wales. Alphasteel's two continuous slab casters were to be transferred to Llanwern to improve the efficiency of that works, but the Alphasteel wide hot strip mill was to be closed and its quotas switched to BSC works. Thus by closing Alphasteel, with a potential capacity of one million tonnes per annum, the government could claim that the UK was continuing to cut its steel capacity, even if none of the BSC works was threatened.

Despite the Ravenscraig announcement, the BSC and government were soon thrown on the defensive as the campaign in Scotland to save Gartcosh gathered steam. Both the Corporation and Ministers were anxious to stress that the closure of Gartcosh did not necessarily entail the demise of Ravenscraig. The BSC claimed that, as a result of the closure of the Linwood car plants and BL's bus and truck works at Bathgate near Edinburgh, only 2.8 per cent of Gartcosh's product was now sold in Scotland. It was now intended that Ravenscraig would supply 70 per cent of Shotton's hot rolled coil, where the chief market for cold reduced sheet, the

Midlands of England, were much closer. At the same time, many in Scotland argued that Shotton was much nearer to Llanwern and Port Talbot than to Ravenscraig, and that these Welsh works could eventually supply all of Shotton's needs. It was also pointed out that the three year guarantee would conveniently last until just after the next General Election!

During the remainder of 1985 the Scottish campaign to save Gartcosh was fought strongly, with many Conservative MPs supporting the case for its retention. Eventually, after much haggling between Conservative and Labour MPs, the Commons Scottish Affairs Select Committee produced a Report which said that the closure of Gartcosh did threaten Ravenscraig (House of Commons Select Committee on Scottish Affairs 1985–86, HC 133, para. 18), but the crucial political element was probably the attitude of George Younger. The Scottish Secretary of State had previously made the closure of Ravenscraig a resigning matter, but now he rejected the Select Committee's case. Thus Gartcosh did close in March 1986, and despite government and BSC assurances inevitably threw doubt on the long-term future of Ravenscraig.

The end of the Davignon era

Notwithstanding the domestic political debate, the future of the BSC was still heavily dependent on policy decisions within the ECSC as a whole. When Davignon resigned as Industry Commissioner at the end of 1984, his Plan was entering its final year. During 1984 the mandatory minimum price regime had held reasonably well, and played a part in the BSC's improved trading performance of 1984–85. Both the BSC and the government were becoming increasingly restive about what they considered to be over-restrictive production quotas imposed on the UK, which reflected a period when the Corporation was less efficient. Indeed, there were even complaints from some customers about steel shortages, and the BSC was forced to spend significant sums of money (the actual amount not specified) on purchasing quotas from other ECSC companies. In 1984, total UK steel output was up 1.79 per cent at 15.14 million tonnes, although the average growth rate for the ECSC as a whole was around 9.7 per cent. Much of this difference can be explained by the loss of the mining industry market for much of the year, but the ossification of trade and output within the ECSC did not allow the BSC much room to exploit its improved productivity.

The new EEC Industry Commissioner, Karl-Heinz Narjes of West Germany, was faced with the task of producing a blueprint for steel in the post-Davignon era. His biggest problem was in the area of subsidies to national steel industries. Under the Davignon Plan, all subsidies would come to an end in December 1985, when in theory the companies would be firmly on the road to viability. In practice, however, it was obvious even in 1984 that a free-standing ECSC steel industry by 1986 was an impracticable goal. The Commission had demanded certain capacity cuts in return for subsidies, but even if these cuts were implemented there was no guarantee that viability would ensue.

As well as Sacilor and Usinor in France, severe problems faced Finsider in Italy, Cokerill-Sambre in Belgium, and Arbed Saarstahl in West Germany, not to mention the BSC itself. Notwithstanding its obvious misgivings about the future viability of these concerns, the Commission had little choice but to continue sanctioning the payment of subsidies, for it was clear that national governments (like that in the UK) would ultimately not allow any of their major steelworks to go under. Operating subsidies (as opposed to those earmarked for the purposes of restructuring) should have ceased at the end of 1984, but in March 1985 the Commission agreed to extend these until the end of 1985, on the grounds that trade had been poor in the second half of 1983. Both the French and Italian governments made it clear that they would be seeking permission to distribute a further round of subsidies, although the Commission insisted that these must be tied to more capacity cuts. This latter condition was demanded by the German government, which was once again becoming worried about the threat to its own industry from what it considered to be unfairly subsidised steel.

The political difficulties in accelerating a restructuring programme were once more illustrated, however, when in July Unimetal of France (a manufacturer jointly owned by Sacilor and Usinor) announced the closure of a plant near Valenciennes. Only six days later pressure from Pierre Mauroy, now Mayor of Lille but Prime Minister at the time of the 1984 restructuring plan announcement, led the government to back down, with the new Prime Minister, Laurent Fabius, declaring that the Valenciennes plant would only close when alternative employment had been found for the 770 workers threatened with redundancy. Thus the Commission could not impose new capacity cuts on either France or Italy, and as early as May 1985 Narjes was admitting the possibility that the 'manifest crisis' might be extended beyond the end of 1985. By July, the Commission was proposing that, although operating and investment subsidies would cease at the end of 1985, they would continue for restructuring, environmental protection, and research and development purposes. In addition, although the minimum price regime would be abolished, output quotas would remain.

ECSC governments generally were reluctant to see an end to output quotas, those such as the French and Italians on the grounds that free competition could undermine their plans for viability, and the West Germans on the grounds that continued subsidies in some countries would in any case destroy hopes of a free market. The Commission, however, proposed that quota controls be removed from certain products, and surprisingly won eventual agreement from the governments. It appeared that Narjes was a less accommodating Industry Commissioner than Davignon, and not so prepared to accept the special pleading of the steelmakers. Thus from the beginning of 1986 quotas were lifted on concrete reinforcing bars and coated sheet, which accounted for 15 per cent of steel products under control (themselves amounting to 85 per cent of total production) (*Financial Times*, 31:10:85). The market in coated sheet was regarded as being more buoyant than in other areas, and so the governments, somewhat reluctantly, agreed to the Commission's suggestion.

In the case of the UK, the government gave its agreement to lifting some of the quotas in return for a regime which would allow British steelworks an improved quota share. In future, a company would be able to arrange a change in its quota for a particular product provided there was not more than a 25 per cent increase in the amount of quota for that product, and provided the tonnage being transferred from one product to another did not exceed 10 per cent of the company's total quota. In addition, the Commission itself would retain a 3 per cent quota in reserve which would be distributed to those companies suffering severe problems in satisfying their market. (*Financial Times*, 31:10:85). This new quota regime was designed to last for two years. Thus the BSC could at last feel that its massive restructuring programme had won a concrete reward.

On subsidies, the governments accepted the Commission's July proposals, with restructuring subsidies now restricted to 50 per cent of redundancy costs, and aid to closures when the companies were leaving the sector altogether. These new subsidies were due to be phased out after three years. By the end of 1985 the Commission claimed that it had exceeded the 1983 objective of 26.3 million tonnes cut in capacity, at a figure of 28.6 million tonnes, with a further 2.5 million tonnes due in 1986. At the same time, the Commission also considered that a further 20 million cut in capacity was due by 1990 if production and consumption were to match. Unlike the Davignon Plan, however, subsidies were now not tied to specific capacity cuts, while the political will of governments to implement these cuts was still problematical. In any case, as Davignon once sagely observed: 'Capacity is anything but what the steel companies say it is' (*Financial Times*, 8:8:85).

For the BSC, the end of 1985 was a time of quiet celebration. For the first six months of the year the Corporation made a profit of £21 million after interest but before taxation and exceptional items – the first half-year profit since 1974. At the same time, as so often happened for the BSC, there were a couple of nasty clouds on the horizon. Firstly, even before the end of the year and the end of controls, the Corporation was becoming concerned about a decline in prices, with accusations of dumping by the French. Secondly, the trading relationship between the EEC and the US was once more deteriorating. The 1982 agreement on EEC exports to the US had expired at the end of the year, but a new four-year agreement had been struck in November 1985. This considerably extended the number of products under trade restraint, but the UK had particular reasons for being the one ECSC country to refuse to ratify the new agreement.

Earlier in the year, the BSC had bought a minority shareholding in the US Tuscaloosa steel company in Alabama. In a plan similar to the Fairless scheme, the BSC was to supply 250,000 tonnes of semi-finished products from Teesside to Tuscaloosa in 1986, rising to 600,000 tonnes per annum by 1989. The US government was now threatening to unilaterally impose a 400,000 tonne maximum on semi-finished imports from the EEC, a move which would have effectively wrecked the Tuscaloosa scheme. The US agreed to ratify the pact only when the US government gave an informal assurance that the BSC would be allowed to honour its Tuscaloosa agreement. This promise appeared to count for nothing, however, when at the end of 1985 the US went ahead with its unilateral action on semi-finished products. This still allowed for a special dispensation of 200,000 tonnes per annum of BSC steel to go to Tuscaloosa each year, but the full scheme appeared to have been squashed. In retaliation for the US action, the EEC placed restrictions on American sales of fertilisers, art paper, and beef grease. The deteriorating relationship between the US and the EEC, taken with continuing doubts on prices and subsidies within the Community itself, emphasised the extent to which the BSC was still dependent for its future viability on widespread and complex political negotiations.

In the event, the US dispute was settled later in the year, when a three-year agreement was signed whereby EEC semi-finished steel was given a quota on the US market of 600,000 short tons in 1986, rising to 670,000 tonnes for the first three quarters of 1989. In return, the assorted EEC trade sanctions, introduced earlier in the year, were withdrawn (*Financial Times*, 10:7:86). This new agreement allowed the BSC some scope, but still represented a severe restraint on the Tuscaloosa arrangement. The BSC's best hope of a substantial improvement in export performance lay within the EEC itself. In Chapter Ten we will examine the BSC's export performance, with particular regard to a significant policy change in the second half of the 1980s which placed considerably greater emphasis on improving exports in tonnage terms.

The BSC in profit

It could be said that in several respects the years 1986 and 1987 heralded a new era for the Corporation. Firstly, for the first time in a decade, it was reporting profits : £38 million in 1985–86, £178 million in 1986–87 and

£410 million in 1987–88. Initially, this upturn was achieved without a significant improvement in domestic market performance, but in the first half of 1987–88 home market sales rose by 16 per cent compared with the corresponding period in the previous year. The greatest improvement was in exports, where in 1986–87 these increased in tonnage terms by 42 per cent.

By the end of 1987 the government felt itself in a position to announce (with considerable encouragement from the BSC) that the Corporation would be privatised as soon as possible, with 1989 the hoped-for date. There was considerable uncertainty about the exact date, given that external conditions were still highly problematical. In particular, within the EEC the political cohesion engendered by Davignon was breaking up, leaving a severe power vacuum which threatened future stability. The lack of leadership and indecision within the Commission left the members of Eurofer in a state of suspense, where despite commitments to free trade as a long-term objective, most wished to see quotas maintained for several years to come. On the other hand, governments and steelmakers were reluctant to implement politically sensitive plant closures. For the moment, the BSC could afford to sound more confident than most producers, but the experiences of the 1970s demonstrated how quickly the mood could change, and so plunge the Corporation into a new crisis.

A power vacuum in Europe

The authority of the Commission rests on the willingness of governments and Eurofer to use it as a totem of power under which a consensus of interests can be arranged. During 1986–87, Narjes attempted to wield authority, but lacked the political skill to bring about the necessary agreement. In fairness, the decisions to be made were of particular sensitivity, but by the end of 1987 there was still no agreed plan for controls and rationalisation.

At the end of 1986 quotas were scrapped only on galvanised sheet, a relatively prosperous product area representing 5 per cent of the market. The Commission had originally sought liberalisation of 20 per cent of the market, but West Germany in particular was opposed to this proposal, and so Industry Ministers postponed any further decisions until Eurofer itself had put forward restructuring plans. Eurofer had proposed cuts of 11.9 million tonnes in annual capacity by the end of 1990, but the Commission was seeking 20 million tonnes. In the now time-honoured tradition, Eurofer commissioned accountants Peat Marwick to draw up proposals for cuts in hot rolled capacity. The Peat Marwick plan, announced in March 1987, proposed cuts of 15.26 million tonnes annual capacity by the end of 1990, in return for the continuation of quotas until that time. Although this figure was not far short of the Commission's target of 20 million tonnes (in any case future capacity requirements are notoriously difficult to forecast), Narjes was concerned that the Eurofer proposals did not include any cuts in hot rolled coil. Closures in these areas involved large strip mills in politically sensitive areas such as Ravenscraig, and throughout the EEC there was continued reluctance to take such radical action. UK Ministers continued the policy that none of the five major plants was in jeopardy, although it was reported that Bob Scholey had offered 3 million tonnes in cuts, but had been over-ruled by the government. (*Guardian*, 17:3:86).

The Commission's rejection of the Eurofer plan led Bob Scholey, in his capacity as President of Eurofer, to write to Narjes suggesting that the Council of Ministers take the initiative. Scholey was in a delicate position at this time, for as Eurofer President he was seeking a continuation of quotas, but as Chairman of the BSC he wished to see a liberalisation of the market. The BSC was exceeding its quota in several products, and paying around £50 per tonne for spare quotas. During 1987 both the UK government and the BSC campaigned for an end to quotas, yet both must have feared the effects of a price war. Usinor-Sacilor (the two companies were merged during the year) Finsider, Cockerill-Sambre and others continued to make losses, and despite the ban on subsidies, governments would not allow their principal companies to collapse. In West Germany, the position was slightly more complex. Although still by far the strongest ECSC steelmaking country, the German companies were seeking state aid for restructuring, and wished for a further period of stability. Unlike the UK government, the West German Ministers also feared that in a price war their companies would be undercut by subsidised producers in France, Italy and Belgium. Only Holland supported the UK position of an immediate end of quotas.

After the Commission rejected the Eurofer proposals, Narjes himself produced a plan which was to cause further controversy. The Industry Commissioner now calculated that there was around 30 million tonnes of surplus annual capacity above the existing figure of 140 million tonnes (one of the Commission's problems in maintaining credibility was that its estimate of surplus capacity varied during the year between 20 and 30 million tonnes). Narjes estimated that a sum of £3.7 billion would need to be spent over the following three years on works closures and job creation, and it was how this money was to be raised which caused widespread dissension. The proposals put forward by Narjes included the retention of quotas on hot rolled coil, cold rolled sheet, heavy plates and heavy sections until 1990, with restrictions on the more prosperous merchant bar and wire rod categories ending in 1987. At the same time, quotas would only be continued if companies promised by the end of the year to make 'a good portion' of the closures needed. Narjes also wished the producers themselves to contribute to the closure and social costs, and so he proposed a complex package in which a steelmaker would pay into the fund a graduated sum for each tonne of steel produced below and above its quota. This proposal was not surprisingly rejected by both the UK government and the BSC, who claimed that efficient producers would be subsidising the less efficient ones.

The Council of Ministers was unable to come to an early agreement on the Narjes plan, which also envisaged that the required closures could mean a loss of 80,000 jobs in the industry. The lack of consensus led to the adoption of a proposal that 'three wise men' be appointed to produce guidelines on how the closures could be achieved (*Independent*, 22:9:87). It was originally suggested that Davignon might make a dramatic return to his former domain as one of the 'three wise men' but in the event they consisted of a former West German Economics Minister, the head of the Italian Atomic Energy Authority and a former Managing Director of Usinor (*Guardian*, 9:10:87). The appointment of the 'three wise men' could not be seen as anything other than a political holding operation, for they lacked either the authority or the experience to resolve the deep differences within the ECSC.

The Commission itself was divided over the remit to be given to the 'three wise men', with Narjes wishing to give them fairly wide powers, but other Commissioners were anxious that the panel should not take too much responsibility (*Financial Times*, 8:10:87). In the event, the 'three wise men' were only given the restricted remit of obtaining information on plant closures required in the next three years. The report produced by the 'three wise men' was thus couched in general terms, and suggested a continuation of quotas on some products until 1990 if closure commitments could be obtained from producers, otherwise quotas should cease at the end of the year.

Thus December 1987 arrived, the month when quotas were due to end, with political agreement as far away as ever. At this point there was a dramatic development domestically when the government announced that early in 1988 it would introduce legislation facilitating the privatisation of the BSC. It was intended that the BSC would be privatised as a single unit, with 1989 as a target date. The political effect of the announcement was heightened by the BSC on the same day declaring a half year profit of £190 million. During the General Election campaign in June 1987 the government had been notably cautious about the prospects for steel privatisation, but now the BSC had been moved up the list.

It was suggested that the Treasury had been particularly instrumental in this policy change, for the government was encountering considerable difficulty in the proposed privatisation of the water and electricity industries, and it was hoped that steel could fill a gap (*Financial Times*, 4:12:87). At the same time, BSC privatisation was still a highly problematical issue. The dissension within the ECSC was not conducive to creating stability, considerable doubts continued about the future of Ravenscraig, and the history of the steel industry indicated that good profits could quickly be turned into heavy losses. On Ravenscraig, the government had obtained a commitment from the BSC that the plant had a future for at least seven years subject to market conditions (a proviso which, as the Beswick review had proved, was not designed to encourage optimism at the Scottish plant). Nevertheless, despite the proposed privatisation, the government was still unwilling to risk further political embarrassment in Scotland by sanctioning the closure of Ravenscraig.

The continuing instability within Europe was emphasised just before the end of the year, when at the eleventh hour a new arrangement on quotas was agreed upon within the Council of Ministers. The UK again sought an immediate end to quotas, but only those for merchant bar and wire rod disappeared. The quotas on hot rolled coil, cold rolled sheet, plates and heavy sections were to remain at least until June 1988. These quotas would be extended until the end of 1990 if concrete commitments could be obtained for about 70 per cent of surplus capacity in hot rolled coil, heavy sections and heavy plates. The political commitment to secure the required cuts in the area of hot rolled coil was still lacking, but the Commission was forced to withdraw its threat of an immediate end to quotas in the face of opposition which could not easily be pacified.

When European industry ministers met in May 1988 there were still no signs of agreement on the 14.5 million tonnes of capacity cuts required by the Commission. In particular, Narjes was particularly upset that no country was prepared to contribute to a proposed 7.5 million tonnes reduction in hot rolled coil capacity. Ministers wishing to maintain quotas, such as those representing France and West Germany, proposed a voluntary scheme, but the Commission was not prepared to countenance this type of private cartel. In contrast to French and West German Ministers, UK Trade and Industry Minister Kenneth Clarke claimed that the end of the quota regime would have a beneficial impact on the BSC, which now would not have to buy quotas, and would also be able to gain at the expense of less efficient EEC producers (*Independent* 28:5:88). After a further meeting of Ministers at the end of June, at which it was finally recognised that no agreement was possible, the eight year old Davignon Plan was laid to rest on 1st July.

Although there was unanimity within the EEC that overcapacity existed within the steel sector, there was not the political will to implement closures at a time of rising demand and relative prosperity for the steel companies. The UK had been in the vanguard of those seeking an end to quotas, and employed the demise of the Davignon Plan as a political asset in the drive towards BSC privatisation. At the same time, there were three other

European developments in 1988 which illustrated that the steel problem within the EEC was not so cut and dried as the government maintained.

First, at the same time as quotas were ending, the Commission was considering a request from the Italian government to pump a further £3.26 billion of aid into Finsider, now renamed Ilva after a reorganisation. Ilva was seeking a 1.2 million tonne cut in annual capacity, but there remained considerable doubt that the Italian government possessed the political will to push through the required closures.

Second, in July 1988, the West German Iron and Steel Federation carried out a threat to take the EEC Commission to the European Court of Justice, over the Commission's failure to act against allegedly illegal subsidies to the BSC. The West Germans claimed that the BSC had received a total of £930 million in illegal aid since 1973. Both the UK government and the BSC took considerable pains to play down this threat during the privatisation process, for it was hardly conducive to encouraging potential investors. Thus it was said by the UK government that the case had little or no chance of success. Nevertheless, the affair had the potential to remind the UK industry of its less than glorious recent past, while also illustrating the West Germans' continued reluctance to accept the BSC's revival at face value.

Third, in October 1988, British Steel was among nine stainless steel producers accused of price fixing by the Commission. This was a political embarrassment to both the Company and the government, who were claiming in support of the case for privatisation that the UK industry was now well capable of standing on its own feet. As with the illegal subsidies case, the government quickly attempted to remove a potential obstacle to privatisation, and claimed that any fine for British Steel would not exceed £10 million (*Independent* 10:10:88).

In the event, none of these political embarrassments had any material effect on the privatisation process itself. At the same time, they each illustrated that the steel problem remained firmly on the EEC political agenda.

The Privatisation Process

As 1988 progressed, it became clear that the government was in something of a cleft stick with regard to BSC privatisation. On the one hand, there was a general enthusiasm to go ahead with the sale in order to fill a potential gap of two or three years in the privatisation programme.

On the other hand, the government was considerably scarred by its experience with the sale of British Petroleum (BP) shares in 1987. Prior to that time, the privatisation programme had built up a considerable head of steam, including the huge sales of shares in British Telecom (51 per cent) in

1984, and British Gas (100 per cent) in 1986. After a low key start, privatisation had thus become one of the chief planks in the Thatcher government's economic and industrial strategy. By 1986–87 the Treasury was budgeting for an annual average of £5 billion in asset sales. Chiefly as a result of these sales, it was estimated by National Opinion Polls that share ownership had trebled since 1979 (*Observer* 29:10:87). Politically, the government could portray the privatisation programme as an exercise in popular capitalism, while at the same time raising revenue which could assist in implementing electorally popular tax cuts.

In Chapter Three, we saw how the Heath government of 1970–74 was largely content to maintain the *status quo* with regard to the Nationalised Industries. Initially, the Thatcher government itself did not appear to be much more adventurous in selling off public assets, although the Prime Minister herself made little attempt to disguise her own dislike of the Nationalised Industries. Once it became clear that popular support could be gained for privatisation, if only through the instant profits to be made on the new share issues, then the programme took on a natural momentum. It therefore came as something of a rude shock for the government when, in October 1987, the sale of BP shares coincided with a catastrophic crash in Stock Market prices. The government planned to sell its remaining 31.5 per cent holding in BP at a total price of £7.2 billion, but the crash put paid to all hopes of gaining another success in popular capitalism. In the event, the sale went ahead by means of a safety-net procedure, wherein the Bank of England agreed to buy-in shares if the price fell below a predetermined level. Nevertheless, the underwriters were left with a loss of £750 million, and the government had received a considerable shock to its nervous system. (*Financial Times*, 18:11:88).

The government thus required the revenue from a BSC sale in 1988, but feared the political consequences of another privatisation failure. Apart from the vagaries of the Stock Exchange, the nature of the business it was selling could not engender any complacency about its sale prospects. Despite its revival, the cyclical nature of the steel business could plunge the BSC back into crisis at any moment. The Corporation itself recognised that it would need to invest at least £300 million *per annum* for several years to come, while the new shareholders would also be looking for their dividends.

This caution on the part of the government could be contrasted with the attitude of the BSC Chairman, Sir Robert Scholey. Given the relatively bouyant state of the steel industry in 1988 (pre-tax profits of at least £550 million were forecast by the BSC for 1988–89), Scholey was convinced that this was the right time to push ahead with privatisation. In the event, as was usual in the government-BSC relationship, it was the Corporation's Chairman who had his way!

The BSC had already gained something of a bloodless victory in ensuring that the organisation would be privatised as an entity, and not split into competing parts. In order to maximise sale receipts, the government did not choose to break up virtual monopolies such as British Telecom and British Gas, and so was not inclined to disturb such a potentially vulnerable organisation as the BSC. The only significant political pressure for BSC fragmentation came from Scotland, where there was continued anxiety about the future of Ravenscraig. Consequently, Motherwell District Council, which contained the Ravenscraig plant within its area, commissioned the consultants Arthur Young to study the feasibility of creating a separate new materials group, which would include Ravenscraig, Shotton (where most of the Scotland plant's steel was rolled), Clydesdale and Dalzell. The Arthur Young report concluded that it would be feasible to form the new group, and there was considerable lobbying in Parliament by Scottish MPs on this point, but with no success. In extracting the promise from the BSC to keep Ravenscraig open for at least seven years, subject to market conditions, the government indicated its political vulnerability on this point, but it could not afford to disturb its relationship with the BSC at such a delicate time.

The Labour Party itself was thrown somewhat on the defensive by the Ravenscraig proposal. Inevitably, Welsh Labour MPs were more concerned with the future of Llanwern and Port Talbot, and were not inclined to see Ravenscraig break away as a potential competitor to these two plants. Officially, Labour supported the new group plan, but in reality the leadership was reluctant to press strongly for BSC fragmentation. Similarly, Labour favoured the public ownership of steel in principle, but refused to commit itself to renationalisation if it were returned to office. In the 22 years after World War Two ownership of the steel industry had been a touchstone of adversarial Party politics. In contrast, by the late 1980s, Labour had at least implicitly recognised that Morrisonian Nationalisation was no longer an electorally popular policy. Thus the 1988 political debate on BSC privatisation was decidedly muted, with Labour still seeking to develop a policy on state ownership which could regain the initiative from the Tories.

One politically vulnerable point which Labour did seek to exploit was the government's proposed capital reconstruction of the BSC. We described in Chapter Eight how the 1981 Iron and Steel Act enabled the government to write-off £4 billion of 'Section 18' capital. Notwithstanding this write-off, in 1988 the BSC retained £3,980 million of this capital. The government recognised that to saddle British Steel with this huge burden, which potentially could be recalled by government, would greatly reduce the sale prospects. On the other hand, Ministers were sensitive to Opposition charges that British Steel was being fattened up for privatisation, while the West German allegation on subsidies could also lead to problems within the EEC.

The government's solution was to, first, use Section 2(1) of the 1988 British Steel Act to reduce the £3,980 million by £642 million in order to eliminate the accumulated defecit on the BSC's profit and loss as at 2 April 1988. Under Section 7(1) of the same Act, the remaining £2,338 million was transferred to a statutory reserve. Out of this sum under Section 7(2) of the Act, £381 million was to be applied as if it were profits available for circulation, while the remaining £1,957 million might be employed for the payment of bonus shares (British Steel Prospectus). In other words, the money which had kept the BSC in business during the 1980s was now to be re-employed to pay off a deficit, supplement profits, pay for bonus shares, and then quietly disappear!

Not surprisingly, Labour claimed that this reconstruction involved a magical process, whereby, at one stroke of the pen, a debt was converted into profits. The government countered by claiming an accumulated deficit was not debt, but a record of past trading losses. In addition, it was argued by the Opposition that employing 'Section 18' capital as a means of making British Steel a more attractive sale prospect stretched commercial credibility to its outer limits. Government sensitivity on the capital reconstruction was perhaps one of the motives in its decision to reduce the amount available to British Steel for tax losses from £1,870 million to £171 million. Even here, however, British Steel's likely tax commitments were to be considerably reduced by unspecified allowances on capital investment (British Steel Prospectus). Despite Opposition complaints, the capital reconstruction did not attract a great deal of public attention, which throughout 1988 was focused on the sale itself rather than the financial machinations which underpinned it.

When the British Steel Act received the Royal Assent in July 1988, the sale process could begin in earnest. The Act transferred all property, rights and liabilities of the BSC to a successor company, with the shares wholly owned by the government. Thus, on 5 September, the BSC passed quietly away just a few weeks after its twenty-first birthday, to be replaced by British Steel plc. As with a number of other privatisations, the British Steel Act also created a 'golden share' which would be held by government in order to restrict the size of individual share holdings in British Steel to 15 per cent of the equity. The share was to be held for five years after privatisation, but could be liquidated before that date with the agreement of both British Steel's Board and the government. In particular, the 'golden share' could be used to protect British Steel from an overseas takeover bid. The existence of a 'golden share' was something of an embarrassment to a government committed to free market principles, but Ministers were also particularly sensitive to charges that privatisation might have the effect of delivering key UK industries into foreign hands.

In the privatisation process itself, the government faced two chief problems. First, to attract sufficient public interest so that all the allocated shares were sold, while also deterring small investors from putting their life savings into a company with an inherently uncertain long-term future. Second, setting the price of the shares at a level which gave a satisfactory revenue for the Treasury, but did not offer investors the prospect of either a heavy early loss (as in the BP case) or huge windfall profits. Early trading on shares such as British Telecom and British Gas had offered a large premium to investors, and left the government open to Opposition charges of seriously undervaluing state assets.

On the first point, Industry and Trade Secretary, Lord Young, summed up the government's position by stating that the British Steel issue was aimed at Sydney rather than Sid (*Financial Times*, 24:11:88). The slogan 'Tell Sid' had been used on a massive scale in advertisements encouraging the public to buy British Gas shares in 1986, but now the message was being conveyed that an investment in British Steel might not be such a guaranteed winner as the earlier sale. The government emphasised this point by eventually setting a minimum £500 investment per person in British Steel, compared with a figure of £50 for BG.

On the second point, the value of the shares was eventually set at 125p each, to be paid in two instalments of 60p and 65p. The price set the value of British Steel at £2.5 billion.

Unfortunately (or perhaps fortunately) for the government, having agreed to Scholey's request for an early privatisation, the offer for sale at the end of November 1988 coincided with a sudden fall in Stock Market prices. Thus although 1.4 million people had registered with the Share Information Office, a day before the offer closed it was reported that the sale had been a flop with the public, as only 200,000 had applied (*The Guardian*, 1:12:88). On the final day, however, there was something of a last minute rush, with about 650,000 individuals applying for shares, including 63,000 British Steel employees and pensioners. (The latter had been given a special offer package). In addition the average size of applications was quite high at £2,300. This had the effect of the public portion of the offer being over-subscribed 3.3 times.

Initially, the share had been allocated on the basis of 42 per cent to UK institutions, 25 per cent to private individuals, and 33 per cent overseas. There was, however, a claw back provision if the shares offered to the public were over-subscribed, and this was now invoked. Consequently, overseas investors retained only a 25 per cent share, the public now received an increase of 17 per cent to 42 per cent, while UK institutions' share fell to 33 per cent. Even, then, larger applications by the public had to be scaled down. It could therefore be said that the government had largely succeeded in its aim of attracting public interest in the shares, without creating an avalanche of applications.

On 5th December, the first day of trading in the new shares, the government's luck continued to hold. The market remained sluggish and the peak price was 62p, which offered only a tiny profit on the partly paid up shares. On this occasion, therefore, the government avoided the embarrassment of being charged with 'giving away' a Nationalised Industry. Ironically, when the market revived early in 1989, the price of British Steel shares quickly rose above 70p, but by then political attention had inevitably turned elsewhere.

Sir Robert Scholey had achieved his wish of seeing British Steel in the private sector during 1988. He himself had been a prominent figure in the sale period, and the old tales of 'Black Bob' were revived and faithfully reported by the media. At the age of 67 Scholey had expressed his intentions of continuing as Chairman for another three years, and his lifelong association with the industry was a positive asset in the general effort by British Steel and the government to convey the impression that the revival in the fortunes of the BSC would be carried over into the new era. At the same time, as we will see in the remaining chapters, many of the problems and dilemmas for both government and the UK steel industry would not disappear as a result of privatisation in itself.

10 The BSC and Its Customers

Introduction

The 1973 White Paper on steel was flawed principally because of its inability to appreciate the link between making steel and selling it. The proposed expansion in BSC capacity was based on the premise that by the 1980s the demand for Corporation steel would require a capacity in the region of 36–40 million tonnes. In the event, in 1980–81 the BSC produced only 11.9 million tonnes, with manned capacity fixed at 14.4 million tonnes. The huge oil price rises of 1973–74 and 1979 could not of course have been foreseen by the BSC, as neither could such factors as radical exchange rate changes, protectionism, high interest rates, and the consequent failure of steel demand to grow in the exponential manner anticipated in the early 1970s. Nevertheless, more down to earth factors, such as the BSC's patchy relationship with its UK customers, the Corporation's once poor reputation on price, quality and delivery (later much improved), and its inability to expand export markets (particularly within the all important EEC), also offer important evidence as to why the policy of the 1970s failed.

On the other hand, there was inevitable disillusionment on the part of the BSC that much of UK manufacturing industry had itself suffered such a catastrophic decline. Such important customers as vehicle manufacturers, shipbuilders, the engineering, capital goods and construction industries all suffered greatly during the 1970s and 80s, with the consequence that even a more efficient and productive steel industry could not have boosted demand to the hoped-for levels.

The long sad road that led to a record visible trade deficit of £20.34 billion in 1988 continues to be one of the most tragically persistent political and economic problems confronting a UK government. Even allowing for increased imports, this decline is reflected in falling UK steel sales. In the first section of the chapter we will briefly examine the range of the BSC's products and customers and their relative importance, together with import levels and the Corporation's export performance.

The second section will look at steel supplies from the point of view of the consumers themselves. The commercial network was obviously of crucial importance to the BSC, even allowing for EEC controls introduced under the Davignon Plan. A significant amount of BSC steel was sold to the private steel sector for reprocessing (see Bryer *et al.* 195–198 for the Corporation's reluctance to release figures with regard to its private sector sales), while stockholders assumed increasing importance in the 1980s. Politically, these interests are represented, respectively, by the British Independent Steel Producers Association (BISPA) and the National Association of Steel Stockholders (NASS). The Phoenix projects and the associated rationalisation in both the private and public steel sectors led to a greater identification of interest between the BSC and BISPA. Indeed, as we noted in Chapter Seven, some companies resigned from BISPA on the grounds that the private sector was in danger of being swamped by BSC interests. With the increase in 'second sourcing', some of the old loyalties between the BSC and its private sector steel customers broke down. Similarly, stockholders are now more liable to look for steel at the price and of the quality required, rather than rely solely on UK steel. At the time of privatisation, stockholders accounted for around 34 per cent of home deliveries to UK consumers, with the result that NASS is a significant actor within the UK steel community (although it should be noted that the BSC itself owned the UK's second largest stockholding company).

The interests of the steel consumers themselves are represented by the British Iron and Steel Consumers' Council (BRISCC). For an organisation with limited resources, BRISCC has gained a relatively high political profile. Its membership consists of both trade associations and individual companies, although over the years there has been a significant turnover of members. Despite BRISCC's political prominence, the BSC itself naturally dealt individually with trade associations such as the British Constructional Steelwork Association, the Society of Motor Manufacturers and Traders and the Metal Packaging Manufacturers' Association. At the same time, the BSC dealt directly with other state owned organisations such as British Shipbuilders, British Coal, British Rail and Rover (privatised in 1988).

together with private sector giants such as Ford, Vauxhall, Peugeot, GKN, Metal Box, Thorn EMI, TI, Shell and BP.

The third section of the chapter will briefly survey the BSC's evolving relationship with some of these interests, while the fourth section will deal specifically with strip steel and the UK vehicle industry. The vehicle manufacturers were one of the BSC's largest customers, and thus give a good indication of commercial relationships which played a large part in determining the fortunes of the Corporation.

The range of steel consumers

The 1986 *Annual Statistics* produced by the UK Iron and Steel Statistics Bureau (ISSB) classified steel consumers into 31 groups (see Appendix B for a table showing net home deliveries to UK steel consumers 1981–86). These statistics were given for production by the UK steel industry as a whole, and were not divided into private and public sectors. Until 1981–82 the BSC produced its own *Annual Statistics for the Corporation*, although these tables only went as far as classifying steel deliveries by product (e.g. ingots, wire rods, plate, sheet, strip, tinplate, tubes) rather than consumers. BSC *Annual Reports* set out broad descriptions of both the state of the market and also the activities of individual groups and businesses, but did not give specific figures on consumers themselves. Nevertheless, by the late 1980s the BSC was directly or indirectly concerned with around 90 per cent of UK steelmaking. The general contraction of the private sector and the emergence of the Phoenix projects left only a relatively few arc and induction steelworks in private hands, e.g. eight special steelmakers in the Sheffield area, Sheerness Steel in Essex and F.H. Lloyd in the West Midlands (ISSB, 1987). The BSC did not give its own figures to tie in with those of the ISSB (ISSB figures are given for a calendar year, while the BSC operated in financial years).

It is also unfortunate that after 1977 the BSC did not give figures for sales of semi-finished steel to the private sector (see Bryer *et al.*, 146), so it must be borne in mind that the statistics given by the ISSB for deliveries of *finished* steel to consumers include a significant tonnage which passed through private sector mills.

A further complicating factor is that from 1986 the ISSB changed its classification of consumers in order to conform with EEC standardisation, with the result that figures given prior to 1981 are not strictly comparable. One notable example is that colliery arches have been moved from the coal category to industrial plant and steelwork. Thus, in these two categories at least, there are considerable variations between the old and new figures. In other categories, such as motor vehicles and tinplate, the disparities are not so significant. In categories with wide disparities we have quoted the figures only after 1981, but with the figures where differences are relatively minor we have used at least some of the earlier figures. Despite these qualifications, the ISSB classification does give us the best available guide to the state of the UK steel market.

In 1986 deliveries of finished steel to the UK market totalled 11,950,400 tonnes. Of this amount imports accounted for 3,613,500 tonnes, leaving 8,336,900 tonnes delivered by UK producers. By far the largest single group of customers for UK produced steel were stockholding merchants, who took 2,835,100 tonnes, with the result that consumers of UK steel took 5,501,800 tonnes during the year.

Notwithstanding the radical improvement in BSC finances during the 1980s, the figures for total finished steel deliveries to UK consumers reflect the decline in UK manufacturing industry which has continued for many years. Thus in 1979 these deliveries totalled 8,829,800 tonnes, slumped to 6,126,100 tonnes in 1981 (excluding the steel strike year of 1980), and henceforth sloped gently downwards to arrive at the figure of 5,501,800 tonnes for 1986. It could therefore be argued that not the least of the BSC's achievements during the 1980s was in winning a position of financial stability despite falling domestic demand (in the next chapter we will see that in the late 1980s cost cutting had become the principal BSC objective, perhaps at the expense of exploiting worthwhile investment opportunities).

Of the 31 classes of UK consumers given by the ISSB, consistently the largest in tonnage terms is wire and wire manufacturers, which in 1986 accounted for 938,200 tonnes (compared with 1,506,200 tonnes in 1979). Nearly 400,000 tonnes of product was used as reinforcement by the construction industry, and indicates the traditionally strong links between the two industries. The large size of this category is also accounted for by wire being a first transformation product, i.e. it may later be turned into a product such as nails.

The second largest consumer of UK steel is industrial and process plant and steelwork, with 667,300 tonnes taken in 1986. This is one area which has shown expansion, as this sector only absorbed 560,100 tonnes of steel in 1981.

Third in the list of consumers is motor vehicle manufacturing (including parts and accessories), which totalled 570,300 tonnes of steel in 1986 (compared with 1,007,900 tonnes in 1979). As we will describe later in the chapter, the BSC's relationship with UK vehicle manufacturers such as Rover and Ford improved considerably during the 1980s. Nevertheless, the problems of these companies have been major factors in placing strip mills such as Llanwern and Ravenscraig under threat of closure.

Fourth highest in the list comes cans and other metal packaging at 521,600 tonnes, compared with 714,800 tonnes in 1979. By far the BSC's largest customer in this category was Metal Box, which by the end of the decade was apparently on a more stable path after several difficult years in the early 1980s.

The final member of the 'big five' is construction (building and engineering) with 364,000 tonnes in 1986, compared with only 297,000 tonnes in 1979. This increase might appear somewhat strange, given that the construction industry was in a depressed state for much of the 1980s. The increase in use of steel can be chiefly attributed to the builders' relative disillusionment with the qualities of reinforced concrete as a construction material, and has allowed steel to make something of a comeback. This boost for steel reflects the extent to which the metal is nowadays in competition with other materials, e.g. aluminium and plastic in the manufacture of motor vehicles, packaging etc.

In 1986, the top five consumers of wire, industrial plant, vehicle manufacture, cans and construction added up to 3,061,400 tonnes, out of total deliveries of 5,501,800 tonnes, or to put it another way, consumed 55.6 per cent of UK produced steel sold to domestic consumers. Thus the BSC was dependent for the majority of its business on a handful of industrial sectors, some of which (particularly vehicle manufacture and cans) were themselves dominated by just a few firms. The crucial importance of the BSC forming an integrated relationship with these customers cannot be stressed too highly, yet it is a subject which politically was neglected during the life of the BSC. Governments tend to look at each sector as an individual entity, when the connections between them are often of equal, if not greater, importance.

Two other categories are particularly worthy of note. Firstly, consumption of UK steel in the coal, coke, petroleum and natural gas category dramatically declined from 289,500 tonnes in 1984 to 134,000 tonnes in 1986. This fall represented not only the decline of the UK coal mining industry, but also the tailing off of investment in the North Sea oil and gas industries. The links between coal and steel run deep, and date back to the early days of both industries, but while the BSC remained British Coal's second largest customer (admittedly a long way behind the CEGB), coal became one of the Corporation's less important consumers. The BSC failed to obtain the full potential benefits from developments in the North Sea, and so this category represented at least partly a lost opportunity.

Secondly, shipbuilding, once one of the Corporation's largest and most valued customers, had slumped to a mere vestige of its former self. By 1986 only 46,800 tonnes of UK steel went to the shipbuilding industry, when even in 1982 the figure was 193,200 tonnes. More than any other sector, the tragic decline of the UK shipbuilding companies epitomised the deeply changed nature of the BSC's operational environment.

Imports

The stark decline in the BSC's domestic market share during the 1970s has been set out succinctly by Pryke:

> During 1968 the BSC accounted for 66 per cent of the finished steel that was delivered to the home market, and in 1970 when imports were expensive, its share reached 70 per cent. Thereafter it declined rapidly to 54 per cent in 1975, a level that the BSC then managed to maintain. The ground that the BSC lost has been taken up by imports. In 1968 they constituted less than 10 per cent of deliveries, but by 1978 the figure had reached 21 per cent (Pryke, 195).

From 1981–1985 the level of steel imports remained relatively static, with the BSC largely recovering the ground lost as a result of the 1980 steel strike. The Davignon Plan was a highly important element in achieving this stability, but by the late 1980s the BSC was able to satisfy a higher proportion of its customers than in the 1970s.

In 1979 UK steel imports stood at 3,784,000 tonnes, rose to 4,620,200 tonnes in the strike year of 1980, but then fell back to 3,306,000 tonnes in 1981. After remaining stable for the next four years there was a rise to 4,118,200 tonnes in 1986. In this year total home deliveries to UK consumers were at their lowest level since 1979 (at 5,501,800 tonnes), suggesting that the improvement in the BSC's financial position at this time was not due to its ability to regain a larger share of the domestic market.

The EEC countries were overwhelmingly the largest importers of steel industry products into the UK, sending 2,895,500 tonnes in 1986. Of these EEC imports, by far the largest proportion was taken by West Germany, with 1,001,400 tonnes, followed, perhaps surprisingly, by the Netherlands, with 458,900 tonnes, and Belgium-Luxembourg with 428,000 tonnes. France and Italy were only fourth and fifth in the list.

In 1986 non-EEC and Eastern European countries accounted for 560,200 tonnes and 202,800 tonnes respectively, leaving only 456,700 tonnes for the rest of the world put together. Of these countries, the only significant importer was Japan, with 149,300 tonnes. The limited non-European imports indicate the success of the EEC third country Voluntary Restraint Arrangements (VRAS), which since the late 1970s have acted as an effective brake on the rush of imports during the early 1970s. Even the figure for Japan in 1986 is something of an abnormality, for in 1985 only 37,700 tonnes of their steel was imported into the UK (tubes and pipes were by far the largest category of Japanese imports, indicating both over-capacity in Japan and the BSC's difficulties in satisfying the demands of the UK oil and gas industries).

The only other importer of any note in 1986 was South Africa, with 76,700 tonnes. The USA only sent 20,500 tonnes (slightly less than Brazil) indicating not only the poor state of the US steel industry, but also the depressingly sterile trade relations between the EEC and the US.

The BSC itself was disappointed that non-EEC countries increased their imports into the UK by 377,800 in 1986, when compared with 1985. The Corporation attributed this rise to increased penetration by countries not bound by VRAs, encouraged by a weakening of the dollar, and pressure created by US protectionism (BSC *Annual Report and Accounts 1986–87*). The Corporation no doubt suspected that some countries were dumping steel in Europe although, as in the case of Japanese tubes and pipes, there was also at least some indication that the BSC was unable to satisfy the domestic market.

From 1984 UK steel customers began to complain to the EEC Commission that certain types of steel were not available from domestic sources, and called for an increase in UK production quotas. The BSC itself and the government were in a dilemma, for although the Corporation's improved performance during the 1980s apparently justified a realignment of quotas, both parties understood the importance of these quotas in maintaining the UK steel industry on a stable path, and so were reluctant to rock the EEC boat by attempting to negotiate a new production regime. Thus, where such a delicate political balance was essential, a restrained approach to the domestic market was a necessary forfeit to be paid.

Exports

The 1970s saw a rapid deterioration in the UK's balance of trade for steel:

> During 1968–70 the UK's exports of steel exceeded its imports by about £375 million and in 1971, when exports were no longer restricted, the figure increased to £475 million. Thereafter net exports fell away sharply, at first because of the availability of cheap imports, but subsequently due to the Corporation's inability to satisfy demand. By 1973 our net exports had fallen to £140 million and during the following year the figure was £250 million. The UK remained an importing nation until 1976 and during 1978 exports only exceeded imports by a paltry £40 million (Pryke, 199).

In the 1980s, UK steel trade was in the red in only 1980 (the strike year, when imports exceeded exports by a value of £512,249) and 1982. In 1986, trade was marginally in the black to the figure of £65,688.

The lack of realism in the BSC's export targets in the early 1970s is well set out by Bryer *et al*, who indicate that in 1971 the Corporation hoped by 1981 to be exporting 10.6 million tonnes of steel per annum (Bryer *et al.* 77). Even after the work of the JSG, the modified 1972 strategy projected exports in the range of 5.5 to 7 million tonnes of steel per annum by 1980 (Bryer *et al.* 128). In the event, BSC exports for 1980–81 totalled 2.3 million tonnes, a figure similar to that for most years during the 1970s. It had been hoped that the UK's entry into the EEC in 1973 would herald an unmatched opportunity for the domestic steel industry to gain a greater share of the European market, but these hopes were unfulfilled. In 1973 the UK imported 3.86 million tonnes of steel, and exported 5.6 million tonnes. By 1974 the trade position had deteriorated to the extent that imports totalled 5.31 million tonnes, and exports 4.32 million tonnes.

As described in Chapter Four, 1974 was a watershed year in several respects. Firstly, it was the last pre-crisis year when steel demand remained relatively buoyant. Secondly, the BSC was attempting to begin implementation of its ambitious development strategy, but instead found itself unable to satisfy the needs of even its domestic customers, with the result that exports were given a low priority. Thirdly, domestic customers began to 'second source' on a large scale, while potential overseas customers gained a poor impression of both the steel quality and the delivery service offered by the BSC.

If the BSC could not take advantage of a buoyant market, then it had little hope of succeeding when the crisis arrived. During the 1970s the BSC argued, with some validity, that a good number of potential export orders were not worth undertaking because general overcapacity had led to a slump in prices, with the result that to fulfil these orders would be an uneconomic proposition. Nevertheless, in tonnage terms, UK steel exports once more exceeded imports by 1977, and by 1979 the balance was 850,000 tonnes. These figures unfortunately disguise the fact that the UK has been in deficit in steel trade with the EEC since we entered the Community. In 1979 the deficit stood at 811,700 tonnes, although by 1986 the gap had narrowed (the Community now including Spain and Portugal) to 592,900 tonnes.

Like much of UK industry, steel had for many years relied on exports to the Commonwealth rather than attempt a greater penetration of the European market. Thus the crises of the 1970s tended to expose the limitations of the BSC's export strategy rather than to create the difficulties themselves.

As we noted in Chapter Six, the nadir of BSC export policy was reached in December 1979 when, in the midst of financial crisis and the threatened strike, the Corporation threatened to withdraw from two-thirds of its export market. In the event, cooler heads soon prevailed and this drastic proposal was withdrawn. Old habits die hard, however, and it was only in 1986 that the BSC finally acknowledged publicly the deficiencies in its EEC commercial network and began to take positive action to remedy the situation. During the early 1980s a combination of the high value of the pound, declining steel demand within the EEC, the output and trade restrictions of the Davignon Plan, protectionism in the US and the growth of steel industries in previously less developed countries, acted as severe constraints on the BSC's export potential.

Despite its earlier reservations about the value of exports, and the continuance of EEC imposed production quotas on a wide range of products, 1986 marked a new watershed year for the BSC, with a more aggressive policy on exports. As we described earlier in the chapter, the Corporation's sales to the domestic market fell during 1986–87, but to counteract this setback exports increased by 42 per cent compared with 1985–86! (these figures exclude results for BSC Special Steels, the greater part of which became part of the Phoenix Two Company, United Engineering Steels in March 1986). This sudden surge from 2.6 million tonnes in 1985–86 to 3.7 million tonnes in 1986–87 included an increase of 0.8 million tonnes in exports of slabs and blooms, semi-finished products which in the past had usually been confined largely to the home market.

Part of the increase could be accounted for by the Tuscaloosa deal (referred to in Chapter Nine), and the reduced value of the Pound against the Deutschmark. Nevertheless, it did appear that the BSC had made a specific commitment to in future offload overseas a significant proportion of its basic production tonnage, rather than rely on demand in the domestic market. This change of policy to tonnage before revenue, rather than *vice versa*, is also indicated by revenue comparisons. Despite the large increase in export quantities, revenues were only slightly up. Thus the BSC earned £1.169 billion from exports in 1985–86, with a marginal increase to £1.185 billion in 1986–87 (BSC *Annual Report and Accounts, 1986–87*). By increasing its export tonnage the BSC was able to maintain the stability of its revenue, but blooms and slabs are not big money earners and, combined with the ending of Davignon price controls together with currency movements, the BSC's export performance in 1986–87 was a good example of how in the steel industry there may be little correlation between production, revenue and profit.

One further manifestation of the BSC's more determined attitude towards exporting, particularly to the EEC, was the acquisition during 1986–87 of various European stockholding and trading companies. These included businesses in France, Holland, West Germany and Belgium. Although production quotas were still in operation, there was no doubt that, thirteen years after the UK entered the EEC, the BSC's new-found enthusiasm for sending steel into Europe was a policy with important implications for the future development of the UK steel industry.

The BSC and private sector steel

In a searching analysis of the relationship between the BSC and its private sector counterparts, Abromeit highlights both the innate tensions of the relationship and the identification of interests which can bring the two sides together (Abromeit, 1986, 218–252). She uses the example of the former West Midlands steelworks, Round Oak, to illustrate the complexity of the relationship. Round Oak was owned jointly by the BSC and Tube Investments (TI), with the result that:

> One of the TI (non-executive) directors, K.G. Webley, was the managing director of TI Weldless Ltd., which had been the major customer of Round Oak; one of the BSC directors, John Pennington, was the managing director of the BSC's Sheffield division, 'a main supplier of TI, in competition with Round Oak'. The BSC is the main supplier of most of TI's steel tube subsidiaries; TI Weldless was not only the main customer of Round Oak (until the latter was closed) but also bought from the BSC, while being in direct competition with the state giant; in the types and sizes of tubes they produce 'there is exact overlap with the BSC tube division'. It has been said also – albeit not specifically – that the TI steel subsidiaries in some instances even supply the BSC, for the latter's downstream production (Abromeit, 219–20).

The changed nature of the relationship between the public and private sectors is illustrated in the sad truth that the Round Oak works has disappeared, together with many other private sector steelworks, while TI itself has now largely withdrawn from the white goods market. The rationalisation of the private sector, the Phoenix projects and other privatisations have combined to consolidate the identification of common interest between the BSC and the private companies.

Bryer *et al.* argue forcefully that a principal explanation for the BSC's problems during the 1970s was the decline in the Corporation's sales of semi-finished steel to the private sector. They point out that, after rising to a peak in 1970, these sales fell away to 1.8 million tonnes in 1977. (Bryer *et al.* 143). Thus from having 80 per cent of the market for ingots, blooms, billets and slabs in 1967, the BSC only supplied about half this market in 1977 (Bryer *et al.* 143). 1977 was the last year in which statistics for the private/public sector split were made available publicly, but by 1987 the BSC and Phoenix companies between them owned over 91.5 per cent of UK steel capacity. The ambivalent status of the Phoenix companies, combined with their dominant position within the private sector, increasingly blurred the traditional private/public sector divide. By 1986, the Phoenix and joint venture companies themselves accounted for about 15.8 per cent of UK steel production, with the BSC responsible for nearly 76 per cent (NEDO, 1986).

In addition to steel manufacture, the BSC also had a significant shareholding in several of the largest steel processing groups, including Bridon, Darlington and Simpson Rolling Mills, Templeborough Rolling Mills and TWIL(NEDO, 1986). Although there were a few remaining large (entirely) private sector companies in the processing sector, such as Glynwed and Arthur Lee, the majority were relatively small companies. The BSC itself still processed around two thirds of UK steel, with the remaining third in the private sector (including joint ventures). Given that joint venture companies such as Allied Steel and Wire, Forgemasters, British Bright Bar and United Engineering Steels supply specific market sectors, the area of product overlap between the BSC and the private sector was reduced sharply. At the same time, the degree of managerial and financial overlap increased substantially. The combined effect of these factors was to lessen competition between the public and private sectors, and to breed a sense of common identity within the UK steel industry as a whole.

In events during the 1980s we can see signs of the fulfilment envisaged by Abromeit in the policy aims underpinning the joint ventures:

- Sorting out all areas of overlap;

- Helping the private sector to concentrate into larger units;

- Accommodating those private investors who wish to rid themselves of their steel interests (Abromeit, 250).

Chapter Seven described the resentment felt by the private sector during the early 1980s at what it considered to be unfair competition created by subsidies given to the BSC. For its part, the BSC found itself during the 1970s losing out to the more flexible investment programmes of the private sector, while also resenting price controls imposed on it by government during these years. By the late 1980s the intense rationalisation on both sides of the industry, together with the growth of joint ventures, had dissipated much of this former rivalry.

Politically, these changes were particularly manifest in the membership of BISPA. The contraction of the private sector had inevitably severely reduced the income of BISPA, with the result that it had been forced to close down its regional offices. Nevertheless, both Sheerness and Glynwed, who had resigned from BISPA during the early 1980s in protest at the Association's willingness to accept the principle of subsidies to the private sector, rejoined the group in 1985 and 1986 respectively. At the same time, BISPA now also included among its membership such joint venture giants as Allied Steel and Wire, United Engineering Steels and Forgemasters (all with BSC representation on their boards). Given the wide range of interests represented, BISPA was placed in a position where its role was more politically ambiguous – a position which reflected the new state of the industry itself.

An even more vivid indication of the new political reality was the creation of a British Steel Producers' Group formed from the BSC and BISPA to help represent British steel interests in Europe. The Davignon Plan, the work of Eurofer and the consequent need to put forward UK steel interests in Brussels, was a more politically imperative task than fighting each other at home.

In concluding her analysis of the public-private sector relationship, Abromeit speculates on the eventual creation of a new BISPA (particularly likely with a privatised British Steel), but with a greater capacity to enforce change than that possessed by the old organisation (Abromeit, 251). The old company structure has now of course disappeared. Indeed, companies such as GKN and TI have left the steelmaking sector altogether and moved into other areas. Steelmaking and processing businesses are not a particularly attractive proposition for new entrants to the industry, and so neither British Steel nor the joint venture companies is likely to encounter any significant new competition in the areas they dominate now. It is thus difficult to escape the conclusion that British Steel will in future occupy the dominant economic and political position within the UK steel industry as a whole.

Stockholders

Steel stockholding is an activity which has gained in importance over the years. In 1968 stockholders took only 27.4 per cent of UK steel deliveries to the home market, but by 1986 this percentage stood at 47 per cent (it should be noted that these National Association of Steel Stockholders' figures are for stockholders' products only, and exclude non-stockholding products such as colliery arches and railway sections). Traditionally, stockholders operate on the basis of offering consumers a flexibility and promptness of service which is often unavailable to the producers. Profits can be assisted by selling stocks which have risen in price since they were purchased originally from the producer. At the same time, when prices fall, as they have done periodically in recent years, then margins can be much tighter. In this respect stockholders had a strong identity of interest with the BSC, for they both wished to maintain prices and were thus prepared to sustain this stability by means of a regulated market.

The National Association of Steel Stockholders (NASS), which represents around 85 per cent of UK stockholders, was placed in a delicate position with regard to its policy towards the BSC. On the one hand, it wished to see BSC prices maintained, but on the other hand some stockholders suspected the BSC of selling steel more cheaply to its own customers than to the stockholding market. A particular crisis occurred in 1981 when NASS complained strongly (and a few of its members resigned in protest) that the BSC's stockholding arm, British Steel Service Centres (BSSC), was undercutting its rivals in order that the Corporation would regain the market share lost during the 1980 strike:

> Had this industry had a long history of letting down the steel consumer there might have been a fair case for that position (BSC undercutting). As it is, for fifty years this industry has given the consumer a first class service and this fact is recognised by both the individual consumers and various Associations to which they belong. Even in the Steel Strike during the first thirteen weeks of last year, there was only one halfhearted complaint of profiteering levelled against this industry and that was after dealing with countless thousands of orders, and at a time when had it not been for the steel stockholder British Industry would undoubtedly have come to a complete stop (House of Commons Industry and Trade Select Committee 1980–81, 217).

Despite these complaints against the BSC, NASS is in a similar position to BISPA in that BSSC is itself a prominent member of the stockholders' group. At the time of privatisation BSSC was the second largest UK stockholding company, with around 15 per cent of the market. BSSC became a limited company in 1981, but remained an integral arm of BSC policy. In the 1980s the company usually turned in a profit, and as with the steelmakers some private stockholders suspected that BSSC was able to undercut its competitors by means of subsidies. Mutual suspicion is endemic in the steel trading business, although there was no doubt that the BSC was in a position to strongly influence the direction of the domestic market.

The development of 'second sourcing' nevertheless prevented the BSC from dictating terms to the stockholders. In its turn, the BSC at times suspected importers of dumping steel on the UK market at prohibitively low prices. In 1981 stockholders selling more than 12,000 tonnes a year (a level later lowered still further) were compelled by the EEC to produce a price list, and this measure did assist in stabilising the market. Margins remained tight, however, with the BSC continuing to suspect that importers were undercutting prices in order to increase sales volume. As we have seen, by the late 1980s the BSC was at last developing its EEC sales network, and the large increase in export volume during 1986–87 can be judged as the development of a policy which had long been common amongst other major EEC steel companies. Unfortunately, figures for UK steel imports by stockholders are not available, despite pressure from NASS that they be produced. The general philosophy on importing was expressed by NASS Executive Director, Robert Rawlins, in his evidence to the Select Committee:

> Nobody in the stockholding industry in my experience ... imports for fun. If you have problems on quality, complaints and stuff, getting stuck in the docks, it takes a long time to unscramble it, so the only reason people import is a defensive measure, primarily based on price. Very occasionally it might have some relationship to quality or something but that is very rare now (House of Commons Trade and Industry Select Committee, 1983–84, Q. 266).

These remarks reflect the position as it stood in the middle to late 1980s, for in the 1970s it was the BSC's poor record on delivery and quality which induced many stockholders to 'second source' their supplies. Mr. Rawlins was at pains to stress to the Committee the extent to which the BSC had improved its record in these two spheres, although there were no indications of a significant reduction in steel imports.

Like BISPA, one of NASS's principal functions is as a lobbying group, not only in the UK but also within the EEC. The EEC association for stockholders is the Club des Marchands de Fer, and this group has joined with Eurofer in putting pressure on the Commission not to reintroduce a free market for steel. When the Davignon Plan came to an end in December 1985, NASS supported the abolition of compulsory price lists but, like the BSC was highly concerned at the consequences of abolishing quotas and third country import controls:

> We appreciate very much that the steel consumer may believe that the abolition of controls affecting quotas and imports into the Common Market might lead to cheaper prices. In the very short term that might be true, but in conversation with dozens of consumers, in addition to those engaged in our own industry, it is abundantly clear that we want one thing above all else – stability (NASS Annual Report, 1986–87).

This strong identity of interest between NASS and the steelmakers is a significant political force within the steel community. As we have seen, for several years the government was inclined to support their case rather than work for a deregulated market. Significantly, there are liaison committees involving NASS, British Steel and BISPA. In addition to specific steel matters, these committees provide a means for the three bodies to discuss steel policy in general. Notwithstanding its role representing prime British Steel customers, NASS can be regarded as a significant political ally of the Company in the campaign to sustain the steel market at a profitable level.

Although UK stockholders control a greater share of the market than in any other major steelmaking country, there are many small companies operating on tight margins. It is estimated that the five largest companies control over half of all stockholders deliveries (*Financial Times*, 20:6:86). Rationalisation has taken place only slowly, although in recent years even large companies have left the market. In 1983 TI sold its stockholding operation to C. Walker, while in 1986 the largest UK stockholder, GKN Steelstock, also sold out to Walker for £50 million. These deals made Walker by far the largest UK stockholder, with BSSC in second place. Nevertheless, if large companies such as TI and GKN see fit to leave the stockholding industry, then the field may be increasingly left open for British Steel to increase its influence within the stockholding sector in a manner similar to that achieved in steelmaking itself (in 1989 British Steel considerably tightened its hold on the UK stockholding market by acquiring Walker for £330 million).

BRISCC

Unlike BISPA and NASS the British Iron and Steel Consumers' Council (BRISCC) does not represent an easily identifiable group of interests. Steel consumers come in many shapes and sizes, and indeed as a whole would encompass nearly all of UK manufacturing industry. Like similar umbrella pressure groups encompassing a range of large and powerful interests (the British Road Federation in the transport sector offers a good example) (see Dudley, 1983 108–9) BRISCC has particular strengths and weaknesses. Its chief strength derives from the quality of its membership. An organisation which includes amongst its members such companies as Ford, Rover, TI, Glynwed, British Coal and Tarmac, together with trade associations such as the Society of Motor Manufacturers and Traders, the Federation of Civil Engineering Contractors and the British Constructional Steelwork Association, must be taken seriously as a political force. Like BISPA and NASS, BRISCC arranges periodic meetings with British Steel executives, officials, Ministers and interested MPs, while together with the other two groups it is also a member of the ECSC Consultative Committee. The Director of BRISCC, John Safford, is also notably successful in being quoted (see the beginning of this book) on matters concerning steel policy!

BRISCC's former Chairmen include Lord Marsh (the Labour government Minister responsible for setting up the BSC and also a member of the panel which settled the steel strike) and Lord Ezra, former Chairman of the NCB.

In comparison with BISPA and NASS, it could be said that BRISCC had a political advantage in that it did not have to accommodate British Steel interests amongst its membership. There is no equivalent of the Phoenix companies or BSSC in BRISCC, but at the same time BRISCC was aware that in order to hold the ear of BSC it could not afford a high degree of confrontation with the Corporation. For example, by supporting the campaign to close Ravenscraig BRISCC could claim that it was encouraging the Corporation toward greater efficiency, but the endorsement of BSC policy also had the political advantage of indicating to the Corporation that it had a friend.

BRISCC's wish to maintain cordial relations with BSC can be seen partly as an indication of the group's particular political weaknesses. Firstly, by representing an amorphous group of interests, it is not easy for BRISCC to articulate 'the steel consumers' view'. Beyond a general wish to see steel producers' efficiency improved, and prices kept under control, BRISCC is not equipped to deal in depth with the disparate needs of consumers. Grant exemplifies how product heterogeneity can inhibit the formation of a business association by using the case of the food industry. Only in 1984 was an organisation formed (the Food and Drinks Federation) for the whole sector in an industry where the product can be differentiated in many ways (Grant, 1987, 142–3).

Secondly, companies such as Rover, Ford, TI and Glynwed, together with a large public corporation such as British Coal, all had their own direct relationship with both BSC and government, and were quite capable of expressing their views individually. In addition, trade associations such as the Society of Motor Manufacturers and Traders and the Metal Packaging Manufacturers Association also had their own co-ordinating committees with the BSC, and were unwilling to sacrifice any of their political status to BRISCC.

Thirdly, although several large steel consumers are members of BRISCC, others have declined to join, e.g. the largest UK motor components group GKN is not a member, while UK car manufacturers such as Vauxhall, Peugeot and Jaguar have declined to join the group. BRISCC has also experienced some significant turnover in membership, with the British Forging Industry Association, METCON, the Process Plant Association and Metal Box all resigning during the 1980s (although Metal Box still had a representative with BRISCC through the Metal Packaging Manufacturing Association).

From 1967–73 the Steel Consumers' Council had a statutory role, but since the UK entered the EEC it has been purely a voluntary body. Given its limited resources (members' subscriptions totalled £32,962 in 1986) and its problems with membership, BRISCC has won a significant niche for itself within the UK steel community. As Abromeit states:

> This stress on more general policy matters and the somewhat careful avoidance of the impression that it is a mere successor to the former BSC's Consumers' Council are highly characteristic; both attitudes allow BRISCC to claim greater importance than it actually has, to pursue wider interests than just passing on the bickering of some BSC customers, and from that comprehensive point of view to make common cause with the BSC (Abromeit, 257–258).

Abromeit also highlights the basically hostile relationship between BRISCC and BISPA, with the latter accusing the former of being superfluous and incompatible with the interests of a free market economy, while also suspecting it of having too cosy a relationship with the BSC (Abromeit, 261–62).

Within the UK steel community, BRISCC is thus a distinctive group in that its function is almost entirely political. It exists as a lobbying group to provide a strategic view of the industry which can be of potential political significance to the interests it serves, but which also could be an ally to the policy aims of BSC.

A new era but old problems

As we have noted earlier, the BSC improved its relationship with many of its customers during the 1980s, yet its market share refused to rise significantly above the 54 per cent it had held (with the exception of the 1980 strike and its aftermath) since the mid-1970s (from a level of 70 per cent in 1970), while domestic steel consumption continued to drift downwards.

Chapter Five described how BSC production problems in the early 1970s, combined with UK entry into the EEC, led consumers increasingly towards 'second sourcing'. The BSC itself commissioned the P-E Consulting Group in 1976 to undertake a survey of 300 customers with the objective of assessing the Corporation's performance and reputation. The survey concluded that continental suppliers produced steel of a markedly superior quality to the BSC, and that this was the chief explanation for the Corporation's loss of market share. A particular example was provided by the Ford Motor Company, which in January 1975 warned the BSC that it would look abroad for supplies still more unless the quality of steel from Llanwern was improved. Ford pointed out that it now obtained nearly 50 per cent of its steel requirements from abroad (*Financial Times*, 29:1:75).

The poor performance of the modern plants was an acute embarrassment to the BSC, and their failure to operate effectively was a large factor in the problems of the 1970s. The BSC was also incapable of taking full advantage of the boom in the pipe market caused by North Sea oil investment. The Corporation did not have the plant capable of producing the quality of steel required for undersea pipelines, and so most of the orders went to Japan! (*The Sunday Times*, 4:8:74). Bryer *et al.* point out that during the 1970s the BSC spent a much lower percentage of its capital investment on finishing facilities than its chief competitors, and as a result was unable to provide steel of the quality required (Bryer *et al.* 161–2).

Until 1975 the chief problem was steel shortages, but after that time the market dipped alarmingly. The BSC introduced price rises in an attempt to recoup revenue, but the crisis years had truly arrived. Evidence given by several steel consumers to the Select Committee on Nationalised Industries during this period indicated the disillusionment of BSC customers. For example, officials of GKN told the Committee that they would prefer domestic supplies because of their geographical proximity, but that the BSC lacked the necessary marketing and price flexibility (House of Commons SCNI 1977–78, HC 26, paras. 190–95).

The Chairman of Metal Box criticised the quality of BSC tinplate, and fiercely condemned the lack of consultation between the BSC's top management and its customers on long-term problems (House of Commons SCNI 1977–78, HC 26, paras. 218–30). Both companies also complained about the delay in improving finishing facilities at Port Talbot – as we noted in Chapter Five a sore political point throughout the 1970s.

The failure of the Davignon Plan to hold prices during the 1970s led to accusations of steel dumping from the BSC, and the Corporation found it impossible to make price rises stick. Periodic increases were invariably followed by heavy discounting, a practice which did nothing to revive the Corporation's weak financial state. Sir Charles Villiers himself was particularly keen to see the BSC become more market orientated, and shortly after becoming Chairman in 1976 highlighted the Corporation's loss of domestic market share. An early personal initiative by Villiers concerned the Murchison North Sea oil rig. It was suggested that the original BSC tender price was some 60 per cent higher than that quoted by Japan. The state owned British National Oil Corporation (BNOC) had a third stake in the majority portion of the Murchison field, and BNOC's Chairman, Lord Kearton, was keen that the BSC should win the contract. Villiers and Kearton were old friends, and after a meeting between them the BSC agreed to cut its price, with the result that it was handed the Murchison contract (*Financial Times*, 11:3:77). Unfortunately for Villiers, this piece of enterprise was not mirrored in general BSC success during his

Chairmanship. Despite his best efforts, it appeared that the Corporation did not possess the materials, expertise or efficiency to either win back lost domestic markets or boost exports.

Ian MacGregor took over at the BSC with a similar zeal for improving the Corporation's market position. He travelled widely, even to the USSR and China, in an attempt to find new customers. Like Villiers, he laid particular emphasis on supplying pipe lines to the North Sea oil industry. By the early 1980s the Corporation was in a better position to satisfy this market, but by then most of the investment had already taken place! Much of this pipe investment was undertaken at the Hartlepool plant. The Corporation was successful in obtaining several orders for domestic oil and gas pipelines and also improved exports to the US, but a severe blow was dealt to Hartlepool in 1984.

In August 1983, the BSC won from Shell an initial £18 million order to supply some 70 per cent (40,000 tonnes) of the steel pipe needed for a gas pipeline from the Fulmar field. The Hartlepool pipe and plate mills had been mothballed earlier in the year, but now they were to open again. In March 1984, however, Shell cancelled the BSC order on the grounds that their pipe was not of sufficient quality to resist corrosive elements in the gas. Instead, the order went to a Japanese consortium. The BSC disputed Shell's claims, but the incident did indicate the BSC's continuing problems in making inroads into markets where a high degree of quality and efficiency was required.

As we have seen, the BSC's improved financial performance was only marginally brought about by reviving its market share domestically. Cost cutting, improved productivity, relatively low raw material prices, debt write-offs by the government and the Davignon Plan, all made a considerably greater contribution to swing round the BSC's fortunes than its domestic market performance. In addition, the BSC was also faced with a severe decline in consumption:

> UK direct consumption of steel has declined by 39 per cent between 1973 and 1984 compared with only 16 per cent in the rest of the EEC. The decline in the UK manufacturing base on top of the structural changes found in other industrialised countries has compounded the problems of the British steel industry's adjustment to changed market conditions. It is not surprising that UK steel makers are cautious about the level of future capacity supportable by the UK domestic market (NEDO, 1986).

The contraction of the market has pushed steelmakers generally towards customers' needs and the production of quality steel:

> The steel industry is having to study users' requirements closely and tailor the steel offering to match needs much more precisely than in the past if steel is to retain existing outlets or displace competing materials in new uses ... Similarly, there is a widespread awareness of the need to put more resources into the research and development of new products to increase the versatility of steel against competitive materials. Finally there is recognition that added value [the difference between sales value and bought-in costs] and quality are now the significant goals for steelmakers rather than tonnage production (NEDO, 1986).

Unfortunately for the UK, the special steels industry has suffered a particularly severe contraction. Between 1958 and 1984 the number of special steel producers shrank from 105 to less than 10, none of them with a significant market share (*Financial Times*, 26:3:84). Imports from France, Austria, West Germany and Sweden have taken out a large part of the market, with the result that the image of Sheffield as the centre of special steel development is sadly a thing of the past. We have previously noted the problems of Sheffield Forgemasters, and although there remains a handful of successful Sheffield steelmakers, the good old days can never return.

On a more general note, the BSC was particularly successful in the development of coated sheet and strip, constructional steel, some types of high quality machine tool steel and tinplate development (NEDO, 1986). The development of high quality steel for specialised requirements (a policy recommended by Redwood, 1980, 83) entailed a great deal of expertise and foresight on the part of the BSC, for as Rhys states:

> If the steel industry is able to give its customers the same scale of technical assistance as that provided by the likes of plastics, steel could meet the challenge of other materials in the future as it has done in the past. However, this will require that the customer is given first-rate service, with the ability and willingness to study the customers' problems... This service has been given, but the problems of the UK industry have reduced the resources available for providing this type of facility at the very time when the need for it was growing. Now a far greater degree of anticipating, and sensitivity to changing market needs is needed. Ideally, this means that equipment should be installed for producing new products ahead of the competition (Technical Change Centre, 1986).

Politically, by the late 1980s, the need for research and development, flexibility in investment, and co-ordination between steel producers and their consumers had begun to replace the dominating concerns of the 1960s and

70s of large scale basic steelmaking development, massive rationalisation, hundreds of thousands of redundancies and the very survival of the UK industry. In the final two chapters we will be examining how government may respond to these less politically cataclysmic but no less important challenges, but as an indication of the long-term problems confronting British Steel, it is interesting to note the views of Professor Nutting, a metallurgist at the University of Leeds. In asking the question as to whether UK educational institutions are capable of meeting the future requirements of the steel industry, Professor Nutting outlines the profound changes in his field:

> The growth of the universities in the sixties also led to the appointment of many new professors. Almost without exception the professors who were appointed in the sixties had a thorough background of physical metallurgy and all had a sound knowledge of ferrous metallurgy. They tried to communicate enthusiasm ... to the newly recruited staff and the students. But since that time many of the professors who were recruited have now retired and others are very shortly to retire. I am one of the last, as it were, of the old school who are left, in the sense of someone who sees himself as a ferrous physical metallurgist. The young men who are coming along to replace them have a knowledge of ferrous metallurgy that is slight. Their interests are now in ceramics, composites, electrical and electronic materials and polymers and if presented with metallographic evidence, they would probably be completely unable to assess how a steel had failed in service. Many would not really know how steel was made (Technical Change Centre, 1986).

These depressing but revealing remarks exemplify how considerations of future steel policy cannot be restricted to the industry itself, or even with manufacturing industry as a whole. The problems are tied in with profound societal attitudes at the grassroots level. NEDO calculated that BSC spending on research and development in 1985 (£18 million) amounted to 0.5 per cent of turnover, a similar percentage to the rest of Europe. The figure for the US was 0.3 per cent, but in Japan it was 1 per cent (NEDO, 1986). At the same time, NEDO acknowledges that it is the effectiveness of research which matters most, and one indicator here is that Japan is way ahead of Europe in the field of foreign patent registration (NEDO, 1986).

Professor Nutting calculates that for £400,000 *per annum* Chairs could be endowed and studentships provided sufficient to revitalise the metallurgical educational process in the UK, and suggests that the steel industry urgently turns its attention to this matter (TCC, 1986). In this respect, at least, steel policy is intrinsically linked with education policy (it should be noted that British Steel's own management training college does have a high reputation).

These problems on training put into context the prospects for British Steel customer relations, although in this field there is no doubting the considerable improvement in recent years. In the final section of this chapter we will look at the BSC and the motor vehicle industry, but in most sectors there was a marked change from the early 1980s onwards. It was not so much that customers abandoned the policy of 'second sourcing', but rather that they were more satisfied with the service they received from the BSC. In 1982 Metal Box said that it would continue to buy the vast majority of its tinplate from the BSC, as the Corporation now had a more realistic approach to the needs of the marketplace (*Financial Times*, 20:12:82). Graham Day, the Chairman of British Shipbuilders (BS) (and later of the Rover Group) compared the approach of the BSC in the early to mid-1970s to that adopted in the early to mid-1980s. He said that in the 1970s there was little regard for the customer, with a quality of steel that would not be accepted in the mid-1980s. Even in the early 1980s he felt that the BSC lacked co-ordination between the mills and the commercial function, with quality unreliable. From 1984, however, there was a considerable improvement, with the result that BS was now a highly satisfied customer (Technical Change Centre, 1986).

As we will see in the final section of this chapter, these testimonials can also be found within the motor vehicle industry, and are a tribute to the BSC's success in improving the quality of its commercial performance. At the same time, as we have seen in the case of tubes, some sectors have improved more than others, while some customers continue to see a much greater need for co-ordination between suppliers and customers. Thus Dr. Browne, of Taylor Woodrow Construction, criticised both the UK steel industry and the government for not giving sufficient attention to the reinforcement of motorway bridges subject to attack from de-icing salts (the BSC replied that the government and local authorities were too concerned with keeping down construction costs), while Professor Nutting highlights his experience with developing a type of special steel for the electricity industry. After hawking his ideas around the UK without success, he received financial support for research from California, the experimental steels were produced in Japan, and finally a large ingot was made in Austria! (Technical Change Centre, 1986).

As with the training of metallurgists, there appeared to be gaps in research and development and the relationship between the BSC and its customers which handicapped the Corporation's progress. Given the establishment of overseas competition, it may be almost impossible for British Steel to regain these lost markets. It is not just a question of price or even of improvement in general quality, but building up a highly integrated relationship with customers over a long period of time, and so producing steels tailor-made to satisfy requirements. Such an ideal state of affairs is, of course, not easily achieved in the real world, but in the final chapter we will be examining the political implications of injecting a higher degree of 'co-ordinated rationality' into the system.

Each of the industries serviced by the BSC e.g. engineering, construction, oil and gas, mining, shipbuilding, metal packaging and white goods, has a distinctive tale to tell, but in a work such as this it is impossible to do justice to all of them. Instead, in the final section of the chapter we will look briefly at the relationship between the BSC strip mills and the domestic motor vehicle industry in order to give just a flavour of how the Corporation improved its dealings with major customers, while at the same time it encountered great difficulty in regaining market share.

The BSC strip mills and the motor vehicle industry

In 1979, despite the BSC's poor performance record in the 1970s, the UK motor vehicle industry still bought 1,097,900 tonnes of steel from domestic producers. In the strike year of 1980 the figure declined by over half to 542,500 tonnes, but in 1981 only rose to 622,400 tonnes. From 1981–85 the figure remained relatively stable, but in 1986 fell once more to 570,300 tonnes (some reclassification slightly deflates this figure when compared with the earlier ones) (ISSB). By 1987, the UK motor vehicle industry demand for flat rolled products represented just under 20 per cent of UK demand for these products. In recent years, the demand is more for lighter steels, and also for zinc coated products (House of Commons Select Committee on Trade and Industry, 1986–87, 238).

Of the 1986 motor industry sales figure, 233,300 tonnes were accounted for by coated and uncoated sheet for vehicle bodies – the steel produced at the BSC strip mills of Port Talbot, Llanwern and Ravenscraig. Other significant components of the total figure included 81,400 tonnes of tubes and pipes, 71,500 tonnes of bright bars, 61,000 tonnes of hot rolled narrow strip, 44,600 tonnes of plates and 44,000 tonnes of cold rolled narrow strip (ISSB). The figure for coated and uncoated steel consisted almost entirely of hot and cold rolled wide strip. This unwieldy category was a deliberate act on the part of the ISSB at the request of the BSC. As the only UK supplier of strip, the Corporation was unwilling to make public its domestic sales in each strip category.

The steep decline in domestic steel consumption after 1979 by UK vehicle manufacturers can be attributed partly to their own problems of falling production levels, but the rise in imports is also significant. Inevitably, there was a rise in coil, strip and sheet imports during the strike year of 1980, but the figure fell sharply in 1981. More significantly, after 1981 there was a steady climb in imports of these categories, with a few startling increases e.g. imports of hot rolled wide strip rose from 403,400 tonnes in 1981 to 883,800 tonnes in 1986, while coated plates and sheets went up from 203,600 tonnes in 1981 to 382,700 tonnes in 1986. Thus 'second sourcing' continued to grow, despite the BSC's improved relationship with the vehicle manufacturers. As we have previously noted, the BSC campaigned for extra quotas from the EEC, while also buying up spare quotas from abroad. Some customers also complained periodically about the unavailability of BSC strip (*Financial Times*, 17:3:84).

These customers may have been looking abroad for their supplies, yet there is something of a paradox here in that the BSC was able to substantially increase its own exports of strip products. Thus UK exports of hot rolled wide strip rose astonishingly from 49,900 tonnes in 1982 to 243,000 tonnes in 1986. In addition, exports of uncoated sheets went up from 279,900 tonnes in 1982 to 519,800 tonnes in 1986, while the figures for coated plates and sheets increased from 187,800 tonnes to 329,800 tonnes respectively. In 1985–86 30 per cent of the BSC's strip mill products were exported (*BSC Annual Report and Accounts*, 1985–86).

It could be argued that the general increase in trade for strip products into and out of the UK is in line with the EEC trends as a whole. Countries such as West Germany, France and Belgium all import a higher percentage of their consumption than in the UK, but all have a much stronger export market. Given that steel consumption in West Germany, France and Italy is considerably higher than in the UK, there was obvious scope for the BSC here, and by the late 1980s the Corporation was at last exploiting some of these opportunities. Nevertheless, it might be noted that in 1986 the BSC exported 11,700 tonnes of hot rolled wide strip to West Germany, while German exports of this product to the UK amounted to 395,500 tonnes (ISSB). The existence of quotas placed restraints on growth, but the extreme imbalance between West Germany and the UK in this important type of strip does exemplify the absence of UK steel from manufactured goods produced by the largest and strongest economy within the EEC.

The steep decline in the UK car market between the mid-1970s and mid-1980s illustrates the mirror image of the BSC's own problems. Between 1974 and 1984 the UK domestic market for new cars grew by 40 per cent to a total of 1.75 million units, but in the same period the number of vehicles produced in the UK fell by 625,000 units. In addition, UK car exports fell by 404,000 (68 per cent) over the same period. On the other hand, car imports' share of the market increased from 27.9 per cent to over 58 per cent (SMMT). The stark figures tell the tale of an industry in similar crisis to that of the BSC during these years. In particular, the Rover Group (formerly BL) was only saved from oblivion by the government taking over responsibility for the company in 1975, while the UK operators of the American Chrysler company were first saved by government loans and were later sold to the French Peugeot-Talbot company in 1979 for the sum of one dollar (the closure of the Linwood plant in Scotland by Peugeot-Talbot was a large factor in placing the future of the Ravenscraig plant in danger).

The Ford and Vauxhall (owned by General Motors) volume car companies, as subsidiaries of powerful US combines, have been in less direct jeopardy, although Vauxhall has consistently made losses. The design and engineering departments for Vauxhall are situated within its sister company Opel in West Germany. The purchase of Vauxhall components is also made in consultation with Opel (House of Commons Select Committee on Trade and Industry 1986–87, Q 371). Vauxhall is thus in something of a 'poor relation' position with regard to Opel, and this is reflected in the number of components sourced in Germany (including steel bodies) and then sent to the UK for assembly (sometimes referred to disparagingly as a screwdriver operation). In contrast, Ford is a more integrated company and sources a higher percentage of its components in the UK. It has become commercially the most successful UK vehicle manufacturer, although it does also import a wide range of models from its sister plants on the Continent. Peugeot-Talbot also relies extensively on imported components. As we will see, this question of imported components became a topic of high political salience in the mid 1980s. Eventually, the government itself expressed disquiet with GM in particular about not only the number of its cars imported into the UK, but also the large amount of foreign sourced components to be found in UK constructed vehicles.

From the mid 1980s the Japanese motor industry began to take a more direct interest in the UK car industry. As with steel, the Japanese had a gentleman's agreement on vehicle imports into the UK (usually about 11 per cent of the market), but by the late 1980s Honda had become a vital collaborative partner with Rover, while Nissan had opened a new factory in the North East of England. In this section we will briefly examine developments in the UK vehicle market during the 1980s as they affected a prime component supplier such as BSC strip steel.

Quality and rival materials

The use of steel itself as a vehicle component has been brought into question in recent years, with the rise to prominence of rival materials such as aluminium and plastics. At the same time, as Rhys points out, the cost of developing and applying new materials is high (Technical Change Centre,

1986), and in the late 1980s there is no hard evidence that steel's traditional importance as a component of vehicle manufacture is about to be replaced. Between 1965 and 1980 the share of steel in the total weight of passenger cars fell by only 2.8 per cent to 60.1 per cent. Aluminium still tends to be expensive, and in any case is hard to weld, while production times for plastic materials are still too slow to be compatible with the production scheduling of assembly plants (International Iron and Steel Institute, 1983).

The difficulties with rival materials were reflected in the decision of General Motors of America during 1986 to abandon a one billion dollar project for a plastic body sports car because the technology required was too expensive (*Financial Times*, 20:10:86). Earlier, the company had stated that it could be producing more than one million plastic body cars by 1990. Steel is thus likely to remain by far the most important vehicle component, although for reasons of economy manufacturers seek lighter vehicles, with a commensurate loss of steel quantity.

Car manufacturers now demand a more specific and consistent quality of steel than in earlier decades, and so close and integrated relationships between producer and customers are at a premium. On the other hand, this collaboration can cause difficulties for a steel producer seeking to invade new territory. Thus UK car manufacturers, having become accustomed to 'second sourcing', would not be persuaded easily to abandon this practice, while the BSC (despite its improved reputation) found it extremely difficult to become a component supplier to such overseas giants as Volkswagon and Fiat. Domestically, the BSC had liaison committees with all the principal UK car companies, while in addition the Society of Motor Manufacturers and Traders had a strip steels sub-committee which met periodically with BSC executives.

The BSC and Rover

As the only UK owned volume car manufacturer, the relationship between Austin Rover (the volume car manufacturer within the Rover business) and the BSC was of particular importance to the Corporation. In 1975, Austin Rover's share of the UK market stood at 29.67 per cent, but by 1980 this figure had slumped to 17.66 per cent. In volume terms the number of cars sold fell from 354,000 in 1975 to 267,000 in 1980. After that time, the company's share of the market remained relatively stable, but the general increase in domestic demand led to 309,000 cars being sold in 1984. Sir Michael Edwardes, Chairman of BL from 1977 to 1982, was brought in by the Labour government as a 'strong leader' figure, who it was hoped would save the company from oblivion. It is worth noting that Ian MacGregor, brought in by the Conservative government to perform a similar function at the BSC, previously held the position of non-executive Deputy Chairman of BL. It might be considered politically ironic that both Edwardes and MacGregor were Labour government discoveries!

Despite the MacGregor-Edwardes connections, a heated row emerged in 1982 over the price BL was to pay for BSC strip steel. The BSC had conducted campaigns for cheaper prices from suppliers such as British Coal, British Gas and the CEGB, but in this case the boot was on the other foot. It could be said that the similarities in character and approach between Edwardes and MacGregor helped to cause the confrontation, for both were strong-willed men. The irony for the government was that, on this occasion, the two Chairmen were fighting each other rather than the unions!

From January 1982 the BSC increased its strip prices by 10–15 per cent. BL immediately objected, and continued to pay only 1981 prices. As the price rises were part of the Davignon Plan to bring stability to the steel industry, Edwardes accused the steelmakers of collusion in a depressed market and blanket protection against cut-price imports (*The Times*, 24:2:82), (although Sir Michael himself complained bitterly about unfair trading practices of car importers into the UK (Edwards, 1983, 283–4)). In 1981 BL had obtained £1 billion in further funding from government, and Edwardes was ill-disposed to using BL resources in order to assist BSC recovery.

In the middle was the government itself which, faced with a position similar to that of BSC and its state-owned suppliers, had to look for a balance of interests. Despite several meetings between Edwardes and MacGregor the dispute rumbled on until June 1982, when the BSC agreed to supply all UK car manufacturers at a discount, but still received most of the proposed increase. Without going into details, Sir Michael accused Patrick Jenkin, Secretary of State at the DoI, of intervening in commercial negotiations with the BSC, which suggested that the government had played a significant part in bringing about the eventual agreement (Edwardes, 259). The BL Chairman was unhappy generally with interventions by Jenkin, and wrote him a letter which pointed out that when government intervened in commercial matters it made it far more difficult to attract and hold first-rate managers (Edwardes, 259). A cynical reply might be that if 'first-rate managers' could not reach agreement in six months over a supplier's prices, then it needed a Secretary of State (who after all was ultimately responsible for the billions of pounds being poured into both organisations) to assist in bringing matters to a conclusion!

Despite the disagreement over steel prices, by the early 1980s the BL-BSC relationship had improved greatly from the low point of the mid-1970s. As we noted earlier, BL had become particularly exasperated with the quality of BSC steel and the service provided by the Corporation. In 1977 the BSC supplied only 58 per cent of BL sheet steel needs, but by 1979 the figure had risen to 74 per cent. In a 1981 memorandum to the Industry and Trade Select Committee, BL stated:

> The efficient operation of press shops is dependent upon the delivery of large volumes of steel on time. BL presses thousands of different parts from its press shops located in various parts of the UK. In 1976 only approximately one third of the tonnage ordered by BL on BSC was delivered on time. BSC were aware of their shortcomings in this respect and succeeded in improving this aspect of their performance so that by the end of 1979, 75 per cent of their tonnage was being delivered on time. BL was able to increase its offtake accordingly and reduce imports (House of Commons Select Committee on Industry and Trade 1980–81, 211).

BL's observations indicate the extremely poor quality of BSC service in the mid 1970s, and the consequent policy change towards a more consumer led service. On quality, in 1981 BL considered that BSC mills were capable of meeting their requirements, but that there remained room for improvement, not only in the introduction of new equipment, but also a new awareness on the part of the workforce (House of Commons Select Committee on Industry and Trade 1980–81, 211). After that time the BSC continued to consolidate its relationship with the new Rover Group, with the company obtaining around 90 per cent of its sheet steel from the BSC. It was unfortunate for the BSC that Rover was unable to secure a higher penetration of the UK market, for other domestic manufacturers were not so wedded to the principle of buying British steel.

Ford and Vauxhall

In 1981, Ford told the Select Committee that the BSC had schedules for approximately 35 per cent of their total car body steels. They believed that the BSC met quality standards in respect of unexposed panel material, but that:

> Their major problem has been to match the high quality standards of exposed panel material achieved by their continental competitors. While this has been the subject of almost continuous discussion between Ford and BSC management, at all levels over a long period, we are not yet satisfied that they can consistently meet the required standards for greater volumes than they currently enjoy (House of Commons Select Committee on Industry and Trade 1980–81, 209–10).

In 1984 Ford told the Trade and Industry Select Committee that it bought body steel from Hoesch, Thyssen, Kloeckner and Krupp in West Germany, Cockerill in Belgium, Sollac in France and also the BSC. (House of Commons Trade and Industry Select Committee, 1983–84, 138). The company also issued figures to the Llanwern Trades Council for the supply of thin panels to the press shop at the Halewood plant in the months January to October 1983. Krupp supplied 6,110.64 tonnes, Hoesch 5,358.96 tonnes, Scottish (Ravenscraig) 4,273.13 tonnes, Llanwern 2,676.46 tonnes and Thyssen 2,513.44 tonnes. These figures confirm that, although Ravenscraig and Llanwern were major suppliers to Ford, the company relied chiefly on German steel for its Halewood produced skin panels.

The sales and production figures for Ford, Vauxhall and Peugeot in the UK are also complicated by the fact that all three companies import parts and complete cars on a major scale. As we have noted, for much of the 1980s nearly 60 per cent of cars sold in the UK were imported (although by 1987 the figure had been reduced to around 51 per cent). In addition, Rhys calculates that UK motor component companies would benefit from an expected increase in domestic car assembly only if the vehicles had an 80 per cent UK content value. Thus although it was estimated that UK car assembly could rise from just over 1 million in 1986 to nearly 1.5 million by 1991 (excluding any contribution from Austin Rover), a figure of 80 per cent local content value was required for the increase to be of significant help to UK component suppliers.

Politically, the question of vehicle components was not (as it surely should have been) a particularly sensitive issue in the 1970s and early 1980s, but in 1985 the subject suddenly rose to the top of the agenda. One reason for this new prominence was the general decline in UK manufacturing industry. The 1985 report from the House of Lords Committee on Overseas Trade (referred to in Chapter One) highlighted this decline, and caused some embarrassment to the government.

Secondly, the Westland helicopters crisis of early 1986, which led to the resignation of two Cabinet ministers, revolved around controversy as to whether a UK company should throw in its lot with a US manufacturer or maintain its independence as part of a European consortium. Although Westland helicopters was not in itself a major business, its fate became a political symbol for an impassioned debate on the relative merits of an official industrial policy or a reliance on market forces.

By coincidence (or perhaps not), concurrent with the Westland crisis potentially major changes were afoot in the UK vehicle industry. Both the Vauxhall and Leyland commercial vehicle operators were in trouble, and it was proposed that General Motors should take charge of both companies, as well as Land Rover and Freight Rover. In the wake of Westland, this proposal caused such a political furore that the government took Land Rover out of the offer with the result that the deal collapsed (in the event, twelve months later, Leyland Trucks was sold to Daf of Holland with barely a whiff of protest, while GM ceased making Bedford heavy and medium trucks). At about the same time there was a proposal that Ford might take over Austin Rover, but strength of protest at this suggestion from MPs of all Parties (together with several members of the Cabinet) again caused the plan to be dropped almost before it had got off the ground.

At the root of all this political emotion was the fear that the UK motor industry could be disappearing entirely into foreign hands. As we have seen, the BSC relied on Rover buying nearly all its steel from domestic sources, and an overseas takeover might have jeopardised this relationship. Notwithstanding Rover's increasingly close collaboration with the Japanese Honda company on a number of models (Honda was building an engine plant at Swindon with the strong suggestion that it might eventually build cars there, while it was already using spare capacity at Austin Rover's Longbridge plants to build its own models), the general political opinion was that further foreign infiltration of the UK vehicle industry must stop. The government encouraged this policy line with its strong support for a takeover of Rover by British Aerospace, a deal which finally went through in July 1988.

This climate was obviously of advantage to the BSC, which could point to its own improvement in quality and service as proof of its determination to improve its performance as a primary component supplier to the UK motor industry. By the late 1980s UK vehicle manufacture was on the increase. Currency changes, particularly the decline in the Pound against the Deutschmark, led both Ford and General Motors to switch a greater share of their production to the UK. In the early 1980s both Ford and Vauxhall imported around 50 per cent of cars they sold in the UK, but by the late 1980s Ford sourced about 75 per cent of its UK cars domestically, while Vauxhall's percentage rose to above 60 per cent. By 1988, with Ford's market share in the UK standing at about 28 per cent, and the decrease in imports, the company was rivalling Austin Rover as the leading domestic vehicle producer.

Nevertheless, the component policies of these companies, particularly Vauxhall, continued to be of high political salience. The complaints against Vauxhall came to a head in 1985. Initially, Professor Dan Jones revealed that the UK content of the cars General Motors sold in Britain had fallen to the extent that these models were almost not British (*Financial Times*, 18:1:85). Vauxhall defended itself by claiming that the age of the Eurocar had now arrived, with manufacturers making variations on a basic model and then assembling these in various European countries to serve the whole market (*The Guardian*, 11:10:85). By October 1985 the Vauxhall controversy had become political property, and the Labour Industry spokesman, John Smith, wasted no opportunity to use it as an example of the decline in UK manufacturing industry under the Conservatives. The Vauxhall plant at Ellesmere Port staged a one day strike in protest at the number of GM cars imported from Germany, and eventually the government reacted.

The Trade and Industry Secretary, Leon Brittan, told the Commons that he had initiated discussions with Vauxhall on the UK content of their cars, but that the company was not yet ready to go further in proving that it really was a British car producer. He contrasted Vauxhall unfavourably with Ford, who had told him that their UK content of British made cars was near to 80 per cent. On the other hand, Vauxhall was committed only to increasing UK content from 46 to 49 per cent during 1985 (Hansard 23:10:85, Col. 285). At this time the government was in a particularly delicate position with regard to Vauxhall. The proposed General Motors takeover of Leyland Trucks was in the pipeline, and so Ministers could not afford to antagonise excessively the world's largest motor manufacturer. The government needed the goodwill of General Motors rather than vice versa. At the same time, the depth of political feeling on the subject of UK content required some kind of reaction. In the event, as we have seen, external conditions enabled both Ford and General Motors to source more vehicles and components in the UK. The government was in no position to enforce this outcome, but a political show of strength at least gave the impression that something was being done.

The arrival of Nissan and the future of the strip mills

A new factor within the UK motor industry arose in the late 1980s as a result of the creation of a UK car operation by the Japanese company Nissan. The plant, situated at Washington in the North East of England, began producing cars in 1986 using kits sent from Japan. In this initial phase, 24,000 cars were to be produced. Under the agreement signed with the government, £35 million in grants and regional development money was to be made available to the company, with the condition that Nissan would work up to a level of 60 per cent local (i.e. EEC) content by 1988. From 1988, however, the factory was to begin a process by which it would have its own sheet metal stamping, full body sub-assembly, engine assembly and engine machining operations. By 1991 it was intended that the plant would produce 100,000 units a year, with an EEC (not necessarily UK) content value of 80 per cent (House of Commons Select Committee on Trade and Industry 1986–87, Qs. 304–5). By moving to this second phase, around £100 million in grants and regional development money was now available to the company. Thus the UK government had a large stake in not only securing the success of the

Nissan development itself, but producing spin-offs in the form of business for UK component manufacturers. The Nissan developments clearly offered a new opportunity to the BSC, particularly as the Company expressed an intention to single source its component suppliers, and that 90 per cent of the 80 per cent EEC content should be UK sourced (House of Commons Select Committee on Trade and Industry 1986–87 Qs. 315–331). This policy suggested that nearly all Nissan's requirements for sheet steel would be supplied by the BSC, and indeed during 1987 the Corporation was holding detailed talks with Nissan (House of Commons Select Committee on Trade and Industry 1986–87, 240).

The problem for the BSC in wider market terms was that the UK production of 100,000 Nissan cars could deflate sales at other manufacturers such as Austin Rover and Vauxhall. On the other hand, a continued expansion of the UK vehicle market, with a higher proportion being domestically sourced, could only be of benefit to the BSC strip mills. The BSC itself continued to campaign for higher EEC production quotas, but remained confident that it possessed the necessary capacity to satisfy its customers (House of Commons Select Committee on Trade and Industry 1986 87, Qs. 1405–6).

Notwithstanding the BSC's confidence on capacity, the problems of forecasting future UK vehicle production highlighted how the Corporation's decisions on plant configuration could be thrown into the melting pot. In particular, throughout the 1980s the BSC had made no secret of its wish to see the closure of Ravenscraig, yet by the end of the 1980s all three of the BSC's strip mills (assisted by the rise in exports) were running close to manned capacity. Although the government's insistence on keeping Ravenscraig open undoubtedly had a strong political ingredient, it could be argued that by the end of the decade the BSC's own commercial performance cast doubt on the management's insistence that the plant was surplus to requirements.

Despite the relative optimism of the late 1980s, it is nevertheless necessary to bear in mind that some UK vehicle manufacturers continued to be highly sceptical about the performance of their suppliers. Particularly relevant are the views of Alan Spencer, Vice-President, Supply, for Ford in Europe. He told the Trade and Industry Select Committee of his experiences with regard to technological development:

> I think that Britain ... has a remarkable ability to come up with very interesting technical inventions ... What is lacking is the follow-through ... What is lacking is somebody who can put down a new product with total reliability so that we feel we can put this into our new model and that the customers will not have the reliability and

> servicing concerns ... Unfortunately, such a new product is rare ... That is something where the German suppliers tend to have an edge (House of Commons Select Committee on Trade and Industry 1986–7, Q. 206).

Similarly, with regard to UK attitudes towards exports, he illustrated his point by noting that UK component suppliers to Ford UK could equally well supply the same parts to Ford plants on the Continent, but that few grasped these opportunities. He went on:

> They have quite satisfactory capacity at the moment, I think. Some of them are a little hungrier than others but I do not think that anybody is short of work in the sense that they actually have empty factory space available. I think that it is also an attitude of mind. I think that they find exporting very difficult (House of Commons Select Committee on Trade and Industry 1986–87, Q. 211).

These views of a senior executive in one of the UK's largest vehicle producers, although not directly aimed at the BSC, are reflected in the Corporation's own difficulties in coming to terms with the EEC market.

At the time of privatisation, the BSC was enjoying the fruits of a boom in steel demand. Steel output in the UK in 1988 totalled 19 million tonnes, the highest annual figure since 1970. Steelmaking at the five BOS works was running near to manned capacity, and the Corporation was making good profits. The level of exports had risen considerably during the late 1980s, in tonnage terms if not in value, while the decade as a whole saw radical improvements in plant efficiency. On the other hand, the BSC had never truly come to terms with new trading patterns brought about by UK membership of the EEC. In addition, major domestic customers, such as Ford, continued to have doubts about the quality of steel provided by the BSC. As we will see in Chapter Eleven, there was also a real danger that, in its anxiety to cut costs and make a profit, the BSC had neglected useful investment opportunities. The journey along the road to viability had compelled the BSC to become a more commercially orientated organisation, but had perhaps also served to resuscitate a conservatism in the UK steel industry which had been evident in pre-nationalisation days.

11 The Politics of Expertise

The persistence of the BSC

The political management of the BSC proved to be a policy problem of the first magnitude. Nationalised amidst great adversarial controversy on the need for such a strategically important industry to be placed in state hands, the chief focus of attention for nearly twenty years was on almost everything but that topic. Given the general political emphasis on privatisation during the 1980s, and indeed the eventual return to the private sector of the BSC itself, this may seem a surprising outcome. Yet even during this latter period, the principal objective was to put the BSC itself on a more stable footing. When it finally arrived in 1988, privatisation was politically a somewhat low-key affair, particularly when compared with the periods of crisis in earlier years. The BSC had nevertheless survived, and the organisation remained by far the most important actor within the UK steel community.

It could be argued that crisis even strengthened the BSC's position, for the more governments pumped resources into it, the more they relied on the organisation itself to pay both a financial and political dividend. Thus when the fortunes of the BSC improved in the mid 1980s, the government claimed at least some of the credit for keeping faith with the Corporation in its darkest days. Throughout its life, the BSC was an extremely powerful organisation, and successive governments challenged its policy-making authority with only spasmodic success (i.e. organisational self-interest has been a key factor in the development of UK steel strategy).

Once the basic strategy of development and rationalisation was adopted, its implementation (despite the work of the JSG) achieved a self-propelled momentum which no external agency could halt (although the government did intervene on prices). Initially, the BSC itself could afford to be relatively relaxed about the closures programme, and the outcome of the Beswick review emphasised that neither government nor BSC perceived the enormity of the impending crisis. When the position began to worsen during the mid-1970s, the government's response was initially slow, with only a slight speeding up of the closures programme. Thus from the autumn of 1977 an incremental programme had started with the closures of Hartlepool and East Moors. This policy was accelerated when BSC losses began to mount, and the Corporation was given its head to achieve closures at any cost.

The new Conservative government hoped to hold the BSC to strict cash limits, and so win almost immediate viability. This policy proved to be totally unworkable, but it did have the effect of forcing the BSC management into a damaging confrontation with the unions, which resulted in the steel strike of 1980. The government could not control the financial position of the BSC, but the organisational behaviour of the BSC was nevertheless greatly influenced by government policy. As we have stated in Chapter One, the size of the EFL, together with the appointment of a Chairman and Board, are two of the few direct policy levers which Ministers can exercise on a public corporation. After the debacle of the strike, the government was left with little choice but to give huge financial aid to the Corporation, while Ministers placed their faith in the ability of successive BSC Chairmen to turn the tide.

The one major issue in the 1980s on which the government successfully prevailed was the future of Ravenscraig. Although the BSC wished to close the plant, the government refused to budge. Even then, the BSC managed to push through the Gartcosh closure. The government could gain some support for its position over Ravenscraig from the integral part played by Ministers in shaping ECSC steel policy. The Davignon Plan depended for its success on a political consensus between national governments. The BSC was involved closely with negotiations through its membership of Eurofer, but the government took the ultimate responsibility. It could thus claim that keeping open Ravenscraig was part of a national strategy designed to show both the Commission and other governments that UK closures had gone far enough, and that any further closures must come from elsewhere. In this way the trend towards international political management within the ECSC strengthened government power in shaping steel policy. The government itself might hope to privatise the BSC, but it had no qualms about playing the game of political negotiation within the ECSC. Thus it could be said that the Commission, at least indirectly, through the negotiating power delegated to the UK government, had become the greatest threat to the policy-making autonomy of the BSC.

The steel unions, by their passivity, have in the past assisted in the development of a restricted steel policy-making community. Moderate by nature, usually they were not consulted with regard to the BSC's corporate plans. The worker director experiment failed, while the 1980 strike was specifically about money, and had no influence over the closures programme. It was the action committees of threatened individual works, particularly well exemplified by the case of Shelton, which came nearest to influencing policy, although even here the BSC had its way in the end.

Parliamentary Select Committees investigated the BSC at regular intervals over the years, and on one notable occasion obtained papers from the unwilling Corporation, but their role was largely one of commenting on past events, rather than genuinely influencing BSC behaviour. Indeed, in June 1981 the relevant Select Committee was still complaining that they secured a copy of BSC's corporate plan only after 'considerable difficulty and delay', and then it was made available to Committee members on a confidential basis (House of Commons Select Committee on Industry and Trade 1980–81, para. 15).

In the early 1970s the private sector of the UK steel industry improved its share of the market, and could see itself as a relatively powerful commercial and political force. From the late 1970s its fortunes declined, and by the 1980s it too was seeking government aid. The various privatisations, including the Phoenix projects, were designed as much to rescue what was left of the private sector as to return parts of the BSC to private ownership. By the late 1980s the BSC had a significant shareholding in nearly all the largest private sector companies, with the result that BISPA had to take account of BSC interests in shaping its policy.

Similarly, the stockholders' group NASS included in its membership the BSC's own stockholding group, and in any case there is a strong identity of interest, particularly over prices, between the steel producers and the stockholders. At the same time, the stockholders do 'second source' extensively, and could afford to operate at least a little distance politically from the BSC. The scrap merchants, through their success in exporting, could also operate with some autonomy. As we saw in Chapter Seven, the government was at times prepared to back the claims of the British Scrap Federation against those of the BSC.

Notwithstanding the political influence of stockholders and scrap merchants, the power of the BSC itself strongly determined the nature of strategic policy-making on steel within the UK. In this chapter we will examine briefly the characteristics of the BSC as an organisation, how these changed over the years, and their influence over the course of UK steel policy. Given the predominance of the BSC in policy-making, we will also see how the generally reactive role of government resulted from a lack of expertise, confidence and political will.

The traditional but ill-defined Morrisonian concept of an arm's-length relationship between a Nationalised Industry and government has been of little assistance in clarifying either the authority of government or the accountability of the BSC. Even the Benn-Finniston debate of 1975, designed to throw light on this matter, was predictably messy and inconclusive. The nature of the relationship leads us to the conclusion that a Minister taking direct charge of a state-owned BSC would have been the best solution for steel. If not a guarantee of good decision-making, this arrangement would at least have clarified the lines of responsibility. The potential quality of decision-making could also have been improved if the Minister had at least some alternative source of expertise to that provided by the BSC. The chapter concludes with an evaluation of how UK membership of the ECSC and EEC has affected the domestic political culture.

The BSC as an organised interest

Tivey highlights the key to the chief political strength of the public corporation:

> The explanation seems to lie in the high advantages of knowledge and understanding of their own industries which the corporations build up. The constitutional (i.e., political and administrative) position of the Morrisonian public corporation was weak, and was intended to be weak. Its autonomy was managerial – that is, was intended to relate to its internal affairs. In practice, the corporations have rarely been as weak as this theory expected ... Whatever the supposed constitutional position, the residual political strength of the industrial public corporations lies in their deep and specific reserves of knowledge (Tivey, 1982, 54).

With regard to the relationship with government, this view is endorsed and expanded on by Foster:

> To some extent, what a portsman thinks about ports or a railwayman about railways is something he has absorbed on the job and never learned explicitly. Thus, he is less likely to be impressed by logic, or dismayed when the illogicality or incompleteness of what he has said is pointed out to him. He is more impressed by someone who, he

> believes, shares some of the craft mysteries that he does. Also, quite apart from this, he knows that, though he might not feel as agile intellectually as the civil servant who is questioning him in the end – he may have to play it long – he must win, because of the final lack of self-confidence the administrator has in his ability, or his right to say No (Foster, 1971, 63).

In the case of the BSC, Sir Robert Scholey, Chairman since 1986, could be taken as the epitome of the professional steelman. As part of the so-called United Steel 'mafia' which later came to dominate BSC management, he was Deputy Chairman and then Chief Executive (with a brief gap during the MacGregor Chairmanship) from 1973 until his appointment as Chairman. Even Sir Robert's 1970s nickmane of 'Black Bob' carries with it the connotation of a link with the tough and determined ironmasters of an earlier age. As Abromeit comments:

> Bob Scholey – an 'idealistic man' and emotionally involved in the steel business, seems to have taken great pains always to remain at the very centre of things, hating 'normal routine' (things not going over his own desk), and thus establishing a superior position as the one who knew about everything (Abromeit, 166).

As President of Eurofer from 1985, Sir Robert placed himself in an even better position to 'know about everything', and so was undoubtedly the most knowledgeable and professionally experienced BSC Chairman. It is significant that it was chiefly Scholey's opposition which saw the end of Sir Charles Villiers' scheme for a 'Steel Contract'. As we saw in Chapter One, each of the BSC Chairmen has given the organisation a particular public image, but we should bear in mind that, even in the late 1980s, a significant proportion of BSC management had been with the organisation since its foundation in 1967.

As Abromeit points out, whatever the wording of the Iron and Steel Acts may have implied, the BSC has taken its chief objective as being to operate in a 'commercial' manner (Abromeit, 200). Given the lack of specific guidance from government, it is understandable that the natural inclinations of the steelman should be predominant in determining the course of UK steel policy. From this conclusion, however, we can note that what the steelman considers to be the primary means of achieving commercial objectives does vary from time to time. In the early days, the organisational objective was to break up the old private sector groupings and establish strong central control. From the early 1970s, these changes were linked to the adoption of the ten year strategy involving massive investment in new plant. Behind this plan lay at least the hope that the BSC could play a strategic role in the industrial regeneration of the UK, while profits and market share were largely taken for granted.

Even though the recession began to bite in 1975, this strategy continued until 1976, when it switched to reducing investment to a minimum level and closing the older steelworks as quickly as possible. The *Road to Viability* (and indeed organisational survival) became the principal aim, and general industrial regeneration was put to one side.

When Ian MacGregor became Chairman in 1980 he laid particular emphasis on placing more responsibility at the operational level. The general adoption of Profit Centres (and devolution of pay bargaining) was the device chosen to obtain this aim, although significant control of investment and overall commercial policy was retained at the centre. Behind these changes lay, partly, the long-term political aim of privatisation, but the policy change in itself appeared to signify a loss of confidence at the centre compared with a decade earlier. Increased emphasis was also placed on improving the efficiency of the modern plants, where it had formerly been assumed that new investment in itself would largely solve all problems.

By 1983 much of the closures programme had been completed. From the late 1970s increased attention had been paid to improving relationships with customers, but the more stable financial situation now allowed *The Road to Viability* to develop into a more basic market appraisal. In earlier years the BSC had generally played down the importance of an export strategy, but the low level of domestic demand and the Corporation's failure to win a significantly increased share of the domestic market led to a more aggressive policy on exports (the long term privatisation aim also no doubt played a role in this policy). From the mid 1980s, the BSC for the first time attempted significant penetration of the EEC market, and so this policy aim had apparently become a 'proper' commercial objective. 'Acting commercially', for all its primacy, was thus a highly amorphous subject, and exemplifies how an organisational objective can be altered almost out of recognition as the years go by. Yet at each stage the strategy adopted would be regarded as quite 'normal', 'inevitable' and 'rational' by the members of the organisation.

These changes indicate the dynamics of the BSC's organisational culture. As Foster argues, the steelman is likely to be most impressed by the arguments put forward by other steelmen, although within this enclosed circle there can be significant change. Thus 'Black Bob', the feared Deputy Chairman and hatchet man, can become Sir Robert, Corporation Chairman and President of Eurofer, who quietly builds profits at the BSC while acting as something of a philosopher statesman within the ECSC! Like Sir Robert, the BSC had survived, and was still in a prime position to influence UK steel policy.

Privatisation in itself is unlikely to change substantially the steelman's culture, although one possible policy direction which could change the organisation fundamentally would be diversification. Many privately owned major steel companies have invested in other activities in order to offset the depressed nature of the steel market. The ultimate example of identity change is offered by the largest American steel company, US Steel. In 1986 US steel became US X (it might be said that X literally marked the spot of the disaster), and so officially signified not only the company's diversification, but also an anxiety to totally obliterate its former identity. A British X Company may be an unlikely event, but diversification would possibly place the steelmen in a more restricted and controlled environment than they had known under public ownership.

A further agent of change could be collaboration (or even integration) with other ECSC steelmakers. Sir Robert Scholey has himself suggested that trans-national product exchanges, in order to supply specific regional markets, could be a practical line of investigation (*The Guardian*, 29:9:87). A logical extension of this policy would be inter-company co-ordination, leading to eventual integration. In the UK vehicle industry such activity has taken place already e.g. the sale of Leyland Vehicles to Daf and Austin Rover's collaboration with Honda, but the reluctance of ECSC governments to forfeit control of their national industries could be a formidable political obstacle in the case of steel.

The BSC and government

The generally reactive role of government to BSC policy-making is primarily a result of superior expertise. In addition to the work of the BSC itself, the failure to match plans with performance also highlights the Departmental role in monitoring and evaluating the Corporation's affairs. It is intriguing to note Ian MacGregor's comment to the Select Committee in 1981, when it was put to him that if there were people working in the DoI who knew more about running the steel industry than the BSC management, then they should be working for the Corporation rather than the Department:

> That is a very sage observation. Any time that we see someone in the Department who could be better used in our business, we try to acquire him. In fact, we have done so (House of Commons Industry and Trade Select Committee, 1980–81, Q.31).

In later evidence, this point of view was supported by both the Secretary of State for Industry, and the Under-Secretary in charge of the Department's Iron and Steel Division.

The philosophy behind this consensus was apparently in line with the arm's-length relationship between a public corporation and government, and placed the Department in a similar position to that of a banker's and a shareholder's relationship with a limited company (House of Commons Industry and Trade Select Committee 1980–81, Q. 535). The problem with this argument is that the Secretary of State is obviously more than either a banker or a shareholder. He was answerable to Parliament about the performance of the BSC, appointed the Chairman and Board Members, and also had the prerogative to give final approval to major investment projects and rationalisation plans. Indeed, legislation apparently strengthened the Minister's potential to control the BSC. After 1977 the BSC could basically turn only to the government for its external finance, and unlike a private company could not switch its banker. In addition, as we have seen, the 1981 Iron and Steel Act gave the Secretary of State an increased capacity to exercise direct control over the affairs of the Corporation. Nor is the inter-personal aspect of the relationship necessarily kept at arm's-length, e.g. during a difficult period for the BSC during 1981, MacGregor considered that he was meeting Ministers two or three times a week (and the BSC had had a succession of 'difficult periods'!) (House of Commons Industry and Trade Select Committee 1980–81, Q. 34). Nevertheless, we have seen that although the quality of BSC decision-making frequently left much to be desired, both Ministers and officials only rarely considered themselves sufficiently qualified to challenge the authority of the Corporation.

We have discussed how the BSC had a tendency to act as an organised interest. If it is generally considered that the Department can offer only 'second-rate' advice, then to whom could the Secretary of State turn (other than the BSC itself) in conducting an evaluation of the Corporation's plans? In addition, in the words of its Under-Secretary, the Iron and Steel Division within the DoI had the responsibility: 'for advising on policy with respect to the steel industry in the UK, whether publicly or privately owned' (House of Commons Industry and Trade Select Committee 1980–81, Q. 354). It might be said that the capacity to co-ordinate British steel policy in general required an even greater expertise than running the BSC, and was in no sense subordinate to the latter.

Yet it was clear that the officials within the Iron and Steel Division were extremely diffident about putting forward any point of view which might have suggested they knew more about the steel industry than the employees of the BSC. When asked by the Select Committee about the market prospects for individual sectors within the steel industry, the Under-Secretary in charge of the Iron and Steel division, Ken Binning, stated that:

> We do not evaluate the market prospects for particular product groups independent of BSC. The reason for this, Chairman, very simply is that we do not have the expertise which would be necessary to investigate these matters in detail; whereas BSC in conjunction with their friends on the Continent, for example, certainly have a vast amount of information which, in the past, has proved to be remarkably reliable in relation to these particular product areas (House of Commons Industry and Trade Select Committee 1980–81, Q. 364).

Instead of detailed analysis, the Iron and Steel Division looked at general trends in manufacturing output and gross domestic product, and then translated these into figures for steel consumption.

In March 1981, forty-two people worked within the Iron and Steel Division, but regardless of the expertise which might have existed within it, its own diffidence, and apparently deep-rooted readiness to place itself in a subservient position to that of the BSC, placed it in a poor position to act as a strong policy adviser to the Secretary of State. Binning himself distinguished the Iron and Steel Division's ability to evaluate from its ability to monitor the affairs of the BSC. In the former case he considered that the Department could not possess the expertise to be found in the BSC, but with regard to the latter he claimed that the civil servants played a more positive role.

To the extent that monitoring represents the identification and measurement of a deviation from predicted policy outputs and outcomes, there is clearly a great deal of validity in Binning's statement, but he went on to say that monitoring of the BSC's plan related primarily to understanding what was causing a variation from the planned path :

> This is a matter which does not necessarily ... require a knowledge of the steel industry which is superior to that of the managers of the industry, but which does require a knowledge of, or access to, advice on the economic or political or international background, or whatever it is which is having this effect on the industry. If I may suggest it, this is precisely what we do have (House of Commons Industry and Trade Select Committee 1980–81, Q. 393).

The function of identifying the reasons for a deviation, and then taking steps to correct it, is clearly a more difficult process than only identifying that deviation, but the Iron and Steel Division was confident that it could handle this arduous task. Yet BSC losses had continued to mount for a number of years, often deviating alarmingly from the predicted path, but with little sign that the Iron and Steel Division was equipped to deal with these intense crises. It might be argued that this corrective aspect of monitoring demands an equal (if not greater) expertise than that required to evaluate a strategy, but the Iron and Steel Division acknowledged that this latter type of expertise was lacking amongst the civil servants.

In reality, it appeared that the Department's best hope was to improve its capacity for receiving 'early warning signals' of variations from the forecast path, by means of acquiring information of a more detailed nature than had hitherto been the case, and with a greater degree of frequency. Thus the Select Committee was told by one of the Departmental officials that :

> Monitoring until now (March 1981) has been on the basis of the Corporation as a whole, rather than the businesses within it, and it has concentrated on the financial performance, in particular the cash flow forecasts. We are just about to receive the first batch of information under a new monitoring regime which will break down the commercial and financial information by at least eleven businesses within the Corporation which we will receive on a monthly basis, and which we will discuss with the senior management of the Corporation at a monthly meeting (House of Commons Industry and Trade Select Committee 1980–81, Q. 508).

When questioned further, the official acknowledged that the Department would not now be monitoring just financial targets, but also manpower reductions, levels of orders, and exports.

The increased level of monitoring suggested that the Iron and Steel Division was attempting to keep a closer watch on the Corporation, but significantly it also appeared that officials were hoping to 'educate' themselves to a greater depth about the affairs of the steel industry, and hence play a more relevant part in both evaluating BSC strategy and also correcting variations from the forecast path.

The Iron and Steel Division's anxiety to involve itself more deeply with the affairs of the Corporation also reflected the limitations of the BSC Board as a policy-making body. As we noted in Chapter Seven, after enlargement of the Board in 1978, by 1983 all the worker directors and civil servants had disappeared. It is significant that the Board was enlarged at a time when the finances of the BSC were deteriorating rapidly, and it was hoped that the new Members would assist the Corporation by providing expert guidance. In order to give the Board a more significant role an Audit Committee had been created in March 1977 and given the responsibility of considering the Corporation's Annual Accounts and Half Year Statements of Results before they were approved and published. The Committee was also to review matters arising from the Corporation's annual audit, and also the key indicators of operational performance. The members of the Audit Committee (three in 1987) were all non-executive Members of the Board. The Audit Committee could supplement the work of the DoI, although as non-executive Members it was likely that they could devote only limited time to the affairs of the Corporation.

A further means by which it might be thought that the DoI could develop its expertise in steel was to initiate exchanges of staff with the BSC, but the Department employed this device only sparingly in the 1970s and early 1980s. Between December 1976 and April 1981 there were four temporary exchanges of staff between the DoI and the BSC – two in each direction. Surprisingly, none of these involved staff in the Iron and Steel Division. The Department considered that a general benefit could be gained from these exchanges, but they were not concerned with improving the manner in which it carried out its responsibilities towards the steel industry (House of Commons Industry and Trade Select Committee 1980–81, 236–7). Yet despite its reluctance to become involved in day-to-day management decisions, this was the only way in which the DoI (including the Secretary of State) was likely to be able to fulfil a significant role in evaluating and monitoring the affairs of the Corporation. The arm's-length relationship appears to have had little real value, for it provided government with the pretence that it was exercising real power, when in reality it was usually either subservient to the will of the Corporation, or at other times clumsily attempting to impose unrealistic policies on the industry.

It seems reasonable to argue that had government been *directly* responsible for the steel industry, then it might have been more likely to ensure that decisions were economically and commercially sound. Lack of direct responsibility seems to have encouraged the Department to rely very heavily on the BSC's commercial and technical judgement, notwithstanding the fact that the financial consequences would eventually fall on the government of the day. Thus, despite (or some may say because of) the Conservative government's adherence to a strict monetary policy, it found itself saddled with a virtually open-ended commitment to fund the losses of the Nationalised Industries.

We saw in Chapter Seven that the DTI had not given sufficient attention to monitoring the Phoenix operations. The Department told the Public Accounts Committee that the new companies were free-standing operations, yet the BSC (and thus indirectly the government) holding in these companies was large. It appeared that only questioning by the PAC caused the DTI to discover how it did monitor the Phoenix operations, although by this time the government had already provided £10 million in order to save Sheffield Forgemasters from going out of business!

The BSC itself sometimes displayed sensitivity about the source of its expertise. In 1984 Tory MP Robert McCrindle, as a member of the Trade and Industry Select Committee, asked BSC Chairman Robert Haslam if the Corporation relied wholly on in-house forecasting. The Chairman replied that it did not:

> We are involved in what is called the St. James Group which is a distinguished group of economists and consultants. We clearly are in contact with the Treasury and also with the DTI as to their views. We are in touch with customers and customer representative bodies... Internationally we are very much involved with the Commission, with Eurofer and, on the international front, with the International Iron and Steel Institute (House of Commons Select Committee on Trade and Industry, 1983–84, Q. 367).

In reply to Mr. Haslam's points, the Labour MP Dr. Jeremy Bray (the MP for the constituency that included Ravenscraig) wrote to the Committee and stated that the St. James Group was in fact a group of corporate subscribers established by the Economist Intelligence Unit (EIU) to finance the operation of the Treasury model on their behalf by the EIU. He knew of no member of the St. James Group, apart from the EIU staff, who would claim them to be a distinguished group of economists and consultants. He also claimed that the BSC had not consulted Treasury officials as to the appropriateness of the Treasury model in forecasting steel demand. Also (as we have seen) the DTI would not claim to have the necessary expertise. He concluded by asking the Committee to:

> commission the genuinely independent forcasts of steel demand contingent on different economic scenarios, with the methods and assumptions fully explained, in a manner on which the Committee has had no evidence from either the BSC or the DTI (House of Commons Select Committee on Trade and Industry 1983–84, 128).

Behind Dr. Bray's argument lay his concern for the future of Ravenscraig, but his scepticism does expose the general point that the lack of alternative expertise to that found within the BSC had been a significant gap in UK steel policy-making since the early 1970s.

The BSC itself also subscribed to Oxford Economic Forecasting Ltd., for forecasts of the UK and major world economies. These predictions were used in assisting the BSC to calculate future UK steel demand. From 1979, the BSC also had the benefit of Commodity Flow Accounts, which enabled the Corporation to build up demand equations for the various components of a particular sector, such as mechanical and electrical engineering. The BSC claimed that this technique enabled them to improve the quality of their demand forecasts (see Monopolies and Mergers Commission, 1988, 123–25). At the same time, the greater resources at the BSC's disposal highlighted the lack of expertise within the Department.

The DTI was placed in a difficult position after the 1982 decision to keep open all three strip mills. If the Department continued to claim, as it had in 1981, that it lacked the expertise to evaluate the BSC's plans, then the strip mills decision would be seen as a purely political decision. Indeed, Norman Lamont, Minister of State for Industry, told the Trade and Industry Select Committee:

> Yes. It might be the implication of our comments, that the BSC's forecasts for 1983–84 were unduly pessimistic ... we had concluded ... that it would actually be wrong to take a decision to close an integrated plant, on the basis of the forecasts that were available, looking at it from the trough of recession (House of Commons Trade and Industry Select Committee 1983–84, Q. 166).

In support of his argument, Mr. Lamont cited the expertise in his own Department :

> We have an Economics and Statistics Department, under Mr. Leisner. We have, under Mr. Leisner, a capability to look at these matters, and we do look all the time at other people's forecasts. We make our own judgements on them and make our own views about what is likely to happen to the economy generally and to steel demand (House of Commons Trade and Industry Select Committee 1983–84, Q. 168).

The implication of Mr. Lamont's remarks was that, at least on the strip mills decision, the Department's judgement was more to be trusted than that of the BSC. Lamont was clearly more optimistic about the expertise to be found in the Department than his predecessors (Sir Keith Joseph and Mr. Binning) three years earlier. Nevertheless, although the DTI might have improved its monitoring of the BSC, there was no hard evidence that the Department had radically improved its capacity for evaluating the work of the BSC. Indeed, as Abromeit comments, from 1983 the Iron and Steel Division (ISD) was transformed into a wider Mineral and Metals Division in which steel formed only one part of its work, but even at the ISD only three people had been concerned directly with the BSC, with considerable job rotation (Abromeit, 203).

In the Treasury there was more continuity, with the Assistant Secretary concerned with the BSC within the Public Enterprise Group having changed only once in a four year period (Abromeit, 203). Abromeit also makes it clear that civil servants see themselves as providing at least some degree of continuity in dealing with the public corporations, and so act as a 'moderating influence' on Ministers and politicians (Abromeit, 203. For a detailed analysis of the BSC-government relationship see Abromeit, 145–217). In this sense, it is not so much that the civil servants wished to challenge BSC expertise, but rather would hope to explain the Corporation's position to Ministers, and so act as mediators between the two sides. In the case of Norman Lamont, there was a political dilemma which caused him on this occasion to place the Department in an adversarial position *vis à vis* its relationship with the BSC.

Expertise and intervention

The dilemma for the Thatcher government was personified in the plight of Sir Keith Joseph. An arch apostle of non-intervention by government, and a firm believer in the power of the market, he was compelled to finance all the Corporation's losses and also to introduce almost day-to-day monitoring of its affairs. Sir Keith wished to discount the expertise to be found in his own Department compared with that of the BSC, yet he could not ignore how this 'expertise' had placed the Corporation on the edge of oblivion. Similarly, the strip mills decision demanded support for Departmental expertise, while evidence to the Select Committee caused the Department to look at how it actually monitored the Phoenix projects.

In spite of its beliefs, the Thatcher government found itself increasing the official supervision and monitoring of BSC affairs. The Audit Committee was already in place, and the BSC crisis led to increased monitoring by the DTI. On a more general level, the 1978 White Paper on the Nationalised Industries (HM Government, Cmnd. 7131) recommended that each Nationalised Industry should select or publish a number of key performance indicators. The BSC produced these from 1978–79 although, as Abromeit comments, there were no performance *targets* against which actual performance could be measured (Abromeit, 174).

Later, the 1980 Competition Act made provision for the Monopolies and Mergers Commission (MMC) to undertake regular efficiency audits of selected Nationalised Industries. The MMC conducted an enquiry into the BSC during 1985–86. The Report was handed to the DTI in August 1986, but inexcusably was only published in July 1988. The BSC had complained that the Report contained commercially sensitive information, and after much wrangling succeeded in having most of these contentious figures removed (*Financial Times*, 28:7:88). Garner argues that, although the MMC is capable of doing good work, the job of auditing the efficiency of the

Nationalised Industries would have been better performed by the Comptroller and Auditor General's National Audit Office, who as paid servants of Parliament would have been better placed to undertake an independent analysis. The government's delay in publishing the MMC Report on the BSC suggests that Garner's case has much validity (Garner, 1982, 425). The MMC Report itself made forty-eight detailed recommendations, and also observed that the BSC's projected level of required investment was too low (Monopolies and Mergers Commission, 1988).

In 1982 a Private Members' Bill attempted to give the National Audit Office power to inspect the accounts of the Nationalised Industries themselves. This proposal was defeated by a combination of interests, notable among them being the Naitonalised Industries Chairmen's Group. The NICG had been formed in 1976, and allowed the BSC another outlet for lobbying the government on matters of interest it shared with other public corporations.

In addition to the responsibility placed on the MMC, the government commissioned its Central Policy Review Staff (CPRS) (popularly known as the Think Tank and set up by the Heath government, but abolished by the Prime Minister in 1983) to investigate the relationship betweeen the government and the Nationalised Industries. The CPRS recommendations were based on the premise that the expertise to be found in Whitehall with regard to the Nationalised Industries required considerable strengthening. Thus new Cabinet Committees on the Nationalised Industries were set up, the Treasury created a Public Enterprise Analytic Unit to monitor the Industries, and more accountants and business executives were hired to bolster the various Departments dealing with the Nationalised Industries (although this last recommendation was not implemented with respect to the BSC).

Thus the government, while seeking the long-term aim of distancing itself from the steel industry in thc form of BSC privatisation, also sought a more detailed oversight of the Corporation's affairs. The Thatcher government, with its deep political commitment to privatisation and generally putting the Nationalised Industries on a more commercial footing, was compelled to seek means of checking that the right job was being done. At the same time, commercial objectives and the essentially monitoring role largely precluded the government from taking a wider look at BSC strategy as a whole. A Report by the Public Accounts Committee highlighted how emphasis on one objective could lead to neglect of others.

In 1985 the Committee of Public Accounts conducted an investigation on the basis of a Report by the Comptroller and Auditor General. The PAC's Report looked into the arrangements employed by Departments for: (a) evaluating capital programmes proposed by certain Nationalised Industries (including the BSC); (b) controlling capital expenditure; and (c) monitoring the progress of capital projects. The evidence given by the DTI, together with the Report's conclusions, provide a revealing insight into the DTI-BSC relationship in the mid to late 1980s. In the case of the BSC, the Committee concluded that:

> it was not clear how the scale of BSC investment, which was concentrated on projects to improve plant efficiency and extend product range, was related to the Corporation's strategy for reduction of capacity and manning (House of Commons Committee of Public Accounts 1985–86, para. 5).

At the root of the problem lay the BSC's policy priorities and the DTI's reluctance to examine too closely the Corporation's investment plans. As the Committee remarked:

> the DTI had no arrangements for examining any BSC project costing less than £4 million, though these accounted for 30 per cent of the Corporation's programme (House of Commons Committee of Public Accounts 1985–86 para. 15).

The 1967 White Paper on Nationalised Industries (HM Government, Cmnd. 1343) set out the need for all important projects to use discounted cash flow techniques based on a test discount rate set at 8 per cent (later 10 per cent). The 1978 White Paper on Nationalised Industries concluded that this system had failed, and instead proposed a more general required rate of return set at 5 per cent. The DTI claimed that all the BSC investment appraisals of recent years suggested a rate well above 5 per cent on all projects, but the key factor, acknowledged by DTI Permanent Secretary, Sir Brian Hayes, was that the BSC now gave priority to cost-saving projects and the achievement of financial targets set by the government, rather than any required rate of return. Sir Brian summed up: 'In the circumstances of the BSC it has been inevitable that the top priority has been given to the short-term financial target, not merely inevitable but right' (House of Commons Committee of Public Accounts 198586, A. 1826).

Thus, in the eyes of both the BSC and government, the long-term quality of the Corporation's investment strategy had become a relatively minor consideration. Both bodies were no doubt deeply scarred from the experiences of the 1970s, and the motto had become 'safety first'. The privatisation aim was also likely to have been a factor in this minimum risk strategy. The difficulty is that this cautious policy may preclude investments with a potential for success in the 1990s, particularly with regard to producing high quality steels. The CPRS report on the Nationalised

Industries had called for corporate plans to include radical policy alternatives, but the Comptroller and Auditor General could find no evidence of these proposals. Elsewhere, we have called for an alternative source of expertise to be 'a power base for pessimism' (a term suggested to us by Grant Jordan) (Richardson and Dudley, 1986, 365), but on reflection it might be that a radical alternative to the accepted wisdom is required i.e. a power base for pessimism in the 1970s and a power base for optimism in the 1980s.

There is an obvious danger that if the Department itself acquired a high degree of expertise, then the industries would become increasingly suspicious and resentful of any intervention, but as we have argued elsewhere, there appears to be a recognition by the NIs that they can 'learn to live with' genuinely *expert* external (and perhaps even internal) review bodies (see Dudley and Richardson, 1984, 132). At the moment, the work of the MMC is conducted on an *ad hoc* basis, but the review process could be institutionalised through the creation of a range of permanent, independent review bodies, modelled on the MMC or Price Commission (for a discussion of an example of such a body, the Norwegian Petroleum Directorate, see Richardson, 1981, 35–50). If such bodies were genuinely independent of both government and industries (British public administration has many such independent bodies), they would earn the respect and co-operation of the industries and would act as an alternative source of expertise to whom the government could turn for advice.

In the case of the Ravenscraig decision the government took direct control over a basic strategic decision, but in other respects the Corporation was still in a strong position in its relationship with government. Sir Keith Joseph's 'blank cheque' policy in itself put the government in the hands of BSC management, while the government also made a public show of placing its faith in Ian MacGregor. Although Sir Robert Haslam and Sir Robert Scholey were less politically prominent figures, there is no evidence to suggest that they had less freedom for manoeuvre than that given to Sir Ian. The greatest political change has come about as a result of the UK's increasing integration into ECSC steel policy, and it is this factor which has the potential to be the most powerful catalyst of political change.

Policy co-ordination within the ECSC

When compared with the UK, the ECSC steel community has a greatly more organised and formal structure. Apart from statutory consultation processes, the Davignon Plan itself could only have been constructed and implemented within a well ordered sphere of activity (for a detailed analysis of the ECSC community see Grunert, 1986, 222–307). On the other hand, there are obvious similarities with the UK in terms of power relationships.

On structure, where the UK steel community has no formal status, and comes together only on an *ad hoc* basis or in the Iron and Steel EDC of the NEDC, the Consultative Committee of the ECSC brings together all the established interests on a regular quarterly basis. Thus British Steel, BISPA, BRISCC, NASS, BSF and the unions are all represented on the Consultative Committee. The UK has the second largest national representation on this Committee, eighteen delegates to nineteen from West Germany (Grunert, 262), although it should be noted that coal and steel interests are both represented. Under the Treaty of Paris, the Commission is compelled to hear the views of the Committee on major issues, and also to keep the Committee informed about its aims, programmes and activities (Grunert, 263). The Committee forms a useful function in ensuring that none of the chief interests within the ECSC steel community can claim to have been excluded from the policy process.

In this respect, the system bears a close relationship to the typical UK policy style described by Jordan and Richardson, in which if all the interests concur, then everyone must take the blame if things go wrong (Jordan and Richardson, 1982, 93). Despite its formal status, the Consultative Committee does not play a central role in sector policy-making. As in the UK, it is the producers' interests which predominate, and although they are represented on the Consultative Committee, it is through Eurofer that they prefer to operate as a political force. As in the UK, the unions, stockholders and consumers etc can have their say, but it is the steel companies which call the tune on steel strategy.

In representing the largest steel producers within the ECSC, Eurofer has established a close relationship with the Commission, and played a crucial part in shaping policy. The Commission-Eurofer relationship inevitably invites comparison with the government-BSC relationship, although there are clear differences. Firstly, unlike the UK government, which appointed the Chairman and Board of the BSC, the Commission has no power to hire and fire Eurofer representatives. Secondly, although the Commission was given wide powers under the Davignon Plan, it did not have direct control over the steel companies. Thirdly, the Commission is not ultimately responsible for the fate of national steel industries, either in terms of being in the public or private sector, or even existing at all.

The great similarity between the BSC-government, and Commission-Eurofer relationships is that in both cases the two parties need each other too much for the implicit understanding between them to break down completely. As Grunert comments:

> In Eurofer the steel producers possess an effective umbrella organisation. With its help, they are able to influence the Commission's steel policy decisively, enabling them to bring their

> interests into the political process. On the other hand, the Commission would find it difficult to formulate and implement its steel policy without its Eurofer partner. It is, therefore, not surprising that over the past few years their co-operation has become more intensive and formalised (Grunert, 267).

As with the BSC, Eurofer has the expertise which is necessary both for collecting information and implementing policy. Unlike the UK government, the Commission has to rely more on its manipulative political skill to win consent for policies which may be highly controversial. It was Davignon's skill in this role which marked out the work of the Commission during the late 1970s and early 1980s. Nevertheless, the Plan only came to fruition when the steel producers realised that voluntary self-imposed controls were not working, and that something more drastic was required. Without Eurofer's consent and active co-operation, the Plan could never have been operated successfully. Since Davignon's departure, the Commission has not been so successful in winning consent for its proposals, while Eurofer itself has adopted a more confrontational approach. Without Davignon, there is something of a political power vacuum, with different national and ideological factions battling for control.

Unlike the UK government, the Commission must also deal with other powerful interests in the shape of the national governments themselves. The UK government has a permanent representative in Brussels, who belongs to COREPER, a permanent committee on the Community level (Grunert, 255). As with the UK government domestically, there is a lack of expertise on steel matters. In 1984, Trade and Industry Select Committee member, Sir Peter Emery, remarked that the permanent representatives in Brussels needed to be better briefed on steel matters than they appeared to be at one meeting the Committee had with them (House of Commons Trade and Industry Select Committee 1983–84, Q. 132). At the same time, it is these representatives who specifically represent the UK interest in Brussels and put the case for higher quotas, the continued existence of BSC strip mills etc. Despite the power of Eurofer, we saw in Chapters Eight and Nine how ECSC plans must ultimately require a political agreement. Although these agreements are eventually ratified by the Council of Ministers, a great deal of preparatory spade work and expert knowledge is needed.

From the point of view of the UK government itself, involvement with the ECSC can serve several useful purposes. Firstly, by means of negotiating in the Council of Ministers it can gain political prestige for any success gained. Thus, if the UK won extra quotas, or was compelled to reduce capacity less than other members, it could be portrayed as a significant victory for Britain. Secondly, as exemplified by the case of Ravenscraig, the government can use the ECSC to buttress its own domestic policy. As we have argued, there were reasonable domestic, economic and political arguments for keeping Ravenscraig open, but it was a plus for the government in that, by claiming other countries should cut back capacity before the UK, it could place its decision on the plant in the broader European context. In this sense, the BSC argument in favour of closing Ravenscraig could be made to look somewhat unpatriotic.

Thirdly, the Davignon Plan itself provided an invaluable breathing space for the ailing UK steel industry. The British government had no hesitation in supporting the declaration of a state of manifest crisis, for it knew that its long term privatisation plans for the BSC could only be fulfilled in a more stable market.

Fourthly, the ECSC allowed the introduction of a system of controls, alien to the philosophy of the Thatcher government, but without any domestic political penalty. Grunert notes that implementation of the full Davignon Plan in 1984 represented a victory for French *dirigisme* over the German philosophy of voluntary regulation within a free market (Grunert, 278). It could also be said that the Davignon Plan by-passed the UK government philosophy, but with the latter's permission! The Plan represented a more rigid regulatory system for the steel industry than had been seen in the UK outside a time of war, and it is hard to envisage that the Thatcher government could easily have introduced these controls by itself. With the help of Davignon, however, it could claim to be active in the spirit of European co-operation, in order to bring about stability in a strategic EEC industry. The introduction of a fresh political culture also indicates that, in steel at least, the UK had become a fully integrated member of the ECSC. This integration might have come about as a result of political and economic necessity, but is none the less genuine. The UK government was happy to cede some of its sovereignty in this area in exchange for a stability which it could not achieve unilaterally. For the BSC, the appointment of Sir Robert Scholey as President of Eurofer also suggested a new departure for that organisation, as all the previous Presidents had represented either French or West German interests. On the commercial front, by the late 1980s the Corporation was finally recognising the importance of exports to Europe, and so had embarked (almost unnoticed) on one of the most important policy changes in its history. Despite the power vacuum in the Commission, and the gradual removal of controls, there continued to be a strong sense, exemplified by the political controversy over strip mill closures, that a UK government (or the privatised British Steel) might never again be in a position to make an autonomous decision on major matters of steel policy. The Commission could not act decisively without the consent of Eurofer, but under Davignon it had proved itself to be more politically skillful and agile than any UK government.

12 Power in the Policy Process

A political paradox

The government's continuing role in representing UK steel interests within the ECSC ensures that even with a privatised British Steel, Ministers will still be integrally involved with the industry. The formal ECSC structure, and the inevitable need for a political negotiation, locks the government into a system where withdrawal is not an option. Privatisation allowed British Steel to appoint its own Chairman and Board and freed it from the constraints of the EFL, but for the purposes of the ECSC it will still be necessary to integrate its policy with the official government line. In addition, any new crisis for steel would be likely to involve the government as a co-ordinator of interests, a guarantor of finance, or a lender of last resort.

In the 1930s the industry was successful in persuading the government to introduce tariff control of imports; the decision in 1958 to divide the strip mill investment was made by the Cabinet; and in the 1980s the government provided finance for restructuring the private sector. Thus, even in a privately owned steel industry, major decisions have a habit of finding themselves on the political agenda. It might be that the contraction of the steel industry has reduced its political prestige and influence, but the strategic position it holds within the economy as a whole ensures the continuance of an official input to policy formulation. As Likierman argues, although the management of Nationalised Industries may have legitimate grievances about the quality of Ministerial interference:

> it is worth considering whether Nationalised Industries (in common with many private sector organisations) delude themselves into believing that a golden age might exist. In such an age, government priorities would never change, or if they did, it would be very slowly and the world would stand still for long enough to allow them to plan free of the hurly-burly of short-term constraints (Likierman, 1984, 99).

It is something of a paradox to argue, on the one hand, that the BSC has acted largely autonomously, yet on the other that the political element is of crucial importance in understanding the development of both the BSC and the UK steel industry as a whole. In this Chapter we will try to identify the chief elements which have shaped the steel policy process, and thereby find an explanation of how these conflicting outcomes can be fitted into the whole.

The BSC as beneficiary

We have noted previously that government intervention has been limited in scale and effect on BSC policy, while in addition there have been few discontinuities in the policies of successive governments. The two clear cut examples of discontinuity on Party lines in the 1970s were the Beswick Review and the new Conservative government's strict enforcement of cash limits in 1979. Both of these changes were quickly overturned, to the extent that it was almost as if they had never happened. In the 1980s, privatisation became the most significant discontinuity, but even here the main core of the BSC remained intact, while the Phoenix projects enabled the BSC to gain a firm hand on the industry as a whole.

Thus it could be argued that the chief beneficiary of state ownership was the BSC itself, which employed the political system to its own advantage. 'Political intervention' enabled the Corporation to dismantle the old company structure, obtain state finance for its investment programme, and later draw on an apparently bottomless purse in order to ensure survival. As Mény and Wright comment:

> the steel industry, like every other Nationalised Industry, has become a 'political quasi-autonomous actor', and this is inherent in the nature of its relationship with the State which has to leave many important decisions to the industry (Mény and Wright, 1986, 87).

By the late 1980s the BSC was quite happy to contemplate a return to the private sector, provided that the conditions were right. In some senses, the period of state ownership could be regarded as mainly an unintended 'fire brigade' operation (in the manner of Rolls Royce, which was saved by the government in 1972, and then sold back to the private sector in 1987), even if the fire took twenty years to be extinguished!

At the time of privatisation, the government's insistence that the Ravenscraig plant should remain in operation stood as the one monument to effective official intervention. From the original 1960s objective of steel nationalisation as a key pillar in the framework of a strategic plan for national growth, public ownership conferred on the BSC a freedom and scope for policy-making which no government could halt. In becoming an integral part of the political process, the BSC could manipulate the system in order to dominate it, and so become the ruler of its own private kingdom.

We have described the catalogue of BSC weaknesses and disasters, including the over-ambitious investment strategy, the loss of domestic market share, the poor record on exports, the problems in bringing on stream the new plants, the inability to appreciate the value of mini works such as Shelton, the lack of an integrated modern investment strategy in the 1980s, the Sidbec-Normines débacle, the mishandling of the 1980 strike, the lack of attention to quality and poor service to customers (at least until the mid 1980s), and the need for more trained ferrous metallurgists.

At the same time, we should also note its genuine successes. The BSC did inherit a private sector industry in a depressed state. Lord Melchett and Monty Finniston might have been over-ambitious and even arrogant at times about the future of the BSC, but they did set in motion a much needed process of modernisation, and also made the industry aware of its investment and productivity deficiencies when judged by international standards. The onset of crisis made the BSC a more humble organisation. More attention was paid to the needs of customers, while the efficiency and productivity of the remaining plants were greatly improved. Sir Charles Villiers was the most beleaguered BSC Chairman, but under him there did at least appear a glimmer of how the Corporation could find a way through the crisis. In the early 1980s, Ian MacGregor highlighted the problem of energy prices and the cost of coal, and under him the political system was used to obtain a better deal for the Corporation. By the mid 1980s, the BSC had achieved a relatively stable financial position, and for the first time there was a genuine attempt to win a share of ECSC markets.

If state ownership assisted the BSC in becoming over-ambitious in the late 1960s and early 1970s, then in the crisis years it saved the UK steel industry from oblivion, and also gave valuable breathing space in allowing the

Corporation to restructure itself. Given the severity of the crisis from 1975 onwards, it is impossible to believe that a private sector industry could have survived without state help.

The potential of government as mediator

Once we move outside the BSC's immediate sphere, particularly with regard to strategic policy co-ordination and implementation space, then a different picture emerges. We have noted that the BSC's relationships with other interests became politicised over the years, and the Corporation inevitably found difficulty in gaining its own way against these strong countervailing interests. As a mediator and/or arbiter between these interests, government comes into its own as a seat of power in the policy process. In the area of policy co-ordination and implementation, official decisions can potentially construct a framework and guide outcomes in a way which government apparently finds extremely difficult when dealing with a single organisation. If we examine these inter-relationships, then it is possible to identify the principal means by which government can guide and control the affairs of the UK steel industry.

After twenty years of turbulence and crisis, it was perhaps understandable that the vision of the BSC as a key element in a strategy for industrial regeneration should have fallen by the wayside. Nevertheless, it is these cross-sectoral weaknesses which lay at the root of many BSC problems. If governments (and ultimately the taxpayer and workers in the industries) are to foot the bill for failure, then governments must develop the necessary expertise to see that plans are soundly based, and encourage an awareness of the number of interdependencies involved in implementing a strategy for any one industry. As Hugh Heclo has noted in the American context:

> partitions which were previously assumed to separate policy areas are more often being called into question ... The interplay of what were once thought to be exogonous factors for a given program is increasingly seen as integral to its very substance (Heclo, 1975, 404).

In the steel community itself there are important interdependencies, and the BSC politically did not always see events go its own way. Both stock-holders and scrap merchants to some extent succeeded in decoupling themselves from BSC interests, while for a period in the 1970s the trade unions were successful in delaying closures and the operation of new plant. Thus if the BSC was King of strategic policy-making in its own sphere, it was not nearly so powerful in the policy co-ordination and implementation processes, even within its own sector. It is these spheres of policy co-ordination and implementation which give a clue as to how government can potentially gain greater control (both direct and indirect) of the system. As a mediator and arbiter of interests, it could shape the BSC's (and British Steel's) operational framework. In some areas, such as the price of iron ore, even the government may have little influence, but in others it has considerable potential power.

In Chapter One we identified areas of BSC activity where it could control its environment and so achieve stability, and other areas where it was more at the mercy of factors it could not control. Thus even where the BSC had scope for political activity, powerful countervailing interests were always present. This interplay of interests, identifiable at both national and ECSC levels, and invariably taking place in a highly charged political atmosphere, epitomised the environment in which the BSC was compelled to operate. It is an area where governments of all colours can continue to have the final say.

Policy co-ordination and implementation in the steel community

The BSC's most effective area of policy implementation was the sweeping plant closures programme of the late 1970s and early 1980s. The BSC quickly realised that by picking off the plants one by one, and offering lucrative redundancy payments, it could proceed smoothly with the closures. The action committees could protest, but were powerless to intervene. The trade unions at national level had no official alternative strategy, and in any case could not prevent their members from voluntarily giving up their jobs. In this area, notwithstanding the Beswick review and the Ravenscraig interventions, successive governments were largely content to remain on the sidelines (while picking up the bill for this costly process). Responsibility for a politically messy and unpopular task could safely be left to the BSC! By setting up BSC (Industry), the Corporation even took some of the responsibility for creating jobs in closure areas.

Opening up new plant at established works was a more difficult matter, particularly in the 1970s. Blast furnacemen could impose conditions on operating a new furnace at Port Talbot, while dockers at Hunterston delayed the opening of that terminal (a dispute settled by the Thatcher government's acceptance of the Dock Labour Scheme). Later, when the closures began to bite, it was easier to convince employees at strip mills such as Llanwern and Ravenscraig that, if productivity did not improve, then their works might be threatened with closure. In the 1980s, this implicit competition between the strip mills was an effective instrument in securing productivity improvements, and achieved its apotheosis during the miners' strike of

1984–85, when both plants stayed in operation despite the mixture of pleas and threats from the NUM.

The one period of direct government intervention in the BSC's relationship with the unions came towards the climax of the steel strike. Having attempted to distance itself from the issues in dispute, the government was instrumental in setting up the machinery which finally settled the strike.

The BSC was largely successful in bringing the private sector within its ambit. During the 1970s the private sector remained in a relatively prosperous position compared with the BSC. After 1980, many of the private sector companies fell into crisis, and were given financial assistance from government. Implementation of Phoenix projects required a great deal of private sector rationalisation, while the BSC had a large financial stake in the Phoenix companies. Although the BSC gained from the Phoenix projects, it should nevertheless also be noted that they offered a convenient solution to a policy problem for the Conservative government. The scheme assisted the private sector, apparently furthered the privatisation programme, but also maintained the strength of the BSC. In setting the future framework for the industry, the government had secured a consensus of interests, while also satisfying political objectives.

Stockholders had a strong identification of interest with the BSC through the need for maintenance of prices. Nevertheless, the stockholders have improved their market share over the years, and by 'second sourcing' partly succeeded in decoupling themselves from the influence of the BSC. At the same time, stockholders do need to maintain prices, and so must seek to secure a close relationship with government in EEC negotiations. Similarly, scrap merchants have increased exports, and although still reliant on the BSC for much of their business, they succeeded in winning concessions from government in the face of BSC opposition. In this area again, government mediation and arbitration gave it potential power over the UK steel community.

In contrast, both the UK steel industry and government have only marginal influence over supplies of iron ore. The BSC's one attempt to secure stability in its environment, through Sidbec-Normines, ended in disaster. During the 1980s, given an excess of supply over demand and the depreciation of the dollar, the BSC was able to make significant savings in the cost of its key raw material, but it was a gain which came about largely through autonomous events.

Government as a cross-sectoral mediator

The BSC had a stormy relationship of long-standing with British Coal. This state of affairs was brought to a head in 1981 when Ian MacGregor threatened that, if it did not receive a better deal from the then NCB, then the

BSC would buy all its coal from abroad. The BSC increased its coal imports, but the key role was played by the government in deciding to subsidise coal used by the Corporation. Similarly, the government also encouraged British Coal to develop the Margam mine as a new supplier of coal to the BSC. In acting as mediator, the government followed what it saw to be the national (and its own political) interest in maintaining an interdependence between the UK coal and steel industries.

Energy prices offer a similar example of a battle between strategic national industries. The Energy Intensive Users' Group evolved out of a common interest to secure lower prices, but despite intensive pressure on government only limited success was achieved. The gas and electricity industries were themselves extremely powerful, while the government was reluctant to lose revenues from industries which made a large profit. Even when British Gas was privatised, the government still retained indirect control through its potential power to impose a regulatory pricing framework for industrial consumers.

In Chapter Ten, our study indicated the role of government in the relationship between the BSC and some of its principal customers. In a few instances the role was explicit, for example when Patrick Jenkin intervened in the MacGregor-Edwardes row over steel prices. In others, the role was more indirect. Two examples were Leon Brittan's pleas to Vauxhall to increase its percentage of locally sourced components, and (not so indirectly in that over £100 million of public money was being put into the project) the stipulation that Nissan should eventually achieve 80 per cent EEC content of its vehicles before permission was given for the company to develop its UK operations.

These examples all indicate the government's potential power to influence the operational environment of the BSC. In creating a framework which pushes rival interests in a particular direction, the government manipulates the system by means of a series of levers.

The pattern of the steel policy process

No government can free itself from this system of control, for even *not* operating a lever in a particular way at a particular time will lead to a particular outcome. The framework of the steel policy process which emerges is one where the BSC had control over its own domain of special expertise, while using its intimate relationship with central government as a means of ensuring its organisational survival. For its part, government appeared to be enfeebled and non-plussed at the prospect of controlling or even entering the inner sanctums of the BSC, but where the Corporation's activities touched those of other organisations, then Ministers could act as mediators and arbiters, and thereby hold the balance of power. Thus we could explain the paradox of apparent BSC autonomy and the primacy of the political element by arguing that the BSC effectively played the political game within its own sphere of potential direct control, but that outside this sphere it was only one of many inputs to a policy process ultimately under the arbitrating and mediating control of government. Within the ECSC, it is the Commission which acts as the fulcrum of the see-saw holding rival interests at each end, and when placed in the hands of a skilful political operator such as Davignon, it too becomes a seat of real power.

In his study of petrochemicals, an industry with similar problems to those confronting steel in the 1970s and 1980s in that supply outstripped demand, Bower concludes that the most appropriate metaphor which can be applied to the industry is one of sports, but not the romantic idea of contest:

> it is the practical one of negotiation over the regulations. In a very real sense it is the rules of play that the companies are able to establish for themselves, as well as their relations with unions and government, that determine the potential profitability of their industry (Bower, 1986, 51).

From this analysis, Bower sets out the essential relationship between companies and government:

> Companies involved in restructuring need the government to give their strategies legitimacy and to make sure that political forces do not lead authority to be used to sidetrack a program ... The problem is that restructuring by major corporations technically involves power shifts that governments regard as part of their domain. Their interest is unavoidable ... There are economic, administrative and political imperatives and all must be met, usually by some kind of compromise (Bower, 222–3).

Adversity in the steel industry increased the pressure on governments and industrial interests throughout the world to maintain stability and minimise suffering. Within the ECSC, the Davignon Plan was the most rational approach yet devised for dealing with these severe problems. Similarly, in the UK, political leadership is required if the steel industry, whether privately or publicly owned, is to secure a position of relative stability. This leadership will be even more urgently required if there continues to be a power vacuum within the Commission. Domestically, government has the power to mediate and arbitrate on (and thus shape) the industry's key relationships. The BSC would clearly have benefited from a more co-ordinated relationship with its chief suppliers and customers, as well as interests such as stockholders within its own community. Fragmentation led not only to the BSC losing nearly a third of its domestic market share in the early 1970s, but also cost British Coal a significant slice of its business with the BSC.

The process of strategy formulation and implementation has been bedevilled by the consequences of segmentation within the policy process, which has been identified as a common feature throughout Western Europe. When the fortunes of key industries are so inter-connected, then much greater effort must be made to achieve what can be termed 'co-ordinated rationality' (see Gustaffson and Richardson, 1979). If some form of 'co-ordinated rationality' is to be achieved, then the concept of relatively autonomous public and private corporations will need to be substantially modified.

Political leadership and 'co-ordinated rationality'

Given the political environment in which the steel industry has operated throughout its history, need political judgement be incompatible with rational decision-making? Does privatisation of the BSC automatically remove any government responsibility for intra and inter-sectoral co-ordination? As a Nationalised Industry, we have seen that the Morrisonian concept of government arm's-length control over the BSC hardly fits with reality, but as a mediator and arbiter of interests within the implementation process the government can make a positive contribution.

We will re-examine briefly three specific areas in which the government could possibly assist in steering British Steel towards the objective of 'co-ordinated rationality'.

Firstly, there is the potential of the NEDC. In Chapter Seven we described how the Iron and Steel EDC had contributed to the debate on energy prices, and also how the group assisted in implementing the 'nuts and bolts' of steel policy. In 1987 the government, which funds the NEDC, decided that henceforth the NEDC council would meet only four times a year instead of once a month. In addition, despite pressure from the CBI, TUC and NEDO itself which caused him to slightly modify his initial proposals, Chancellor Nigel Lawson decided that from 1988 the number of EDCs would be reduced from 37 to 14 sector groups and 3 working parties, with NEDO cutting its staff by 40 per cent. Amongst the EDCs to disappear was that for iron and steel. The political fashion of the 1980s dictated that a tripartite consultative group such as the NEDC could not work effectively in a deregulated economy.

The Iron and Steel EDC acted as a useful informal forum where representatives of government, BSC, BISPA and unions produced studies on UK demand for steel, opportunities for exports, energy prices etc while at the same time also meeting other EDCs. The group was chaired by an industrialist, not a Minister, while government was represented on the Committee by civil servants (surprisingly the Executive Director of NASS received only monitoring reports from the Committee, while the British Scrap Federation was not represented at all). A Minister at the head of the Committee could have given it more authority – Michael Heseltine states that he was the first Minister to attend a sector working party of NEDO, twenty years after the organisation was set up (Heseltine, 1987, 120).

This political neglect of NEDO indicates the lack of official leadership from successive governments. The staff of NEDO could provide a valuable source of expertise, particularly on cross-sectoral issues, while the EDC itself offered the one opportunity for the steel community as a whole to work regularly on policy co-ordination. The work of a body such as the Iron and Steel EDC may not be the essence of political glamour, but it can at least oil the wheels where interests collide and connect.

Secondly, the government's key role in determining energy prices to industry. As we saw in Chapter Seven, the government's decision to allow British Gas to continue its existing commercial contracts after privatisation rebounded on it in 1987 when the nature of these contracts was referred to the MMC. A similar situation could arise with a privatised CEGB, while we also described in Chapter Seven how, by subsidising CEGB coal, there was a knock-on effect resulting in lower electricity prices to the BSC. By tightening or loosening this regulatory framework, government can push costs onto the energy industries or a high energy user such as British Steel. Thus, if price rises within BG and the CEGB are restricted, then some pressure will be taken off British Steel. On the other hand, if the energy companies are allowed more latitude in raising their prices, then British Steel will bear the cost. This regulatory framework can be used as an indirect tool of control, for a government which wished to tighten the pressure for cost efficiency on BG or the CEGB could restrict prices, while if it was thought that an organisation such as British Steel was living too much on its fat, then energy price control could be loosened (a thorough analysis of this control concept can be found in Dunsire, 1978, 207–18).

Thirdly, the government stipulation to Nissan that as a condition of developing a UK operation they must guarantee an eventual 80 per cent EEC content of their vehicles. Although this stipulation by no means precludes Nissan from importing steel, such a requirement is of substantial potential benefit to British Steel. Thus it is of close interest to the government that this agreement be adhered to, and the British Steel-Nissan relationship should be carefully monitored, particularly with over £100 million of public money invested in the project.

These three specific cases illustrate how an apparently loose British Steel-government relationship can still involve a significant degree of official mediation, arbitration and indirect control. The ties which bind the UK steel industry and government are of long standing and are not about to be broken, notwithstanding privatisation. The industry, although diminished, remains in a strategic position, and thus subject to a strong official policy input. The potential sources of an effective political leadership include greater Ministerial responsibility, alternative pools of expertise, a co-ordinated and statesmanlike approach to ECSC policy. and a strong framework supporting the industry's inter-relationships. Our analysis suggests that this latter element is the most effective source of official power, and is persuasive to the extent that the need for action will exist with either a publicly or privately owned steel industry.

The relationship between the conjuror and his assistant may well be entering a new phase, but the fundamentals of their alliance are little changed. Privatisation and the continuance of better times may see the assistant apparently going solo and holding centre stage alone, but the conjuror himself can still have a large influence over certain of the act's ingredients, and in any case will be standing in the wings ready to leap onto the stage at a moment's notice should something go wrong.

Meanwhile, back in the real world, both the government and British Steel face the challenge of reconciling greater ECSC integration (both in policy and perhaps even real industrial terms), with the need to maintain a significant UK steel industry. It would certainly be a tragedy of the first order if the sacrifices made by hundreds of thousands of steelmen during the lifetime of the BSC should prove to have been in vain.

Bibliography

Abromeit, H. (1986), *British Steel*, Leamington: Berg.

Aylen, J. (1980), 'Innovation in the British Steel Industry', in Pavitt, K. (ed.) *Technical Innovation and British Economic Performance*, London: Macmillan.

Barlow, W. (1981), 'The Problems of Managing Nationalised Industries' in *Allies or Adversaries?*, London: RIPA.

Bower, J. L. (1986), *When Markets Quake*, Boston: Harvard Business School Press.

Brannen, P., Batstone, E., Fatchett, D. and White, W. (1976), *The Worker Directors: A Sociology of Participation*, London: Hutchinson.

Brannen, P. (1983), 'Worker Directors: An Approach to Analysis. The Case of the British Steel Corporation' in C. Crouch and F. A. Heller (eds.) *International Yearbook of Organisational Democracy* (Vol. 1), London: John Wiley & Sons.

British Steel Corporation (1967) *First Report on Organisation,* Cmnd. 3362, HMSO.

British Steel Corporation (1969) *Second Report on Organisation*, HC 163, HMSO.

British Steel Corporation (1969), *Third Report on Organisation*, HC 60, HMSO.

British Steel Corporation (1969), *Finance for Steel.*

British Steel Corporation (1975), *Organisation Review*, HMSO.

British Steel Corporation (1978), *Prospects for Steel*.

Bryer, R. A., Brignall, T. J., and Maunders, A. R. (1982), *Accounting for British Steel*, London: Gower.

Burn, D. (1961), *The Steel Industry 1939–1959*, Cambridge: Cambridge University Press.

Centre for Policy Studies (1984), *Which Direction? Board Appointments in Nationalised Industries*.

Darnell, H., Miles, T. P., and Morrison, H. C. (1984), *Steel: Future of the UK Industry*, London: Technical Change Centre.

Downs, A. (1967), *Inside Bureaucracy*, Boston: Little, Brown and Co.

Dror, Y. (1973), *Public Policymaking Re-examined*, Bedfordshire: Leonard Hill Books.

Dudley, G.F. and Richardson, J. J. (1984), 'The Political Framework' in J. Grieve Smith (ed.), *Strategic Planning in Nationalised Industries*, London: Macmillan.

Dudley, G. F. (1976), *Lobbying in North Staffordshire: A Study of Two Local Pressure Groups*, M.A. Thesis, University of Keele.

Dudley, G. F. (1979), 'Pluralism, Policy-making and Implementation: The Evolution of the British Steel Corporation's Development Strategy with Reference to the Activity of the Shelton Action Committee', *Public Administration*, 453–70.

Dudley, G. F. (1983), 'The Road Lobby: A Declining Force?' in D. Marsh (ed.), *Pressure Politics*, London: Junction Books.

Dudley, G. F. (1984), 'The British Steel Corporation and Problems of Political Management', *The Political Quarterly* 427–37.

Edwardes, M. (1983), *Back from the Brink*, London: Collins.

Fight for Shelton Bar (1977), Victoria Theatre transcript, London: Eyre Methuen.

Finer, S. E. (1975) *Adversary Politics*, London: Wigram.

Foster, C. D. (1971), *Politics Finance and the Role of Economics*, London: Allen and Unwin.

Garner, M. R. (1982), 'Auditing the Efficiency of the Nationalised Industries: Enter the Monopolies and Mergers Commission', *Public Administration* 409–28.

Grant, W. (1978), 'Insider Groups, Outsider Groups and Interest Group Strategies', *Working Paper*, No. 19, Warwick: Department of Politics, University of Warwick.

Grant, W. (with Sargent, J.) (1987), *Business and Politics in Britain*, London: Macmillan.

Grieve Smith, J. (1981), 'Strategy – The Key to Planning in the Public Corporation', *Long Range Planning* 24–31.

Grunert, T. (1986), 'Decision-Making Processes in the Steel Crisis Policy of the EEC: Neocorporatist or Integrationist Tendencies?' in Y. Mény and V. Wright (eds.), *The Politics of Steel: Western Europe and the Steel Industry in the Crisis Years (1974–1984)*, Berlin: de Gruyter.

Gustafsson, G. and Richardson, J. J. (1979), 'Concepts of Rationality and the Policy Process', *European Journal of Political Research*, 415–36.

Hannah, L. (1982), *Engineers, Managers and Politicians*, London: Macmillan.

Hayward, J. E. S. (1974), 'National Aptitudes for Planning in Britain, France and Italy', *Government and Opposition*, 397–410.

Headey, B. (1974), *British Cabinet Ministers*, London: Allen and Unwin.

Heal, D. W. (1974), *The Steel Industry in Post War Britain*, London: David and Charles.

Heclo, H. (1975), 'Frontiers of Social Policy in Europe and America' *Policy Sciences*, Vol. 6, pp. 403–21.

Heseltine, M. (1987), *Where There's a Will*, London: Hutchinson.

H. M. Government (1967), *Nationalised Industries: A Review of Economic and Financial Objectives*, Cmnd. 3437, HMSO.

H. M. Government (1973), *Steel. British Steel Corporation: Ten Year Development Strategy*, Cmnd. 5226, HMSO.

H. M. Government (1978), *British Steel Corporation: The Road to Viability*, Cmnd 7149, HMSO.

H. M. Government (1978), *The Nationalised Industries*, Cmnd 7131, HMSO.

Hogwood, B. W. (1979), *Government and Shipbuilding*, London: Saxon House.

Hood, C. (1976), *The Limits of Administration*, London: John Wiley and Sons.

House of Commons Select Committee on Nationalised Industries. Session 1972–73 *British Steel Corporation*, HC 141, HMSO.

House of Commons SCNI, 1977–78 *British Steel Corporation*, HC 26, HMSO.

House of Commons SCNI 1977–78, *British Steel Corporation*, HC 127, HMSO.

House of Commons SCNI 1977–78, *Financial Forecasts of the British Steel Corporation*, HC 238, HMSO.

House of Commons Industry and Trade Committee 1980–81, *Effects of BSC's Corporate Plan*, HC 336, HMSO.

House of Commons Industry and Trade Committee 1982–83, *The British Steel Corporation's Prospects*, HC 212, HMSO.

House of Commons Trade and Industry Committee 1983–84, *The British Steel Corporation's Prospects*, HC 344, HMSO.

House of Commons Trade and Industry Committee 1984–85, *The British Steel Corporation,* HC 474, HMSO.

House of Commons Trade and Industry Committee 1986–87, *The UK Motor Components Industry*, HC 407, HMSO.

House of Commons Committee of Public Accounts 1984–85, *Control and Monitoring of Investment by British Steel Corporation in Private Sector Companies – The Phoenix Operations*, HC 307, HMSO.

House of Commons Committee of Public Accounts 1985–86, *Monitoring and Control of Investment by the Nationalised Industries in Fixed Assets*, HC 37, HMSO.

House of Commons Select Committee on Scottish Affairs 1985–86, HC 133, HMSO.

House of Lords Select Committee on Overseas Trade 1984–85, HL 238, HMSO.

International Iron and Steel Institute (1983), *Steel and the Automative Sector.*

Iron and Steel Trades Confederation (1980), *New Deal for Steel.*

Jones, K. (1983), *Impasse and Crisis in Steel Trade Policy*, London: Trade Policy Research Centre.

Jordan, A. G., Richardson, J. J., and Kimber, R. H. (1977), 'The Origins of the Water Act 1973', *Public Administration*, 317–34.

Jordan, A. G. and Richardson, J. J. (1982) 'The British Policy Style: Or the Logic of Negotiation?' in J. J. Richardson (ed.), *Policy Styles in Western Europe*, London: Allen and Unwin.

Likierman, A. (1979), 'The Financial and Economic Framework for Nationalised Industries', *Lloyds Bank Review*.

Likierman, A. (1984), 'The Impact of Financial Constraints' in J. Grieve Smith (ed.), *Strategic Planning in Nationalised Industries*, London: Macmillan.

Lundquist, L. (1972), 'The Control Process: Steering and Review in Large Organisations', *Scandinavian Political Studies*.

MacGregor, I. (1986), *The Enemies Within*, London: Collins.

McEachern, D. (1980), *A Class Against Itself: Power in the Nationalisation of the British Steel Industry*, Cambridge: Cambridge University Press.

Mény, Y. and Wright V. (1986), 'State and Steel in Western Europe' in Y. Mény and V. Wright (eds.), *The Politics of Steel: Western Europe and the Steel Industry in the Crisis Years (1974–84)*, Berlin: deGruyter.

Middlemas, K. (1983), *Industry, Unions and Government*, London: Macmillan.

Monopolies and Mergers Commission (1988), *British Steel Corporation*, Cmnd. 437, HMSO.

Monopolies and Mergers Commission (1988), *Gas*, Cmnd. 500, HMSO.

Montjoy, R. S. and O'Toole, J. (1979), 'Toward a Theory of Policy Implementation: An Organisational Perspective', *Public Administration Review*, 465–76.

Moon, J. (1983), 'The Spectrum of Issue Salience: Britain and European Economic Integration 1950–63', *Strathclyde Papers on Government and Politics* No. 14.

National Economic Development Office (1986), *Steel: The World Market and the UK Steel Industry*.

Olson, M. (1965), *The Logic of Collective Action*, Cambridge Mass.: Harvard University Press.

Ovenden, K. (1978), *The Politics of Steel*, London: Macmillan.

Prior, J. (1986), *A Balance of Power*, London: Hamish Hamilton,

Pryke, R. (1981), *The Nationalised Industries: Policies and Performance since 1968*, Oxford: Martin Robertson.

Rawson, G. E. (1980), 'Organisational Goals and their Impact on the Policy Implementation Process', *Policy Studies Journal*, 1108.

Redwood, J. (1980), *Public Enterprise in Crisis*, Oxford: Blackwell.

Richardson, J. J. (1971), 'The Administration of Denationalisation: The Case of Road Haulage', *Public Administration*, 385–402.

Richardson, J. J. (1977), 'The Environmental Issue and the Public', in *Decision Making in Britain*. Milton Keynes: Open University Press. Block V: 'Pollution and the Environment'.

Richardson, J. J. and Jordan, A. G. (1979), *Governing Under Pressure*, Oxford: Martin Robertson.

Richardson, J. J. (1981), 'Problems of Controlling Public Sector Agencies: The Case of Norwegian Oil Policy', *Political Studies* 35–50.

Richardson, J. J. and Dudley, G. F. (1986), 'Steel Policy in the UK: The Politics of Industrial Decline', in Y. Mény and V. Wright (eds.), *The Politics of Steel: Western Europe and the Steel Industry in the Crisis Years (1974–84)*, Berlin: de Gruyter.

Ross, G. W. (1965), *The Nationalisation of Steel*, London: MacGibbon and Kegan Paul.

Sirs, W. (1985), *Hard Labour*, London: Sidgwick and Jackson.

Technical Change Centre (1986), *UK Steel – There is a Future*, (H. Darnell and N. J. Read eds.).

Tivey, L. (1982), 'Nationalised Industries as Organised Interests', *Public Administration*, 42–55.

Tsoukalis, L. and Strauss, R. (1986), 'Community Policies on Steel 1974–82: A Case of Collective Management', in Y. Mény and V. Wright (eds.), *The Politics of Steel: Western Europe and the Steel Industry in the Crisis Years 1974–84)*, Berlin: de Gruyter.

Vaizey, J. (1974), *The History of British Steel*, London: Weidenfeld and Nicolson.

Villiers, C. (1984), *Start Again, Britain*, London: Quartet.

Wildavsky, A. (1979), *Speaking Truth to Power: The Art and Craft of Policy Analysis*, Boston: Little, Brown and Co.

Wootton, G. (1970), *Interest Groups*, New York: Prentice Hall.

Young Fabian Steel Group (1974), *Crisis in Steel.*

Young, S. (1986), 'The Implementation of Britain's National Steel Strategy at the Local Level' in Y. Mény and V. Wright (eds.), *The Politics of Steel: Western Europe and the Steel Industry in the Crisis Years 1974–1984*, Berlin: de Gruyter.

APPENDIX A
BSC: miscellaneous statistics

	Crude steel production (million ingot tonnes)	steel deliveries HOME (million ingot tonnes)	steel deliveries EXPORTS (million ingot tonnes)	numbers employed	capital invested (£m)	profit/(loss) after taxation, depreciation, interest & exceptional items (£m)
1967–68	22.4	13.3	3.4	253,640	73	(19)
1968–69	23.7	15.6	2.8	254,390	74	(23)
1969–70 (6 months)	12.1	8.3	1.4	255,200	39	12
1970–71	24.2	16.3	2.7	252,400	142	(8)
1971–72	19.9	12.3	3.5	229,700	237	(68)
1972–73	24.2	14.4	3.6	226,600	198	3
1973–74	23.0	15.4	2.7	220,400	187	50
1974–75	20.8	13.0	2.1	228,300	311	73
1975–76	17.2	10.5	2.2	210,200	530	(255)
1976–77	19.7	11.0	2.7	207,900	579	(95)
1977–78	17.4	10.2	3.2	197,100	476	(443)
1978–79	17.3	9.6	2.9	186,000	318	(309)
1979–80	14.1	8.0	2.5	166,400	282	(1,784)
1980–81	11.9	7.2	2.3	120,900	185	(1,020)
1981–82	14.0	8.0	2.7	103,700	148	(504)
1982–83	11.7	6.8	2.5	81,100	122	(869)
1983–84	13.4	7.5	2.9	71,100	164	(256)
1984–85	13.0	7.8	2.8	64,500	182	(409)
1985–86	14.0	7.7	3.0	54,200	220	38
1986–87	11.7	6.6	3.7	52,000	269	178
1987–88	14.7	7.7	4.4	51,600	253	410

Source: BSC Annual Reports and Accounts

APPENDIX B: FINISHED STEEL

Net Home Deliveries To Consuming Industries And Merchants, 1981–1986

Ref No.	Industry Group	Total steel industry products (*Thousand Tonnes*) 1981•	1982	1983	1984	1985	1986
	ALL QUALITIES						
1	Coal, coke, petroleum & natural gas	268.1	288.7	210.8	289.5	167.8	134.0
2	Chemicals and allied industries	15.7	15.0	20.1	26.2	28.8	32.0
3	Steelworks own consumption	225.6	223.2	206.5	204.8	205.1	210.4
4	Foundries	16.3	18.1	10.9	17.6	15.3	22.2
5	Forging and ring rolling	534.9	466.1	408.2	407.2	371.6	331.5
6	Wire and wire manufactures	1,123.4	1,023.2	1,090.3	1,044.3	1,018.5	938.2
7	Bolts, nuts, rivets, screws, etc	77.6	48.7	41.2	45.3	62.4	67.9
8	Metal working machine tools & engineers' tools	15.0	10.1	14.3	13.8	11.5	10.4
9	Agricultural machinery and tractors	38.3	46.5	47.1	37.3	27.9	22.7
10	Construction, excavation & mechanical handling	66.0	60.5	45.1	45.3	48.7	31.8
11	Other non-electrical machines and engines	143.4	129.8	145.2	136.8	99.9	87.9
12	Other non-electrical engineering	153.6	206.6	165.0	93.5	98.0	84.9
13	Industrial & process plant and steelwork	560.1	642.7	523.7	464.3	613.8	667.3
14	Implements, cutlery, hand-tools and instruments	47.3	37.6	35.2	37.0	43.7	41.6
15	Electrical power machinery and equipment	90.6	90.7	90.0	108.2	98.9	104.4
16	Domestic electrical appliances	33.0	33.8	61.3	78.2	102.4	84.0
17	Other electrical and electronic goods & apparatus	43.3	43.4	41.4	46.0	45.4	43.1
18	Shipbuilding	163.4	193.2	102.2	72.3	99.3	46.8
19	Motor vehicle manufacturing (inc. parts & accessories)	595.5	599.3	630.4	600.6	622.3	570.3
20	Motor cycles, cycles, aircraft, etc	30.0	29.3	22.7	24.1	28.1	27.8
21	Railway rolling stock and parts	21.8	20.7	15.3	15.4	12.1	18.6
22	Cans and other metal packaging	504.0	524.6	499.0	530.2	510.5	521.6
23	Metal furniture	43.0	75.0	62.2	76.6	79.3	72.3
24	Metal doors and windows	40.9	48.5	43.2	48.4	39.3	47.9
25	Industrial and domestic hollow-ware	147.8	125.2	128.5	127.1	114.3	105.6
26	Other metal industries	257.4	253.4	250.6	297.9	312.9	267.2
27	Food, drink and tobacco	60.5	59.9	64.8	65.3	61.8	52.2
28	Construction (building and civil engineering)	204.0	203.1	248.2	348.7	327.6	364.0
29	Gas, electricity and water	134.2	67.8	41.2	60.4	40.0	43.7
30	Transport and communication	115.4	95.4	117.7	114.3	104.1	66.6
31	Other UK consumers inc. unallocated	333.0	334.6	369.4	372.7	384.7	382.8
	Total Deliveries to UK Consumers	**6,126.1**	**6,014.7**	**5,751.7**	**5,849.3**	**5,796.0**	**5,501.8**
	Stockholding merchants	3,072.3	2,775.3	3,005.4	3,186.7	3,121.2	2,835.1
	Total Home Deliveries by UK Producers	**9,175.4**	**8,790.0**	**8,757.1**	**9,036.0**	**8,917.2**	**8,336.9**

† Includes own repair and maintenance and further manufacture into steel products not covered by the industry definition, e.g. drawn wire, forgings, etc.

* Some minor changes have been made to the summary classification to eliminate very small tonnage headings or ensure confidentiality.

• 53–week period

Source: UK Iron and Steel Statistics Bureau

Index